Modern Ceramic Engineering

Modern Ceramic Engineering

Edited by **Carl Burt**

WILLFORD PRESS

New York

Published by Willford Press,
118-35 Queens Blvd., Suite 400,
Forest Hills, NY 11375, USA
www.willfordpress.com

Modern Ceramic Engineering
Edited by Carl Burt

International Standard Book Number: 978-1-68285-003-9 (Hardback)

The publisher's policy is to use permanent paper from mills that operate a sustainable forestry policy. Furthermore, the publisher ensures that the text paper and cover boards used have met acceptable environmental accreditation standards.

Trademark Notice: Registered trademark of products or corporate names are used only for explanation and identification without intent to infringe.

Printed in the United States of America.

Contents

Preface

Ceramic engineering is a rapidly emerging field of engineering concerned with manufacturing of ceramics and ceramic composites. It focuses on understanding and analysing processes related to purification of raw materials, production and formation of chemical compounds, their characteristics, etc. The topics covered in this extensive book deal with the core subjects like sintering technologies, hybrid processes, analysis of different structures, mechanical and thermal properties, etc. which are relevant to this field. Some of the modern usage of ceramic materials like high-temperature ceramics, ferromagnetic materials, advanced ceramic coatings, are also discussed in the book. It will help the readers to keep up with the rapid changes in this field.

This book is the end result of constructive efforts and intensive research done by experts in this field. The aim of this book is to enlighten the readers with recent information in this area of research. The information provided in this profound book would serve as a valuable reference to students and researchers in this field.

At the end, I would like to thank all the authors for devoting their precious time and providing their valuable contributions to this book. I would also like to express my gratitude to my fellow colleagues who encouraged me throughout the process.

Editor

Substitution of a fraction of zircon by cristobalite in nano hematite encapsulated pigment and examination of glaze application

Maryam HOSSEINI-ZORI[*]

Department of Inorganic Pigments and Glazes, Institute for Color Science and Technology (ICST),
P.O. Box 1668814811, Tehran, Iran

Abstract: Hematite pigment has a long history, but it cannot be used for ceramic application, because it must be thermally and chemically stable at high firing temperature when using a pigment in a glaze or ceramic body. Recently, through encapsulated systems, a new pigment with suitable thermal and chemical stability can be obtained by encapsulating hematite crystals into selected silica or zircon matrices. It means that nano-sized red hematite has been encapsulated into the protected phases. Transmission electron micrographs of hematite encapsulated into silica and zircon matrices by sol–gel method show spherical single crystals with diameter of about 5–10 nm. In order to optimize ceramic glaze formulations for application of the synthesized red inorganic nanocomposite inclusion pigment by sol–gel method, four different types of glazes (i.e., alkalis, borate, earth alkalis, and leaded glazes) have been tested. The results show that the substitution of a fraction of zircon by cristobalite in hematite–zircon pigment produces acceptable stability with red hue.

Keywords: nano hematite; inclusion pigment; zircon; silica; ceramic glaze

1 Introduction

In the production of colored ceramic materials, coloring ions can play their roles by being dissolved in the material (ceramic body or glaze) which we intend to color [1–3]. However, these alternatives do not appear to be recommendable in industrial production, since the solution and/or reduction processes involved are rather difficult to control and good reproducibility must always be guaranteed in industrial production. Instead of these procedures, synthetic pigments are widely used in ceramic industry; in these pigments, the color agent has been already incorporated into an appropriate host lattice during the calcination stage by some kind of reaction. Of course, the host lattice must have adequate thermal stability and remain insoluble to withstand the aggressive action of the glazes (ceramic frits and/or sintering additives) in which they are formulated. Based on the glaze formulation, some pigments will be useful to apply [4,5].

The inclusion or encapsulation of a reactive, colored or toxic crystal into a highly stable crystalline matrix gives a protection effect to the guest crystal. The guest crystal becomes inactive within the host [6–8]. There is a restricted choice for red/pink and orange colors among synthetic inorganic ceramic pigments to be used in ceramic industry. They are generally easily etched by glasses, and sensitive to the atmosphere and firing temperature; a number of them contain toxic and pollutional elements. Therefore in recent years, there has been a developing interest toward new nontoxic

---
* Corresponding author.
E-mail: mhosseini@icrc.ac.ir

inorganic red pigment and applying them into ceramic glazes [4,9–11].

Hematite encapsulated pigment based on silica and/or zircon crystals that protect the occluded red α-Fe_2O_3 chromophore crystals is usually utilized, even if the traditional preparation leads to powders that negatively affect the glaze composition and sintering temperature. In particular, in order to achieve high efficiency of chromophore agent encapsulation, the matrix sintering and/or crystallization must be synchronized with the nucleation and growth of the occluded chromophore agent. Crystallization, sintering and inclusion process are thus considered to take place simultaneously, and consequently, the control of particle size of the raw powders is one of the main steps [5–9,12–16].

Silica may be considered to have a potential to be used in the occluded pigment as a matrix due to its properties, relatively low price, and particularly, the sufficient thermal and chemical stability towards glassy phases. The high thermal stability of the silicon lattices and their properties such as the sinter-ability at relatively low temperature are well known [8–11], but an important question is why the hematite–silica pigment is not applied into glaze by ceramic industry? The hematite–silica system presents better red shade pigment than hematite–zircon system, and the hematite–zircon pigment is more expensive than hematite–silica system [17–19].

Based on these considerations, the aim of the present work is to compare the synthesized inclusion pigments in hematite–silica, hematite–zircon, and hematite–silica–zircon systems of several ceramic glazes for applications.

2 Experimental procedure

2.1 Materials

Samples of 1 mol SiO_2–0.2 mol Fe_2O_3, 1 mol $ZrSiO_4$–0.2 mol Fe_2O_3, and (0.5 mol SiO_2–0.5 mol $ZrSiO_4$)–0.2 mol Fe_2O_3 compositions were prepared by colloidal sol–gel method. A concentrated aqueous solution was prepared by adding iron sulfate ($FeSO_4 \cdot 7H_2O$, Merck) in deionized water and refluxing at 70 ℃ for 30 min. Then, the required aqueous suspensions of colloidal silica (30 wt% SiO_2, LUDOX) and zirconium chloride (Merck) based on final composition were added to the aqueous solution by drops of the concentrated solution. The systems were

continuously stirred and maintained at 70 ℃ until the pH was stabilized at about 4. The resulting gels were dried at 110 ℃. 0.2 mol NaF as mineralizer or flux agent was applied to all dried gels by milling. The powders were fired at 1000 ℃ in an electrical furnace with a soaking time of 3 h. The fired samples were micronized, wet milled and washed in deionized water. Finally three kinds of inclusion pigments were obtained after drying at 110 ℃ and screening [20].

Constant 5 wt% of the pre-synthesized nano-composite inclusion pigments of the mentioned systems were added to several kinds of transparent ceramic glazes, i.e., earth alkalis (high calcium), borate, alkalis (high sodium), and leaded glazes. All of the colored glazes were formulated by regular amounts of kaolin and additives, then wet milled for 20 min. After air brushing on conventional wall tiles, they were dried at 110 ℃ for 2 h and fired at 1100 ℃ in an electrical furnace with a soaking time of 10 min. The heating rate was 20 ℃/min.

2.2 Characterization methods

In order to determine the effects of the glaze formulation on the stability of the nanocomposite inclusion pigments and to define the color developed by the samples, a Gretag Macbeth Color-Eye 7000 was used, employing a 10° standard observer. L^*, a^* and b^* color parameters were measured following CIE (Commission International de l'Eclairage) colorimetric method. In this method, L^* is the lightness (black (0) → white (100)) axis, a^* is the green (−) → red (+) axis, b^* is the blue (−) → yellow (+) axis, c^* is the concentration of color, and h^* is the hue factor of color. Particle size analysis of the samples was controlled by a sift mesh (No. 400) and equipment of Mastersizer Malvern with water dispersant.

The microstructures were characterized using a scanning electron microscope (SEM, Leo 1455 VP) equipped with secondary electron (SE) detector and Robinson solid-state backscattered electron (BSE) detector (to display the chemical contrast in the observed objects). The glazes were selected from some ceramic industries; in order to identify their compositions, X-ray fluorescence (XRF) and inductively coupled plasma (ICP) analysis were used.

X-ray diffraction (XRD) patterns have been collected using a conventional powder technique by a Siemens Diffractometer (D500 mod) with Cu Kα Ni-filtered radiation to identify the crystalline phases present in the raw and fired powders.

The powders' microstructure characterization and hematite's morphology have been studied by transmission electron microscopy (TEM, Jeol JEM 2010 with GIF gatan Multiscan Camera 794 and Software Digital Micrograph 3.1).

3 Results and discussion

3. 1 Nomenclature of glaze formulations by XRF and ICP analysis

Table 1 shows the formulations and codes of the four types of frits that have been applied in the colored glazes. They are collected among different glazes. The one with the code of A has higher sodium oxide, so it is called alkalis glaze, and similarly, the borate glaze B, earth alkalis (high calcium) E and leaded glaze L.

3. 2 Thermal evolution of crystalline phases by XRD analysis

According to the XRD results in Fig. 1, three phases (i.e., hematite, cristobalite and zircon) crystallize after calcination at 1000 ℃ in (0.5 mol SiO$_2$–0.5 mol ZrSiO$_4$)–0.2 mol Fe$_2$O$_3$ sample. Before heat treating, just an amorphous phase with iron chloride crystals can be detected (Fig. 1(a)).

Figures 2 and 3 are related to binary systems of

Table 1 Formulations and codes of the applied frits by XRF and ICP (%)

Chemical analysis	Leaded glaze L	Borate glaze B	Alkalis glaze A	Earth alkalis glaze E
SiO$_2$	57.7	55.5	51.5	57.4
Al$_2$O$_3$	6.82	5.47	4.30	7.53
Fe$_2$O$_3$	0.23	0.21	0.28	0.28
Na$_2$O	4.36	5.98	**18.4**	5.47
K$_2$O	1.53	1.98	**4.29**	1.16
CaO	5.17	3.21	5.65	**11.5**
MgO	0.23	0.45	0.15	0.22
B$_2$O$_3$	0.14	**20.8**	4.45	6.24
Pb	**22.0**	4.05	0.05	1.73
Zn	1.27	2.01	7.20	**8.51**
Zr	0.40	0.21	0.04	—
Cl	0.06	0.10	3.72	—

hematite–silica and hematite–zircon respectively. XRD results in Fig. 2 detect hematite and cristobalite phases, and hematite and zircon phases are detected in Fig. 3.

3. 3 Colorimetric analysis of the colored glazes by CIELab values

CIELab colorimeter results are shown in Tables 2 and 3, providing easy comparison of the stability and color shades of the nanocomposite pigments and a traditional red pigment with code of K4272 made by Reimbold & Strick Company.

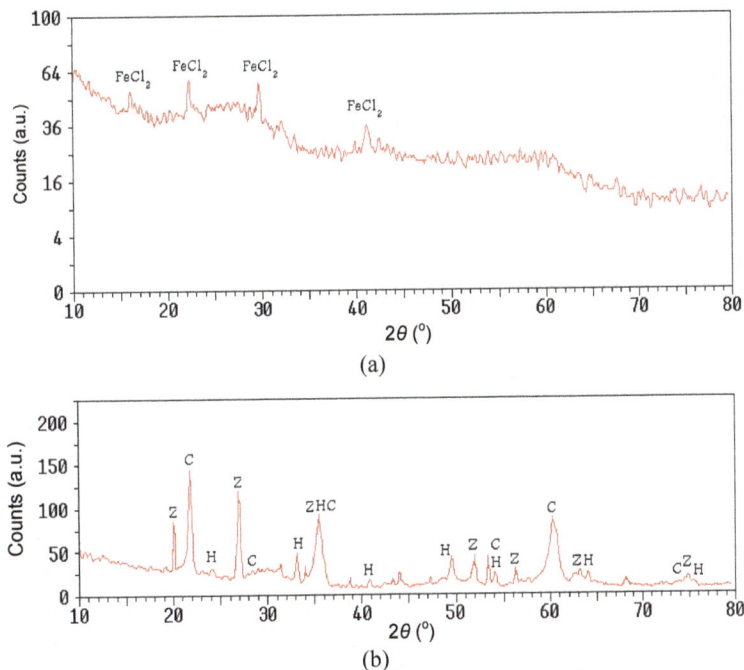

Fig. 1 XRD patterns of (0.5 mol SiO$_2$–0.5 mol ZrSiO$_4$)–0.2 mol Fe$_2$O$_3$ sample: (a) raw; (b) after calcination at 1000 ℃ for 3 h. H: hematite, C: cristobalite, Z: zircon.

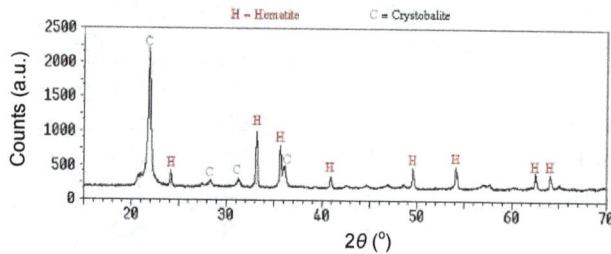

Fig. 2   XRD patterns of the fired sample at 1000 ℃ for 3 h with composition of 1 mol SiO$_2$–0.2 mol Fe$_2$O$_3$. H: hematite, C: cristobalite.

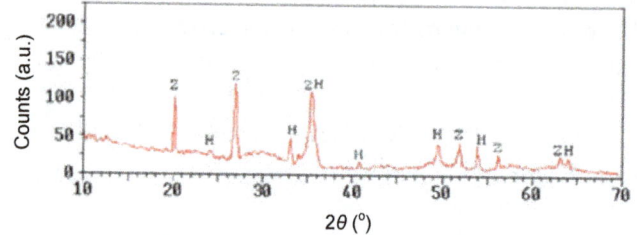

Fig. 3   XRD patterns of the fired sample in 1000 ℃ for 3 h with composition of 1 mol ZrSiO$_4$–0.2 mol Fe$_2$O$_3$. H: hematite, Z: zircon.

Table 2   Color shades and CIELab values of the colored glazes with 5 wt% of the synthesized nanocomposite inclusion pigments in different systems by light source standard of D65

Code of frit	Color	L^*	a^*	b^*	c^*	h^*
Hematite–silica pigment						
E	Red	39.92	26.59	18.40	32.33	34.69
L	Light red	45.21	27.49	18.80	33.30	34.36
A	Brown	39.26	13.99	11.87	18.35	40.33
B	Brown	41.60	11.51	9.23	14.76	38.72
Hematite–silica–zircon (ternary) pigment						
E	Red	39.46	30.62	23.61	38.67	37.63
L	Red	47.40	30.24	27.34	40.77	42.12
A	Red	45.83	27.45	19.58	33.72	35.50
B	Red (dark)	48.50	27.55	24.41	36.80	41.54
Hematite–zircon pigment						
E	Coral	33.13	17.65	9.92	20.25	29.34
L	Coral	36.79	16.67	10.67	19.80	32.63
A	Coral	33.56	14.82	8.45	17.06	29.69
B	Coral	34.03	17.55	9.54	19.98	28.52

Table 3   Color shades and CIELab values of the colored glazes with 5 wt% of a traditional red pigment in system of Sn–Si–Ca–Cr by light source standard of D65

Code of frit	Color	L^*	a^*	b^*	c^*	h^*
E	Coral	34.47	18.00	12.20	21.74	34.13
L	Coral	35.26	14.73	9.95	17.78	34.03
A	Coral	34.69	15.27	10.38	18.46	34.21
B	Coral	35.40	17.97	11.88	21.55	33.47

The color of the glazed tile after firing depends on the system of nanocomposite pigments and glaze formulation. Table 2 presents color shades and CIELab values of the colored glazes with 5 wt% of the synthesized nanocomposite inclusion pigments in systems of hematite–silica, hematite–silica–zircon (ternary) and hematite–zircon, respectively.

From Table 2 about hematite–silica pigment, it can be seen that the hues of red glazes are different. The best red shade has been obtained by the frit of code E with high a^* 26.59 and hue factor 34.69; therefore, the glaze based on earth alkalis is more suitable for the application of binary nanocomposite inclusions with colorant agent of hematite and silica matrix. The hematite–silica system of pigment did not have necessary thermal and chemical stability in the glazes except for the earth alkaline frit. Color shades for other glaze formulations are very different from brown to light red, thus they are not reliable and consistent enough for industrial application. In point of comparison, a traditional hematite–silica red pigment useful for body stain made by INCO Company was not stable in any glaze. In fact, nano hematite encapsulated in cristobalite phases by the mentioned procedure will introduce better inclusion microstructure and therefore better thermal and chemical stability in comparison with an industrial pigment.

The systems of hematite–silica–zircon (ternary) and hematite–zircon (binary) do not dissolve in these selected glazes, and present approximately unique red-orange shade and red-coral glazes respectively. It seems that the color shade of traditional red pigment with code of K4272 is more or less similar to the synthesized hematite–zircon pigment.

The cause of changes in the red hues can be studied by SEM.

3.4 Particle size analysis of the samples

The synthesized pigment was added in glazes after screening. The particle size of the samples is about 10 μm. Figure 4 presents a normal curve of particle distribution with $d(0.5)$: 9.078 μm related to one of the samples (hematite–zircon pigment).

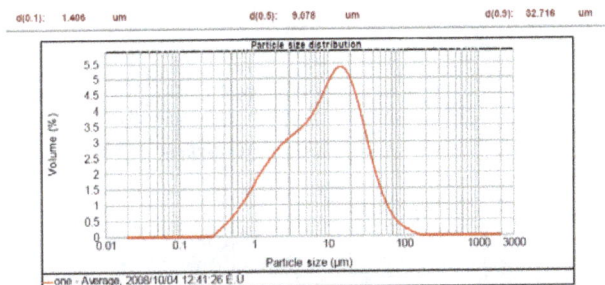

Fig. 4   Particle size analysis of the synthesized hematite–zircon pigment.

3.5 Microstructure analysis of the samples (pigments and glazes) by SEM and TEM techniques

TEM micrographs of hematite crystals encapsulated into silica, zircon and both matrices can be seen in Fig. 5. The spherical nanoparticles have diameter of about 5–10 nm.

Figures 5(a) and 5(b) are high magnifications of nano hematite encapsulated into zircon–silica and zircon crystals, respectively. They show that the morphology and size of nano hematite in different systems is more and less similar as in the

hematite–silica system seen in Fig. 5(c). The morphology of hematite is important for the color of red pigments [21].

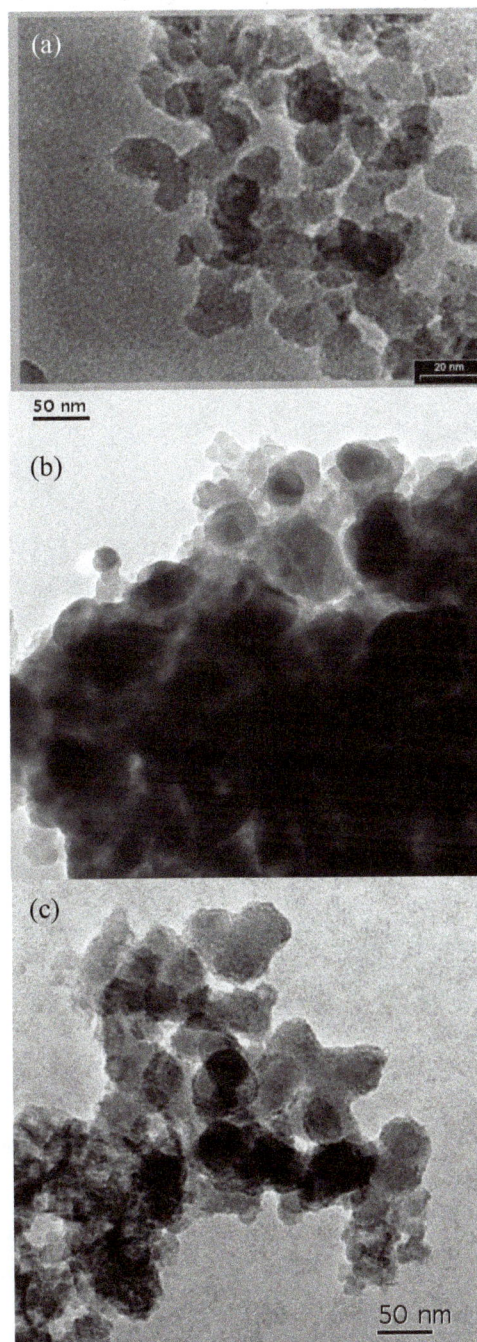

Fig. 5   TEM micrographs of the synthesized (a) hematite–zircon–silica inclusion pigment, (b) hematite–zircon inclusion pigment, and (c) hematite–silica inclusion pigment.

From Fig. 6(a), the microstructure studies of the brown glaze samples related to the colored glaze with 5 wt% of the synthesized inclusion pigments in system of hematite–silica after firing show only some spherical dissolved particles and/or uniformed glassy phase, which indicate that the borate kind of glaze formulation is not useful for the applied system of the nanocomposite pigments, while hematite–zircon pigment in the same glaze show some dispersed particles in the glassy phase (Fig. 6(c)). Borate, leaded and alkalis glazes have the lower viscosity than earth alkalis in the firing of glazes. The viscosity of a molten glaze affects the mobility of atoms and the speed of diffusion that control the reactions. Gualtieri *et al.* [22] reported the same result about a natural pigment based on hematite–silica.

The SEM analysis of the red shade glaze samples containing 5 wt% of hematite–silica nanocomposite inclusion pigment in frit of code E, and/or 5 wt% of the ternary and binary zircon nanocomposite inclusion pigment in all frits, observe some irregular micron-sized particles dispersed in the glass matrix.

The EDX analysis of the stable particles in the fired glazes prove the presence of Si–Fe and/or Zr–Si–Fe elements. It means that these glaze formulations after melting do not attack to nanocomposite pigments. The microstructures of a red shade glaze and point EDX of a stable pigment particle into glaze can be viewed in Fig. 7 (5 wt% of hematite–silica–zircon nanocomposite pigment with frit of code E) with different magnifications by SE detector.

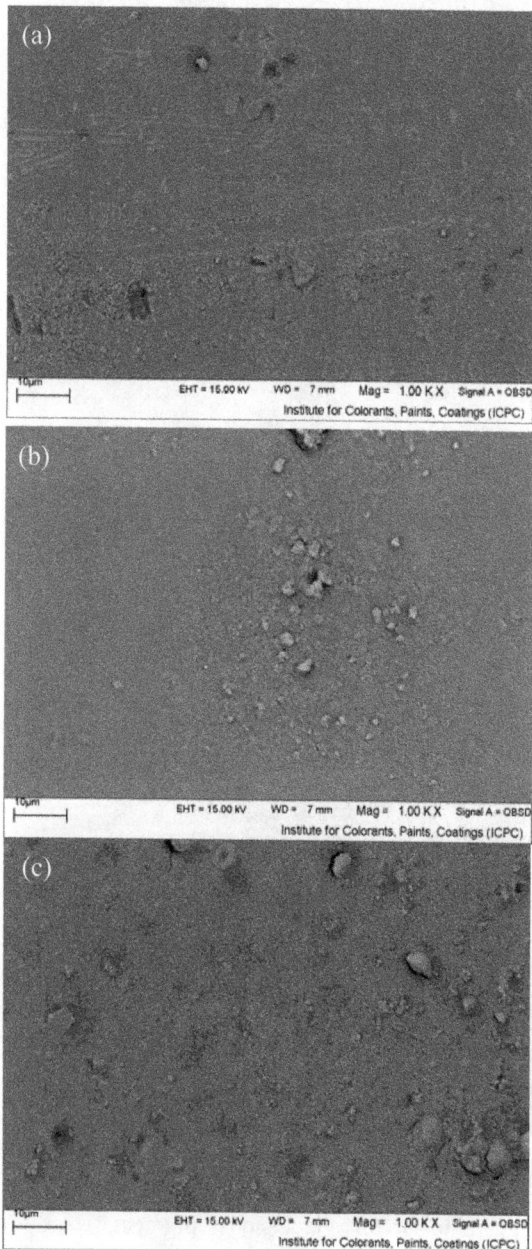

Fig. 6 Microstructures of the brown shade glaze containing 5 wt% of (a) hematite–silica, (b) hematite–silica–zircon, and (c) hematite–zircon nanocomposite pigments with frit of code B by BSE.

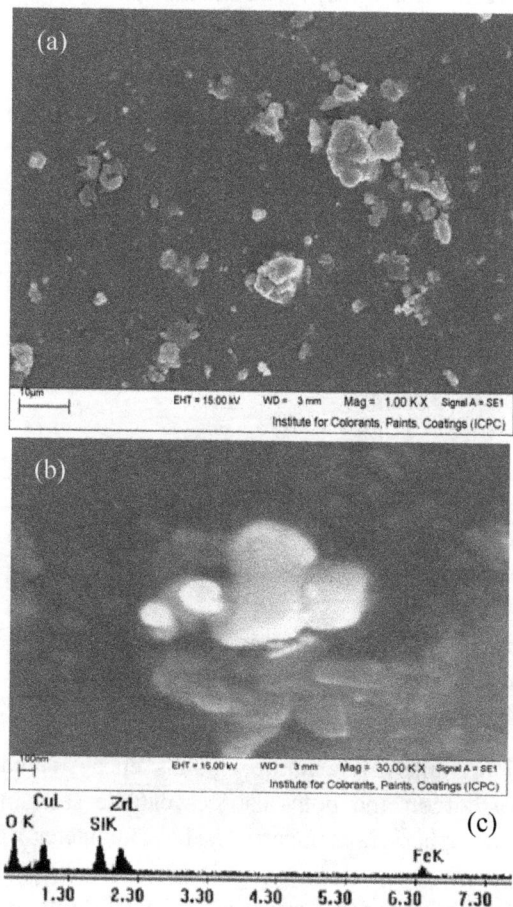

Fig. 7 Microstructures of the red shade glaze (5 wt% of hematite–silica–zircon nanocomposite pigment with frit of code E by SE detector ((a) and (b)), and point EDX of a stable particle (c).

4　Conclusions

In order to optimize the ceramic glaze formulations for application of the synthesized red inorganic nanocomposite inclusion pigments by colloidal sol–gel method, several ceramic glazes have been tested. The results show that the best formulation of the glaze with hematite–silica inclusion pigment is Si–Ca–Zn system (earth alkalis glaze). This system is almost applied for the monoprosa tile glazes in ceramic industry. hematite–silica–zircon and hematite–zircon systems of the nanocomposite inclusion pigments have more stability in ceramic glazes than the other system.

Colorimeter data of hematite–zircon system show that this pigment presents more or less similar shades in different types of glazes, and its color can be predicted in every application, so it is more respectable. By the way, hematite–silica–zircon inclusion pigment in earth alkaline glaze presents more intense red shade. Due to the thermal and chemical stability of ternary nanocomposite pigment in some glazes, it may be considered to be a suitable red pigment for ceramic applications.

A fraction of zircon is substituted by silica in the mentioned pigment; therefore, it can be obtained cheaper than hematite–zircon pigment. On the other hand, it shifts coral shade of hematite–zircon pigment to better or different red hue and performs new red pigment.

Acknowledgements

We are grateful to Mashad Glaze Industrial Group for the provision of the glazes used.

References

[1] Bondioli F, Ferrari AM, Leonelli C, et al. Synthesis of Fe_2O_3/silica red inorganic inclusion pigments for ceramic application. *Mater Res Bull* 1998, **33**: 723–729.

[2] Candeia RA, Bernardi MIB, Longo E, et al. Synthesis and characterization of spinel pigment $CaFe_2O_4$ obtained by the polymeric precursor method. *Mater Lett* 2004, **58**: 569–572.

[3] Sorlí S, Tena MA, Badenes JA, et al. Structure and color of $Ni_xA_{1-3x}B_{2x}O_2$ (A=Ti, Sn; B=Sb, Nb) solid solutions. *J Eur Ceram Soc* 2004, **24**: 2425–2432.

[4] Harisanov V, Pavlov RS, Marinova IT, et al. Influence of crystallinity on chromatic parameters of enamels coloured with malayaite pink pigments. *J Eur Ceram Soc* 2003, **23**: 429–435.

[5] Hosseini-Zori M, Taheri-Nassaj E, Mirhabibi AR. Effective factors on synthesis of the hematite–silica red inclusion pigment. *Ceram Int* 2008, **34**: 491–496.

[6] Bondioli F, Manfredini T, Siligardi C, et al. A new glass–ceramic red pigment. *J Eur Ceram Soc* 2004, **24**: 3593–3601.

[7] Hosseini-Zori M. Synthesis of inclusion nano composite as a non toxic red ceramic pigment by sol–gel method. In *Ceramic Transactions Volume 210: Ceramic Materials Components for Energy and Environmental Applications*. Jiang DL, Zeng YP, Singh M, et al. Eds. Hoboken: John Wiley & Sons, 2010: 60–65.

[8] Hosseini-Zori M, Bondioli F, Manfredini T, et al. Effect of synthesis parameters on hematite–silica red pigment obtained using a coprecipitation route. *Dyes Pigments* 2008, **77**: 53–58.

[9] Bondioli F, Manfredini T. The search for new red pigments. *Am Ceram Soc Bull* 2000, **79**: 68–70.

[10] Vicent JB, Llusar M, Badenes J, et al. Occlusion of chromophore oxides by sol–gel methods: Application to the synthesis of hematite–silica red pigments. *Bol Soc Esp Cerám Vidrio* 2000, **39**: 83–93.

[11] Hosseini-Zori M, Taheri-Nassaj E. Nano encapsulation of hematite into silica matrix as a red inclusion ceramic pigment. *J Alloys Compd* 2012, **510**: 83–86.

[12] Hradil D, Grygar T, Hradilová J, et al. Clay and iron oxide pigments in the history of painting. *Appl Clay Sci* 2003, **22**: 223–236.

[13] Dondi M, Matteucci F, Cruciani G. Zirconium titanate ceramic pigments: Crystal structure, optical spectroscopy and technological properties. *J Solid State Chem* 2006, **179**: 233–246.

[14] Llusar M, Badenes JA, Calbo J, et al. Environmental and colour optimisation of mineraliser addition in synthesis of iron zircon ceramic pigment. *Br Ceram Trans* 2000, **99**: 14–22.

[15] Bondioli F, Manfredini T, Siligardi C, et al. New glass–ceramic inclusion pigment. *J Am Ceram Soc* 2005, **88**: 1070–1071.

[16] Ricceri R, Ardizzone S, Baldi G, *et al*. Ceramic pigments obtained by sol–gel techniques and by mechanochemical insertion of color centers in Al_2O_3 host matrix. *J Eur Ceram Soc* 2002, **22**: 629–637.

[17] Garcia A, Llusar M, Badenes J, *et al*. Encapsulation of hematite in zircon by microemulsion and sol–gel methods. *J Sol–Gel Sci Technol* 2003, **27**: 267–275.

[18] Nero GD, Cappelletti G, Ardizzone S, *et al*. Yellow Pr–zircon pigments the role of praseodymium and of the mineralizer. *J Eur Ceram Soc* 2004, **24**: 3603–3611.

[19] García A, Llusar M, Sorlí S, *et al*. Effect of the surfactant and precipitant on the synthesis of pink coral by a microemulsion method. *J Eur Ceram Soc* 2003, **23**: 1829–1838.

[20] Hosseini-Zori M. Synthesis of the new ternary nano composite pigment of hematite–zircon–silica by sol–gel method. Iran Patent PV 49877, 2008.

[21] Katsuki H, Komarneni S. Role of α-Fe_2O_3 morphology on the color of red pigment for porcelain. *J Am Ceram Soc* 2003, **86**: 183–185.

[22] Gualtieri AF. Natural red pigment for single-fired ceramic glaze. *Ceram Bull* 2002, **81**: 48–52.

2

Microstructure and corrosion resistance of ultrasonic micro-arc oxidation biocoatings on magnesium alloy

Lijie QU^{a,b}, Muqin LI^{a,b,*}, Miao LIU^{c}, Erlin ZHANG^{b}, Chen MA^{b}

^{a}State Key Laboratory of Advanced Welding and Joining, Harbin Institute of Technology, Harbin 150001, China
^{b}Department of Materials Science and Engineering, Jiamusi University, Jiamusi 154007, China
^{c}Department of Stomatology, Jiamusi University, Jiamusi 154007, China

Abstract: The ultrasonic micro-arc oxidation (UMAO) was used to fabricate ceramic coatings on magnesium alloy. UMAO coatings were produced at 60 W input ultrasonic. The effects of the ultrasound on the microstructure, phase composition, elemental distribution and corrosion resistance of the coatings were extensively investigated by scanning electron microscopy (SEM), X-ray diffraction (XRD), energy-dispersive X-ray spectrometry (EDX) and electrochemical workstation. The results showed that ultrasound improved the homogeneous distribution of micro-porous structure. The coatings were mainly composed of MgO ceramic and small amount of calcium and phosphorus with porous structure. The Ca/P ratio of the coatings increased when 60 W ultrasonic was used. The corrosion potential in simulated body fluid (SBF) changed from −1.583 V of bare magnesium alloy to −0.353 V of magnesium alloy coated under 60 W ultrasonic. The corrosion resistance of UMAO coatings was better than that of MAO coatings.

Keywords: micro-arc oxidation (MAO); ultrasonic treatment; magnesium alloy; microstructure; corrosion resistance

1 Introduction

Magnesium and its alloys have been investigated as implants for almost two centuries due to the advantages and obvious benefits from biodegradable metal implants [1,2]; however, commercial implants containing magnesium and its alloys are still not available until now. Magnesium is present in high concentration in sea water and is the eighth most abundant element on earth. Furthermore, it is the fourth most abundant cation in human body. It also has excellent specific strength and low density, inherent biocompatibility, and adequate mechanical properties [3,4]. Unfortunately, magnesium is too reactive and generally exhibits a poor corrosion resistance because of high dissolution tendency in biological environments. Protective coating is an effective way to improve the corrosion resistance of magnesium and its alloys [5]. Many technologies have been used to obtain protective coatings, such as electroplating, thermal spraying, chemical conversion coatings, bio-mimetic approach, electrochemical deposition and anodization, and so on [6–9]. Recently, micro-arc oxidation (MAO) treatment, a common technique for the corrosion protection of magnesium alloys in industrial sector, has been used for the surface modification of magnesium and its alloys for biomedical applications due to their

* Corresponding author.
E-mail: jmsdxlimuqin@163.com

low cost and simplicity in operation.

The porous microstructure of MAO ceramic coatings would also result in an increase in corrosion rate of magnesium alloys. So many efforts have been done to improve the corrosion resistance of magnesium alloys during the MAO process. Zheng et al. [10] changed the applied voltage to increase the corrosion resistance of Mg–Ca alloy. Liang et al. [11] chose silicate and phosphate electrolytes to improve the corrosion resistance of magnesium alloy. Additives were used to improve corrosion resistance of the coatings [12–14]. The corrosion resistance was improved greatly; however, MAO coatings still exhibited poor corrosion resistance [15]. The porous microstructure might offer a beneficial surrounding for cells [16–19], but MgO exhibited poor biological activity. Calcium and phosphorus coatings on magnesium alloy were potential biomaterials [20,21]. So MAO combined with other methods was used to improve the corrosion resistance of magnesium alloys and the biocompatibility of MAO coatings. Hu et al. [22] used MAO with chemical deposition to prepare calcium phosphate coating on the surface of micro-arc oxidized magnesium alloy. Electro-deposition was also used to fabricate a top layer of DCPD on AZ80 Mg alloy coated by MAO to improve the corrosion resistance and bioactivity [23]. These composite methods need two steps, though the corrosion resistance of magnesium alloy and the bioactivity of MgO layer were improved evidently.

Recently, ultrasound technology in the synthesis of new materials has played a very significant effect [24]. Mechanical, thermal and active effects generated by ultrasonic and liquid medium make ultrasonic wave applied in surface engineering and electrochemistry [25]. At the same time, ultrasound is beneficial to inducing chemical modification on many materials [26,27]. Ultrasound would have cavitation effect when the ultrasonic power density is equal to or greater than 0.3 W/cm^2. In the present study, the minimum ultrasonic power density was applied during the MAO process to fabricate ultrasonic micro-arc oxidation (UMAO) coatings on magnesium alloy to develop a one-step method to improve corrosion resistance of

magnesium alloy and make the excellent biological performance of UMAO coatings. Furthermore, the microstructure and corrosion resistance properties of UMAO coatings were studied.

2 Experiment

2.1 Sample preparation

Magnesium bar with a diameter of 15 mm was sliced into thin samples with 2 mm in thickness and 15 mm in diameter. The samples were ground with SiC papers progressively up to 1500 grits followed by ultrasonic cleaning in acetone and ethanol for 5 min, respectively. The composition and mechanical properties of the magnesium alloy are shown in Table 1. UMAO treatment was carried out at 300 V pulse voltage with a duty cycle of 10% and a frequency of 700 Hz. The electrolyte, comprising 20 g/L $Ca(H_2PO_4)_2 \cdot H_2O$ and 13 g/L NaOH, was prepared using distilled water and used continuously during the treatment with 60 W (the ultrasonic power density was 0.3 W/cm^2), 40 kHz ultrasound. A sheet of 316L stainless steel with dimensions of 30 mm × 10 mm was used as the cathode. Samples treated in the electrolyte with and without ultrasound process for 5 min were rinsed in distilled water and dried in air, respectively.

2.2 Characterization

The specimens were examined by scanning electron microscope (SEM, JSM-6360LV) equipped with energy dispersive X-ray (EDX) analysis facility. The phase composition of the coatings was investigated by X-ray diffraction (XRD, Rigaku D/max-rB) using Cu Kα radiation, with a step size of 1° and a scan range from 10° to 90° (in 2θ). Polarization curves tested in simulated body fluid (SBF) were conducted using CHI660C potentiostat with a scan rate of 0.01 mV/s, from −2.5 V to 1 V. SBF was composed of 7.996 g/L NaCl, 0.35 g/L $NaHCO_3$, 0.224 g/L KCl, 0.228 g/L $K_2HPO_4 \cdot 3H_2O$, 0.305 g/L $MgCl_2 \cdot 6H_2O$, 0.278 g/L $CaCl_2$, and 0.071 g/L Na_2SO_4.

Table 1   Composition and mechanical properties of the magnesium alloy

Composition (wt%)					Mechanical property		
Zn	Zr	Mn	Ca	Mg	Tensile strength (MPa)	Yield strength (MPa)	Percentage elongation (%)
5.9	0.59	0.08	0.2	Bar	244.8	105.7	10.6

3　Results and discussion

3. 1　Morphologies and elemental distribution of the coatings

The surface and cross morphologies of coatings on magnesium alloy are shown in Fig. 1. The MAO coating without the ultrasonic treatment is shown in Figs. 1(a) and 1(c). Nonhomogeneous distribution of the micro-porous structure and uneven surface are seen and there is local cracking on the surface of coating in Fig. 1(a). The cross morphology shows inter-connective porous structure, and there is no obvious dense layer in Fig. 1(c). In theory, all of these structures do not offer good protective effect on the magnesium alloy substrate. The MAO coating with the ultrasonic treatment is shown in Figs. 1(b) and 1(d). Homogeneous distribution of micro-porous structure and even surface are seen in Fig. 1(b). What is more, local cracking on the surface of UMAO coatings disappears and the hole sizes increase compared with that of MAO coatings. The thickness of dense layer increases and independent hole is formed in Fig. 1(d).

The atomic force microscopy (AFM) analyses, performed with scanning area of $25\ \mu m^2$, show the cross topographies of MAO and UMAO coatings in Fig. 2. They are all composed of nano-particles, but the three-dimensional morphologies are different. There is island structure in MAO coating and column layer structure in UMAO coating, which indicates that the ultrasonic changes the growth pattern of coatings. The system of micro-arc oxidation and magnesium matrix with ultrasound obtains the energy ΔU from ultrasonic vibration. The energy provides the condition for the activation of magnesium matrix. So the initial reaction in MAO process happens as follows [24]:

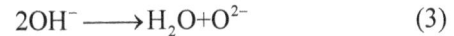

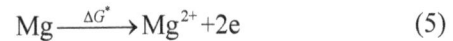

$$H_2O \xrightarrow{\Delta U} H^+ + OH^- \tag{1}$$

$$2H^+/H^+ \longrightarrow H_2/H_2O_2/H_2O \tag{2}$$

$$2OH^- \longrightarrow H_2O + O^{2-} \tag{3}$$

$$H_2O + Mg \longrightarrow MgO + H_2O \tag{4}$$

$$Mg \xrightarrow{\Delta G^*} Mg^{2+} + 2e \tag{5}$$

$$Mg^{2+} + O^{2-} \longrightarrow MgO \tag{6}$$

Magnesium matrix rapidly achieves the activation status when MAO system absorbs energy ΔG^* obtained from ultrasound, which makes electron transfer from magnesium matrix and form Mg^{2+} [28]. Mg^{2+} quickly passes through the solid phase MgO and arrives at the interface of magnesium and electrolyte. However, MgO is an insulation layer which hinders electron transfer. The effect of ultrasonic cavitation makes the movable carrier O^{2-} move to the interface and react with Mg^{2+} to form the dense layer [29]. Therefore, the thickness of dense layer of UMAO coating is greater

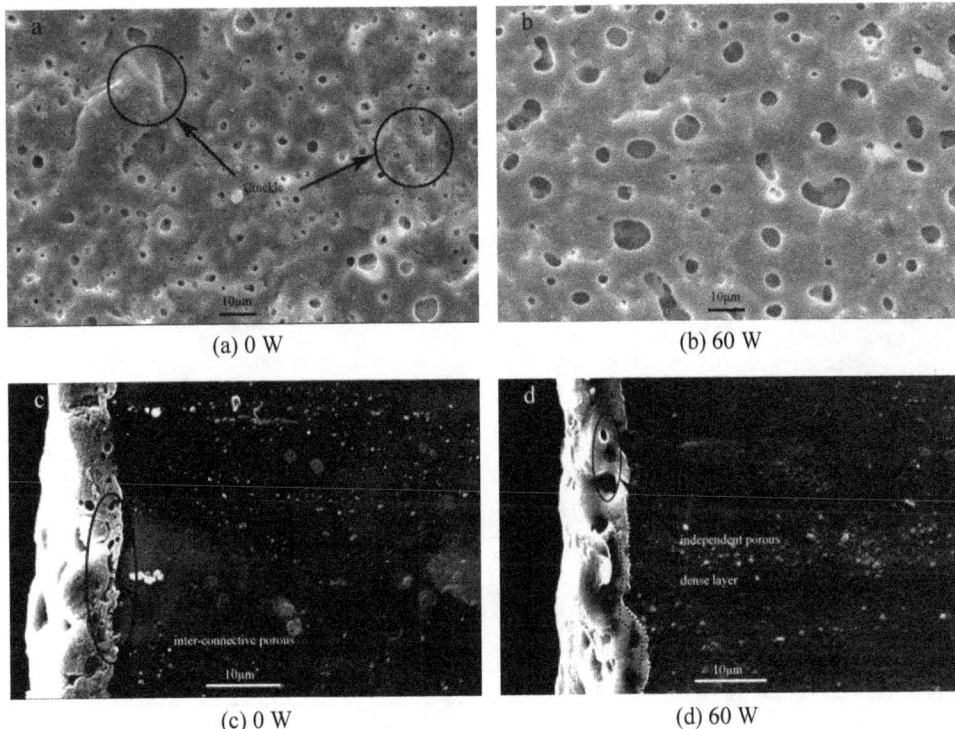

(a) 0 W　　　　　　　　　　　　　　　　(b) 60 W

(c) 0 W　　　　　　　　　　　　　　　　(d) 60 W

Fig. 1　Surface and cross morphologies of MAO and UMAO coatings.

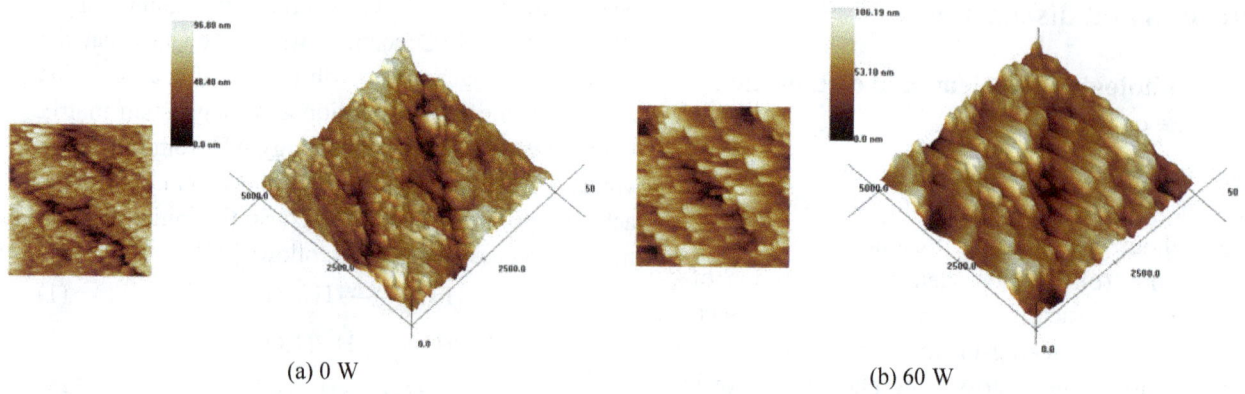

(a) 0 W   (b) 60 W

Fig. 2   AFM topographies of the cross coatings coated under different ultrasonic powers.

than that of MAO coating. At the same time, mechanical and cavitation effects of ultrasound cause damage to outward growth layer. The results of cross dimension show that the total thickness of UMAO coated under 60 W ultrasonic power is accordant with that of MAO coating (shown in Figs. 1(c) and 1(d)). MgO film is a necessary condition for micro-arc oxidation, so ultrasound accelerates the process of MAO to make the crystal energy increase. The higher crystal energy promotes crystal orientation growth, which shows different crystal morphology and size. The maximum crystal sizes of UMAO and MAO coatings are 106.19 nm and 96.80 nm, respectively.

The ratio of calcium and phosphorus (Ca/P) is calculated according to Fig. 3. Mg and O elements are detected and the electrolyte elements are also the constituents of the coatings. It can be found that the relative Ca/P ratio increases from 0.252 to 0.380 when

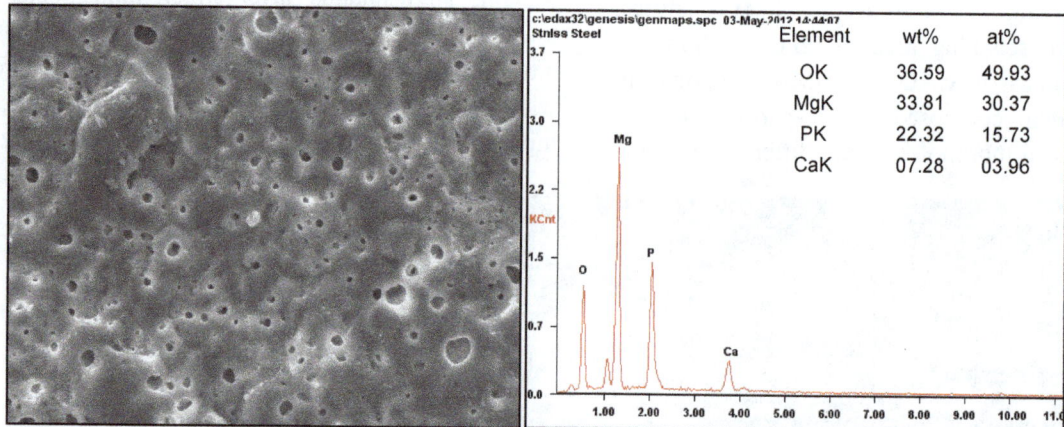

Element	wt%	at%
OK	36.59	49.93
MgK	33.81	30.37
PK	22.32	15.73
CaK	07.28	03.96

(a) 0 W

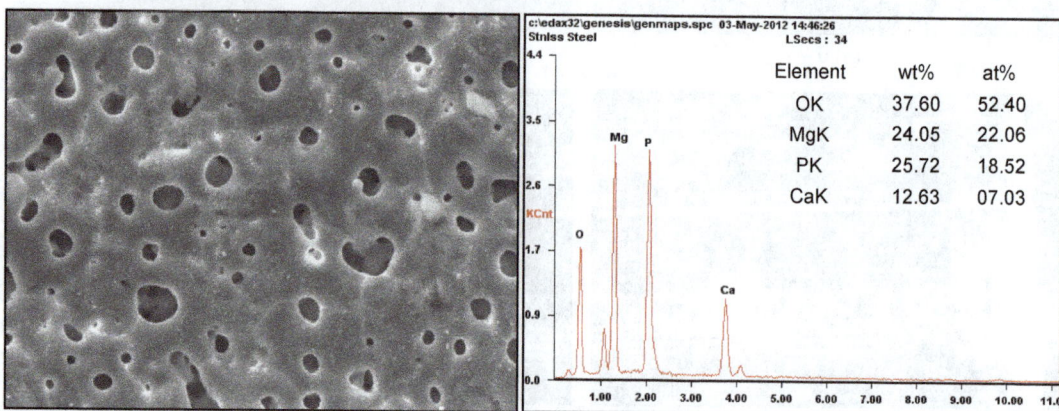

Element	wt%	at%
OK	37.60	52.40
MgK	24.05	22.06
PK	25.72	18.52
CaK	12.63	07.03

(b) 60 W

Fig. 3   EDX analysis of coatings coated under different ultrasonic powers.

the ultrasonic power is input. At the same time, the contents of calcium and phosphorus also increase. When ultrasound passes through an electrolyte, it produces rapidly fluctuating pressures to make rapid movement of the fluid and phosphate ions will be introduced into the coatings [30]. A thin oxidation film forms on the base metal at the beginning of oxidation process. Meanwhile, $Ca_3(PO_4)_2$ sol particles and PO_4^{3-} anions in the electrolyte are incorporated into oxidation film. Ultrasound in the MAO process increases the electrolyte temperature which makes $Ca_3(PO_4)_2$ sol particles generate violent collision with each other to form relative big sol particles [31]. The released energy of ultrasound by mechanical and cavitation effects lowers the critical voltage value of MAO, which makes the response of MAO be facilitated. In contrast, the forming high energy of ultrasound warms the anode and more electrical sparks take place on the vicinity of the anode. Under such a high energetic field, big sol particles $Ca_3(PO_4)_2$ and PO_4^{3-} ions in the electrolyte move toward the anode faster and enter into the coatings, which makes the quantity of $Ca_3(PO_4)_2$ increase in the coatings. The calcium and phosphorus film is deposited on the magnesium substrate and the Ca/P ratio of UMAO coating increases compared with that of MAO coating, which indicates that the reaction is activated more easily by the ultrasound agitation.

3.2 Composition of coatings

The phase composition of MAO coatings formed with and without ultrasonic power is examined, as shown in Fig. 4. This reveals that the coatings are mainly MgO. UMAO mainly includes two stages, namely the

ultrasound exciting stage and interaction of ultrasound and micro-arc oxidation. At the first stage, ultrasound plays the main role in the forming process of movable carrier O^{2-}, which accelerates the MAO reaction process. The phase of coatings is not changed though ultrasound promotes the coating in growth. At the same time, increased voltage in the second stage also makes the following reaction happen [32]:

$$H_2O \longrightarrow H^+ + OH^- \qquad (6)$$

$$2OH^- \longrightarrow H_2O + O^{2-} \qquad (7)$$

$$Mg^{2+} + O^{2-} \longrightarrow MgO \qquad (8)$$

No peak corresponding to calcium and phosphorus phases are detected by XRD. However, EDX analysis results indicate that there exist calcium and phosphorus elements in the coatings, as seen in Fig. 5. The reasons for the phenomenon can be the following two aspects: calcium and phosphorus exist in an amorphous phase due to the fast cooling rate of the meltdown thing, and the amount of calcium and phosphorus is too low to be

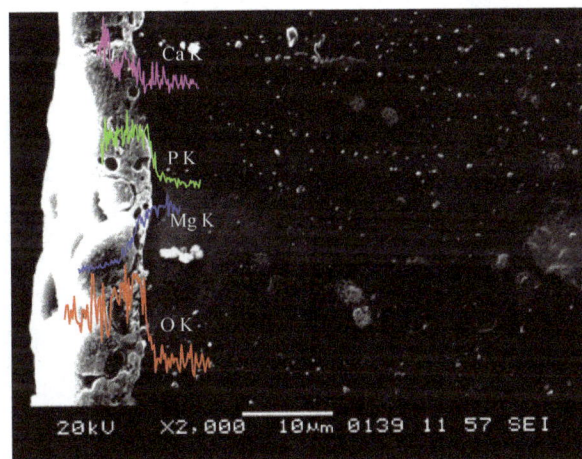

(a) 0 W

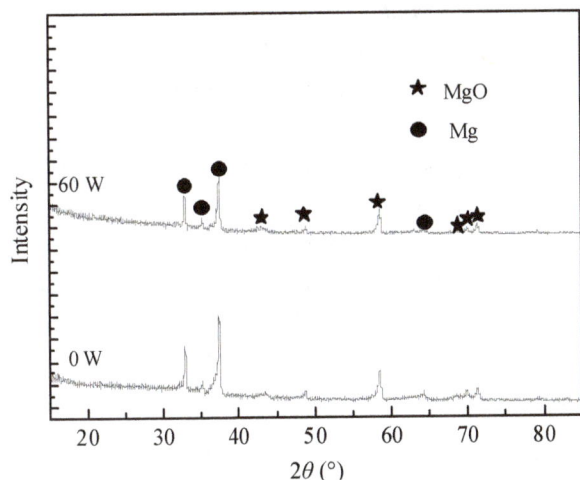

Fig. 4 XRD patterns of magnesium alloy coated under different ultrasonic powers.

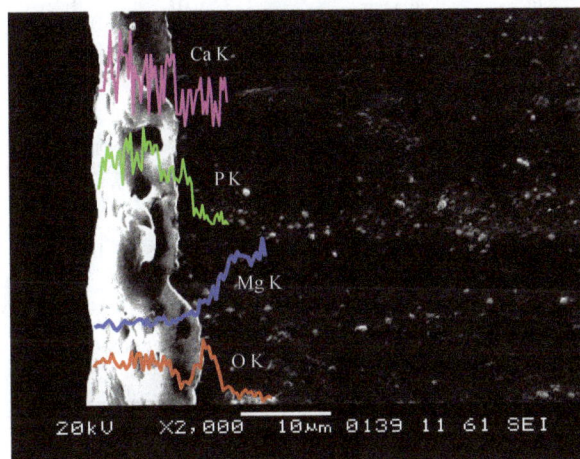

(b) 60 W

Fig. 5 EDX of coatings line scanning analysis.

detected by XRD.

3.3 Corrosion resistance

The polarization curves of different coatings under the electrochemical corrosion are examined compared to that of the substrate as shown in Fig. 6. All data are listed in Table 2. After the polarization tests, corrosion craters could be observed evidently on the surface of magnesium alloy substrate by low-magnification microscopy observation. Smooth MAO sample disappears, but there are no obvious changes. The flat UMAO coatings are still being though there is local corrosion pit (shown in Fig. 7). The E_{corr} and i_{corr} of the bare magnesium alloy was −1.583 V and 2.529× 10^{-5} A/cm^2, respectively. In comparison, the E_{corr} of MAO magnesium alloy obviously increases. The corrosion resistance of magnesium alloy is greatly improved by the MAO process. Furthermore, the E_{corr} of UMAO coated sample is −0.353 V and the i_{corr} is $1.560×10^{-6}$A/cm^2. From above results, it could be seen that the corrosion rates of the coated magnesium alloys by ultrasound are greatly reduced. It also could be concluded that UMAO coatings perform the best corrosion resistance. The anticorrosion property of the ceramic coatings is decided by microstructure and chemical composition, etc. [14]. As shown in Fig. 1, the microstructure of coatings is porous, so the corrosive medium could easily enter into the porous

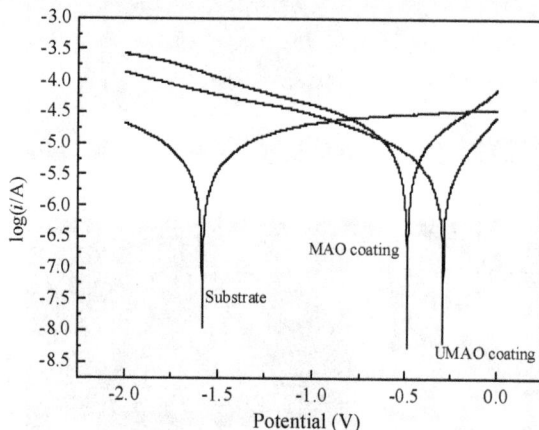

Fig. 6 Potentiodynamic polarization curves of the bare and coated magnesium alloy.

Table 2 Results of potentiodynamic corrosion tests in SBF

Film	E_{corr} (V)	i_{corr} (A/cm^2)	R_p (kΩ)
Bare magnesium	−1.583	2.529×10^{-5}	1.020
MAO magnesium	−1.098	5.188×10^{-6}	1.945
UMAO magnesium	−0.353	1.560×10^{-6}	4.888

(a) substrate

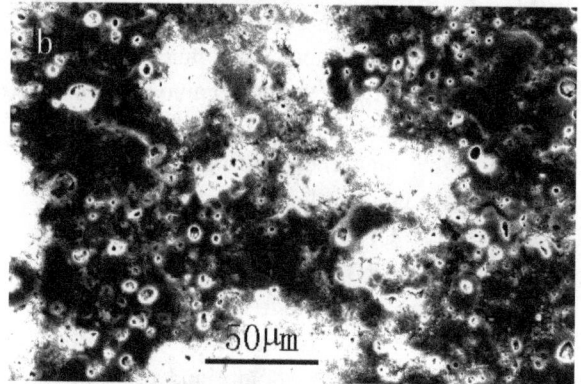

(b) MAO coating

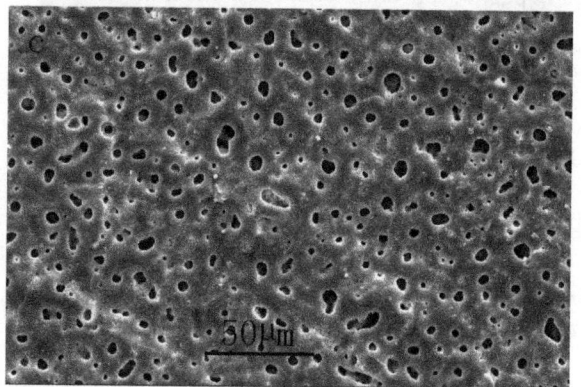

(c) UMAO coating

Fig. 7 Electrochemical corrosion morphologies of the bare and coated magnesium alloy.

film. However, ultrasound makes the movable carrier O^{2-} in electrolyte react with Mg^{2+} transferred from magnesium matrix to form the thicker dense layer, which offers more protection layer for magnesium substrate. Therefore, ultrasound in MAO could effectively improve the corrosion resistance.

4 Conclusions

(i) Ultrasonic improved the homogeneous distribution

of micro-porous structure, but did not change the phase of coatings. The coatings coated with and without ultrasonic were all porous MgO ceramic containing small amount of calcium and phosphorus.

(ii) Ultrasonic could increase Ca/P ratio of the ceramic coatings.

(iii) Ultrasonic could increase the corrosion resistance of magnesium alloy. Corrosion resistance of UMAO coatings was better than that of MAO coatings, and ultrasonic treatment during the process of MAO effectively decreased the corrosion rate of magnesium alloy.

Acknowledgements

The authors are grateful for the supports from the National Natural Science Foundation of China (No. 31070859) and the Natural Science Foundation of Heilongjiang Province (No. ZD201008).

References

[1] Witte F. The history of biodegradable magnesium implants: A review. *Acta Biomater* 2010, **6**: 1680–1692.

[2] Hermawan H, Dubé D, Mantovani D. Developments in metallic biodegradable stents. *Acta Biomater* 2010, **6**: 1693–1697.

[3] Alvarez-Lopez M, Pereda MD, del Valle JA, *et al.* Corrosion behaviour of AZ31 magnesium alloy with different grain sizes in simulated biological fluids. *Acta Biomater* 2010, **6**: 1763–1771.

[4] Li JN, Cao P, Zhang XN, *et al.* In vitro degradation and cell attachment of a PLGA coated biodegradable Mg–6Zn based alloy. *J Mater Sci* 2010, **45**: 6038–6045.

[5] Mandelli A, Bestetti M, Da Forno A, *et al.* A composite coating for corrosion protection of AM60B magnesium alloy. *Surf Coat Technol* 2011, **205**: 4459–4465.

[6] Chen H, Lv GH, Zhang GL, *et al.* Corrosion performance of plasma electrolytic oxidized AZ31 magnesium alloy in silicate solutions with different

additives. *Surf Coat Technol* 2010, **205**: S32–S35.

[7] Cai J, Cao F, Chang L, *et al.* The preparation and corrosion behaviors of MAO coating on AZ91D with rare earth conversion precursor film. *Appl Surf Sci* 2011, **257**: 3804–3811.

[8] Yang Y, Wu H. Effects of current frequency on the microstructure and wear resistance of ceramic coatings embedded with SiC nano-particles produced by micro-arc oxidation on AZ91D magnesium alloy. *J Mater Sci Technol* 2010, **26**: 865–871.

[9] Guo HF, An MZ. Growth of ceramic coatings on AZ91D magnesium alloys by micro-arc oxidation in aluminate–fluoride solutions and evaluation of corrosion resistance. *Appl Surf Sci* 2005, **246**: 229–238.

[10] Gu XN, Li N, Zhou WR, *et al.* Corrosion resistance and surface biocompatibility of a microarc oxidation coating on a Mg–Ca alloy. *Acta Biomater* 2011, **7**: 1880–1889.

[11] Liang J, Hu L, Hao J. Characterization of microarc oxidation coatings formed on AM60B magnesium alloy in silicate and phosphate electrolytes. *Appl Surf Sci* 2007, **253**: 4490–4496.

[12] Shi L, Xu Y, Li K, *et al.* Effect of additives on structure and corrosion resistance of ceramic coatings on Mg–Li alloy by micro-arc oxidation. *Curr Appl Phys* 2010, **10**: 719–723.

[13] Lin P, Zhou H, Li W, *et al.* Interactive effect of cerium and aluminum on the ignition point and the oxidation resistance of magnesium alloy. *Corros Sci* 2008, **50**: 2669–2675.

[14] Liang J, Guo B, Tian J, *et al.* Effects of NaAlO$_2$ on structure and corrosion resistance of microarc oxidation coatings formed on AM60B magnesium alloy in phosphate–KOH electrolyte. *Surf Coat Technol* 2005, **199**: 121–126.

[15] Chen J, Zeng R, Huang W, *et al.* Characterization and wear resistance of micro-arc oxidation coating on magnesium alloy AZ91 in simulated body fluids. *Trans Nonferrous Met Soc China* 2008, **18**: s361–s364.

[16] Zhong L, Cao F. Shi Y, *et al.* Preparation and corrosion resistance of cerium-based chemical conversion coating on AZ91 magnesium alloy. *Acta Metall Sin* 2008, **44**: 979–985 (in Chinese).

[17] Wu CS, Zhang Z, Cao FH, *et al.* Study on the anodizing of AZ31 magnesium alloys in alkaline borate solutions. *Appl Surf Sci* 2007, **253**: 3893–3898.

[18] Chang L, Cao F, Cai J, *et al.* Formation and transformation of Mg(OH)$_2$ in anodic coating using FTIR mapping. *Electrochem Commun* 2009, **11**:

2245–2248.

[19] Khaselev O, Yahalom J. The anodic behavior of binary Mg–Al alloys in KOH–aluminate solutions. *Corros Sci* 1998, **40**: 1149–1160.

[20] Srinivasan PB, Liang J, Blawert C. Characterization of calcium containing plasma electrolytic oxidation coatings on AM50 magnesium alloy. *Appl Surf Sci* 2010, **256**: 4017–4022.

[21] Xu L, Pan F, Yu G, *et al*. In vitro and in vivo evaluation of the surface bioactivity of a calcium phosphate coated magnesium alloy. *Biomaterials* 2009, **30**: 1512–1523.

[22] Liu GY, Hu J, Ding ZK, *et al*. Bioactive calcium phosphate coating formed on micro-arc oxidized magnesium by chemical deposition. *Appl Surf Sci* 2011, **257**: 2051–2057.

[23] Yang S, Qi M, Chen Y, *et al*. MAO–DCPD composite coating on Mg alloy for degradable implant applications. *Mater Lett* 2011, **65**: 2201–2204.

[24] Zhang Y, Lin SY, Fang Y. New developments in sonochemistry—Preparation of nanomaterials by ultrasound. *Physics* 2002, **31**: 80–83 (in Chinese).

[25] Mo RY, Lin SY, Wang CH. Methods of study on sound cavitation. *Appl Acoust* 2009, **28**: 389–400 (in Chinese).

[26] Cravotto G, Tagliapietra S, Robaldo B, *et al*. Chemical modification of chitosan under high-intensity ultrasound. *Ultrason Sonochem* 2005, **12**: 95–98.

[27] Aurousseau M, Pham NT, Ozil P. Effects of ultrasound on the electrochemical cementation of cadmium by zinc powder. *Ultrason Sonochem* 2004, **11**: 23–26.

[28] Morison SR. Wu HH, translator. *Electrochemistry of Semiconductor and Metal Oxide Film*. Beijing: Science Press, 1988: 95–96 (in Chinese).

[29] Wu HH. *Electrochemistry*. Beijing: Chemical Industry Press, 2004: 218 (in Chinese).

[30] Pandey AK, Kalsi PC, Iyer RH. Effects of high intensity ultrasound in chemical etching of particleyer tracks in solid state nuclear track detectors. *Nucl Instrum Meth B* 1998, **134**: 393–399.

[31] Wang XM, Zhu LQ, Liu HC, *et al*. Influence of surface pretreatment on the anodizing film of Mg alloy and the mechanism of the ultrasound during the pretreatment. *Surf Coat Technol* 2008, **202**: 4210–4217.

[32] Abbasi S, Bayati MR, Golestani-Fard F, *et al*. Micro arc oxidized HAp–TiO$_2$ nanostructured hybrid layers-part I: Effect of voltage and growth time. *Appl Surf Sci* 2011, **257**: 5944–5949.

3

Synthetic process and spark plasma sintering of SrIrO$_3$ composite oxide

Yongshang TIAN[a,b], Yansheng GONG[a,b,*], Zhaoying LI[a],
Feng JIANG[a], Hongyun JIN[a,b]

[a]*Faculty of Material Science and Chemistry, China University of Geosciences, Wuhan 430074,
People's Republic of China*
[b]*Engineering Research Center and Application of Nano-Geomaterials of Ministry of Education,
China University of Geosciences, Wuhan 430074, People's Republic of China*

Abstract: Single phase of SrIrO$_3$ powders and ceramics were obtained by solid-state chemical reaction method and spark plasma sintering (SPS) technique, respectively. Phase evolutions, characteristics, morphology and resistivity of the samples were studied by using thermogravimetric analysis–differential scanning calorimetry (TG–DSC), X-ray diffractometry (XRD), field emission scanning electron microscopy (FESEM) and four-point probe method, respectively. The results showed that the reaction process to form SrIrO$_3$ phase occurred between SrCO$_3$ and IrO$_2$ directly during the heating process. By using optimum fabrication conditions established from the TG–DSC results, single phase of SrIrO$_3$ powders was synthesized at 800–1000 ℃. SrIrO$_3$ ceramics were sintered by SPS technique at 1000–1100 ℃ with a pressure of 30 MPa, showing a high relative density of 92%–96% and dense microstructure. The room-temperature resistivity of SrIrO$_3$ ceramics was about 2×10^{-4} Ω·m. The present study can provide high-quality ceramic target for the preparation of SrIrO$_3$ films in traditional physical vapor deposition (PVD) method.

Keywords: SrIrO$_3$; powder; controllable synthesis; spark plasma sintering (SPS)

1 Introduction

Iridium oxide (IrO$_2$) has been applied as electrode and electrical conducting paste due to its excellent electrical conductivity. However, IrO$_2$ has a high volatility in air atmosphere due to the formation of volatile IrO$_3$ phase at high temperature (above 800 ℃), resulting in the degeneration of electrode [1,2]. In recent years, the 4d- and 5d-electron transition metal oxides (TMOs), e.g., the ruthenates and iridates, have received growing attention for their potential application in catalysis, electrochemistry and microelectronic devices [3–6]. The alkaline-earth iridates AIrO$_3$ (where A is the alkaline-earth element Ca, Sr or Ba) is an important system in oxide iridates [2,7], which can suppress the volatile nature of IrO$_2$ at high temperature.

* Corresponding author.
E-mail: gongyansheng@hotmail.com

According to McDaniel and Schneider's study on the equilibrium phase diagram of Sr–Ir–O in 1971, Sr–Ir–O compounds have three stable phases: Sr_4IrO_6, Sr_2IrO_4 and $SrIrO_3$ [8], whereas $SrIrO_3$ decomposes to Sr_2IrO_4 and Ir at 1205 ℃; Sr_2IrO_4 decomposes to Sr_4IrO_6 and Ir at 1445 ℃; Sr_4IrO_6 decomposes to SrO and Ir above 1540 ℃ [9]. $SrIrO_3$ has a monoclinic distorted hexagonal $BaTiO_3$ structure with a space group $C2/c$ and lattice parameters of $a = 0.5604$ nm, $b = 0.9618$ nm, $c = 1.4170$ nm and $\beta = 93.26°$ [10] at room temperature under atmospheric pressure.

Most studies among $SrIrO_3$ compound have been focused on thin film preparation by sol–gel, sputtering, pulse laser ablation, and so on [1,6,11–13]. Orthorhombic $SrIrO_3$ perovskite has also been synthesized at a high pressure (5 GPa), and its unusual magnetic characteristics are reported [7]. However, the synthesized processes of $SrIrO_3$ powders and its intrinsic properties, such as the electrical properties of $SrIrO_3$ sintered bulk bodies, have rarely been reported.

So far, dense sintered bodies of alkaline-earth iridates are still difficult to be obtained due to the evaporation of volatile oxide species (IrO_4/IrO_3) in conventional sintering process. For the purpose of application, a very dense ceramic is required. Spark plasma sintering (SPS) technique may consolidate alkaline-earth ruthenates or iridates bodies because the fast heating rate in SPS process may avoid the evaporation of volatile oxide species. Keawprak *et al.* [14] reported the thermoelectric properties of Sr–Ir–O compounds by SPS technique at 1100 ℃, showing a 81.5% relative density of $SrIrO_3$ body, indicating the feasibility of SPS in alkaline-earth iridates compound sintering. However, the relative density of $SrIrO_3$ body still needs to be improved and the SPS process of $SrIrO_3$ ceramics also needs to be further studied.

In the present work, the synthesized conditions of pure $SrIrO_3$ powders derived by solid-state chemical reaction method were studied in detail. In addition, SPS technique was employed to increase the ceramic density, and the effect of sintering temperature on crystal structure, microstructure and resistivity of $SrIrO_3$ ceramics were reported.

2 Experiment

$SrIrO_3$ powders were synthesized by using conventional solid-state chemical reaction method.

Strontium carbonate ($SrCO_3$, 99.9%) and iridium oxide (IrO_2, 86.0%) were used as the raw materials. Stoichiometric mixed powders were ground more than half an hour in an agate mortar, then placed into a muffle furnace (JML-5.4-1.6), and calcined for 9 h at a specific temperature. The synthesized powders were removed into a graphite die (Φ10 mm) and sintered in an SPS equipment (SPS-1050). The sintering temperature was from 900 ℃ to 1050 ℃ for 10 min with a uniaxial pressure of 30 MPa and a heating rate of 150 ℃/min. The carbon diffusion layers on the $SrIrO_3$ ceramic surface were removed to avoid any contamination before different detection.

The crystal structure and phase composition were examined by X-ray diffraction (XRD, X'Pert PRO) with Cu Kα radiation at room temperature. Field emission scanning electron microscopy (FESEM, SUV-1080) was used to characterize the morphology of $SrIrO_3$ powders and ceramics. The cumulative distribution of $SrIrO_3$ particles was tested by laser particle size analyzer (JL-1155). The density of ceramics was measured by the Archimedes immersion method. The room-temperature electrical resistivity of the $SrIrO_3$ ceramics was measured by Hall test system (Accent HL 5500 PC).

3 Results and discussion

Figure 1 shows the $SrCO_3$ powders' thermogravimetric (TG) curve and thermogravimetric analysis–differential scanning calorimetry (TG–DSC) curves (at the heating rate of 20 ℃/min in atmosphere) of $SrCO_3$ and IrO_2 mixed powders with mole ratio ($R_{Ir/Sr}$) of 1. From TG–DSC curves of the as-mixed powders, it could be seen that evaporation of free water occurrs below 150 ℃, corresponding to a 0.97% weight loss; the reaction between $SrCO_3$ and IrO_2 to form single-phase $SrIrO_3$ is observed from three exothermic peaks at about 150 ℃ to 850 ℃ with a 12.62% weight loss. No apparent peak and weight loss can be observed at the temperature over 800 ℃. In addition, from the $SrCO_3$ powders' TG curve, it can be deduced that the reaction process to form $SrIrO_3$ phase occurs between $SrCO_3$ and IrO_2 directly, since there is no endothermic peak of the decomposition of $SrCO_3$ during the heating process. From the TG–DSC curves, the proper synthesized temperature of $SrIrO_3$ powders is about 800 ℃.

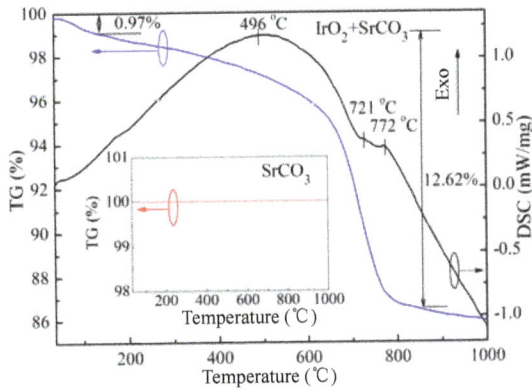

Fig. 1   TG–DSC curves for mixed powders of SrCO$_3$ and IrO$_2$ at $R_{Ir/Sr}$=1. The inset shows the SrCO$_3$ powders' TG curve.

Figure 2 shows the XRD patterns of synthesized SrIrO$_3$ powders at a calcination temperature of 800–1000 °C with $R_{Ir/Sr}$ = 1. The standard XRD pattern of SrIrO$_3$ is also shown in Fig. 2 for comparison. Every peak in the patterns at the temperature from 800 °C to 1000 °C can be attributed to SrIrO$_3$ phase. This indicates that the powders are single phase of monoclinic SrIrO$_3$, which starts to form from about 800 °C and is in accordance with the TG–DSC results. Some of the peaks are not obvious at the temperature of 800 °C, while all the characteristic peaks appear and the intensity of the diffraction peaks improves with increasing calcination temperature. The appropriate temperature of synthesizing SrIrO$_3$ powders is identified at 850 °C and the lattice parameters of synthesized SrIrO$_3$ are a = 0.5617 nm, b = 0.9621 nm, c = 1.4089 nm, and β = 93.30°, which is calculated according to the XRD results of synthesized SrIrO$_3$ powders at 850 °C.

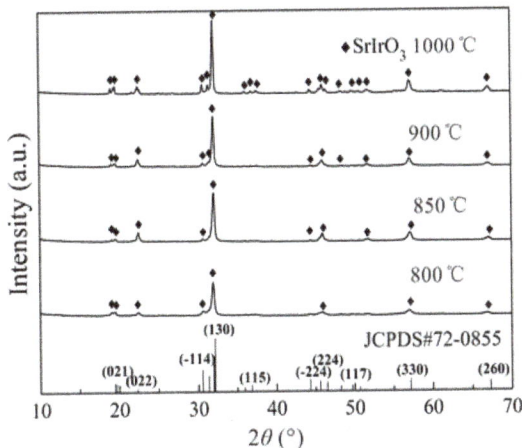

Fig. 2   XRD patterns of synthesized SrIrO$_3$ powders at different calcination temperatures.

Figure 3 presents the FESEM image of SrIrO$_3$ powders, which was synthesized at 850 °C and dispersed by ultrasound in alcohol for 30 min. It could be seen that the particle size of SrIrO$_3$ powders is almost uniform, showing a particle size of about 0.4 μm. The cumulative distribution of SrIrO$_3$ particles with different size is shown in Fig. 4. The particle size distribution obeys normal distribution through differential calculation, and the calculated particle sizes of D$_{50}$ and D$_{AV}$ are 0.75 μm and 1.21 μm, respectively, which are a little higher than the FESEM image.

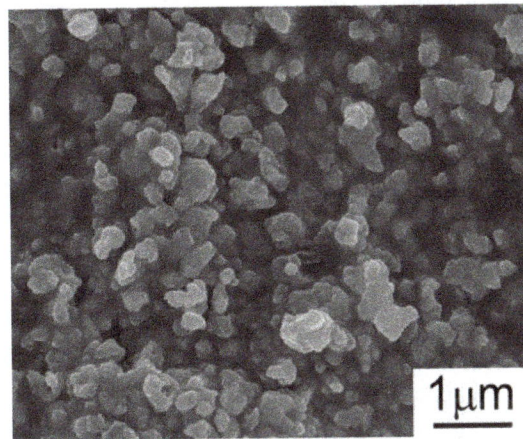

Fig. 3   FESEM image of synthesized SrIrO$_3$ powders at 850 °C.

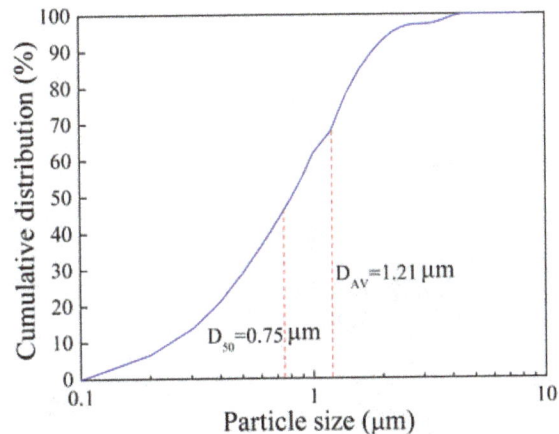

Fig. 4   The cumulative distribution of synthesized SrIrO$_3$ powders at 850 °C.

Figure 5 shows the XRD patterns of SrIrO$_3$ ceramics by SPS technique at the sintering temperature of 1000–1100 °C under a uniaxial pressure of 30 MPa. From the patterns, it can be found that the main diffraction peaks are consistent well with JCPDS card 72-0855 for SrIrO$_3$. Compared with SrIrO$_3$ powders,

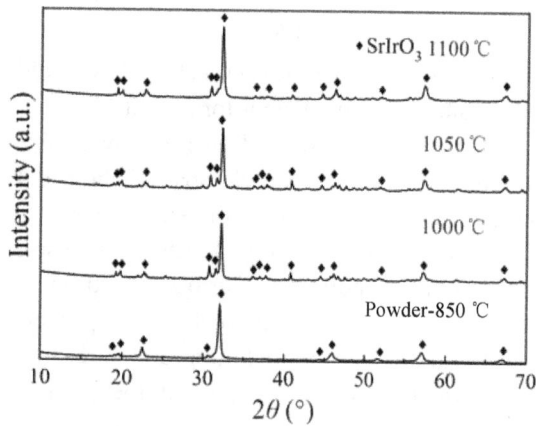

Fig. 5　XRD patterns of SrIrO$_3$ ceramics by SPS technique at different conditions.

the crystallinity of SrIrO$_3$ ceramics are significantly strengthened.

Figure 6 presents the FESEM images of SrIrO$_3$ ceramic fracture surface sintered at 1000–1100 ℃ by SPS technique. The relative densities of SrIrO$_3$ ceramics sintered at 1000 ℃, 1050 ℃ and 1100 ℃ are 93.9%, 96.2% and 92.7%, respectively, which are much denser than the literature [14]. From the FESEM images, it could be seen that a dense structure forms and the average grain size is about 0.6–1.0 μm in length, indicating that the grain growth is slight in comparison with the grain size of the SrIrO$_3$ powders. However, there still exist some nanoscale grains in the sintered bodies. The proper sintering temperature of SrIrO$_3$ ceramics is fixed at 1050 ℃ according to the relative densities.

Fig. 6　FESEM images of SrIrO$_3$ ceramic fracture surface sintered at 30 MPa and (a) 1000 ℃, (b) 1050 ℃, (c) 1100 ℃ by SPS technique.

Figure 7 shows the typical sintering displacement and heating curves of the SrIrO$_3$ ceramics with sintering time variation. At the initial stage, the powders are swelled with increasing temperature, while the powders are contracted rapidly with the intervention of pressure. In the heating preservation stage, the densification of SrIrO$_3$ ceramics is almost finished above 900 ℃. Figure 8 shows the shrinking rate of SrIrO$_3$ ceramics with the variation of sintering temperature in SPS process. The low or high speed of powder expansion and ceramic contraction could be found from the shrinking rate curve. The variation of shrinking rate below 600 ℃ is due to the powder expansion, which is caused by the increasing temperature and the powder inherent expansion properties. When the pressure gradually adds to 30 MPa, the shrinking rate increases significantly with increasing sintering temperature, which starts from 700 ℃ and ends at 920 ℃. After that, the densification process reaches to the balance. All the processes are consistent with sintering displacement and heating

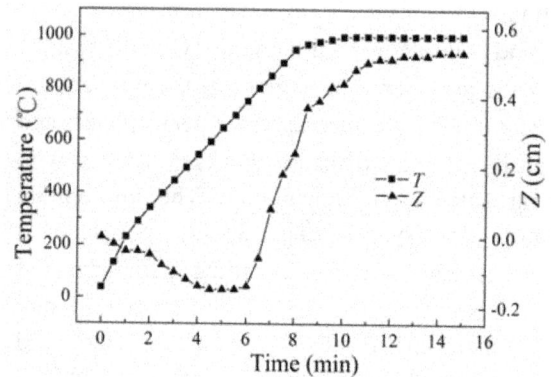

Fig. 7　Typical sintering displacement and heating curves of the SrIrO$_3$ ceramics by SPS technique.

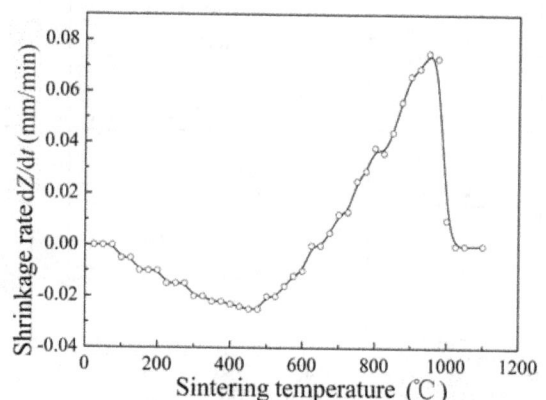

Fig. 8　Shrinkage rate of SrIrO$_3$ ceramics sintered by SPS with increasing sintering temperature.

curves of sintering $SrIrO_3$ ceramics (Fig. 7), and the density of $SrIrO_3$ ceramics reaches the maximum at the setting condition.

The room-temperature electrical resistivity and Ir/Sr ratio ($R_{Ir/Sr}$) of $SrIrO_3$ ceramics are shown in Table 1. The results of $R_{Ir/Sr}$ were obtained by energy-dispersive spectrometer (EDS), showing a nearly stoichiometric ratio at the sintering temperature of 1000–1100 ℃. The nonstoichiometric ratio may due to the volatility of iridium oxide at high sintering temperature. The electrical resistivity is about 2×10^{-4} Ω·m, showing a little higher value than the literature [14,15], which is probably due to the coulomb-scattering mechanism on carrier at the boundary causing by the smaller grains (Fig. 6) [16]. In addition, the SPS $SrIrO_3$ ceramics in the present study show a high relative density and dense microstructure, which could be used as a ceramic target to satisfy the demand of preparation of $SrIrO_3$ films in physical vapor deposition (PVD) method.

Table 1   Relative density, $R_{Ir/Sr}$ and electrical resistivity of $SrIrO_3$ ceramics

Temperature (℃)	Relative density (%)	$R_{Ir/Sr}$	Bulk resistivity (10^{-4}Ω·m)
1000	93.9	0.953	2.049
1050	96.2	0.989	2.032
1100	92.7	0.946	2.194

4   Conclusions

In the current work, nearly stoichiometric $SrIrO_3$ powders and ceramics were obtained by solid-state chemical reaction method and SPS technique, respectively. $SrIrO_3$ powders could be synthesized through the reaction of $SrCO_3$ and IrO_2 in a wide range of temperature (800–1000 ℃). With increasing calcination temperature, the crystallinity of the as-prepared $SrIrO_3$ powders was improved significantly. The optimum SPS condition for the dense $SrIrO_3$ ceramics was 1050 ℃ with a pressure of 30 MPa, showing a relative density of about 96.2% and electrical resistivity of about 2×10^{-4} Ω·m.

Acknowledgements

This work was financially supported by Hubei Provincial Nature Science Found of China (2011CDB331), State Key Laboratory of Advanced Technology for Materials Synthesis Processing (Wuhan University of Technology, 2012-KF-3), and the Fundamental Research Founds for National University, China University of Geosciences (Wuhan) (CUG120118).

References

[1] PauPorté T, Aberdam D, Hazemann J-L, *et al*. X-ray absorption in relation to valency of iridium in sputtered iridium oxide films. *J Electroanal Chem* 1999, **465**: 88–95.

[2] Keawprak N, Tu R, Goto T. Thermoelectricity of $CaIrO_3$ ceramics prepared by spark plasma sintering. *J Ceram Soc Jpn* 2009, **117**: 466–469.

[3] Xu W, Zheng L, Xin H, *et al*. $BaRuO_3$ thin films prepared by pulsed laser deposition. *Mater Lett* 1995, **25**: 175–178.

[4] Choi KJ, Baek SH, Jang HW, *et al*. Phase-transition temperatures of strained single-crystal $SrRuO_3$ thin films. *Adv Mater* 2010, **22**: 759–762.

[5] Cao G, Durairaj V, Chikara S, *et al*. Non-Fermi-liquid behavior in nearly ferromagnetic $SrIrO_3$ single crystals. *Phys Rev B* 2007, **76**: 100402.

[6] Liu Y, Masumoto H, Goto T. Structural, electrical and optical characterization of $SrIrO_3$ thin films prepared by laser-ablation. *Mater Trans* 2005, **46**: 100–104.

[7] Zhao JG, Yang LX, Yu Y, *et al*. High-pressure synthesis of orthorhombic $SrIrO_3$ perovskite and its positive magnetoresistance. *J Appl Phys* 2008, **103**: 103706.

[8] McDaniel CL, Schneider SJ. Phase relation in the $SrO–IrO_2–Ir$ system in air. *J Res NBS A Phys Ch* 1971, **75A**: 185–196.

[9] Jacob KT, Okabe TH, Uda T, *et al*. Phase relations in the system Sr–Ir–O and thermodynamic measurements on $SrIrO_3$, Sr_2IrO_4 and Sr_4IrO_6 using solid-state cells with buffer electrodes. *J Alloys Compd* 1999, **288**: 188–196.

[10] Longo JM, Kafalas JA, Arnott RJ. Structure and properties of the high and low pressure forms of $SrIrO_3$. *J Solid State Chem* 1971, **3**: 174–179.

[11] Sumi A, Kim YK, Oshima N, *et al*. MOCVD growth of epitaxial $SrIrO_3$ films on (111) $SrTiO_3$ substrates. *Thin Solid Films* 2005, **486**: 182–185.

[12] Jang SY, Kim H, Moon SJ, *et al*. The electronic structure of epitaxially stabilized $5d$ perovskite $Ca_{1-x}Sr_xIrO_3$ ($x = 0$, 0.5, and 1) thin films: The role of strong spin-orbit coupling. *J Phys: Condens Matter* 2010, **22**: 485602.

[13] Jang SY, Moon SJ, Jeon BC. PLD growth of epitaxially-stabilized $5d$ perovskite $SrIrO_3$ thin films. *J Korean Phys Soc* 2010, **56**: 1814–1817.

[14] Keawprak N, Tu R, Goto T. Thermoelectric properties of Sr–Ir–O compounds prepared by spark plasma sintering. *J Alloys Compd* 2010, **491**: 441–446.

[15] Qasim I, Kennedy BJ, Avdeev M. Synthesis, structures and properties of transition metal doped $SrIrO_3$. *J Mater Chem A* 2013, **1**: 3127–3132.

[16] Zhang J, Tu R, Goto T. Fabrication of transparent SiO_2 glass by pressureless sintering and spark plasma sintering. *Ceram Int* 2012, **38**: 2673–2678.

4

Structural and photocatalytic characteristics of TiO$_2$ coatings produced by various thermal spray techniques

Pavel CTIBOR[a,*], Vaclav STENGL[b], Zdenek PALA[a]

[a]Institute of Plasma Physics, ASCR, v.v.i., Za Slovankou 3, Prague, Czech Republic
[b]Institute of Inorganic Chemistry, ASCR, v.v.i., Husinec-Rez, Czech Republic

Abstract: Titanium dioxide (TiO$_2$) was elaborated by four different thermal spray techniques—(i) plasma spraying using a water-stabilized torch, (ii) plasma spraying using a gas-stabilized torch, (iii) high velocity oxy–fuel gun, and (iv) oxy–acetylene flame. The porosity of the coatings was studied by optical microscopy, nano-structural features by scanning electron microscopy (SEM), phase composition by X-ray diffraction (XRD); the microhardness, surface roughness and wear resistance were evaluated. The diffuse reflectance was measured by ultra-violet/visible/near-infrared (UV/Vis/NIR) scanning spectrophotometer. The kinetics of photocatalytic degradation of gaseous acetone was measured under a UV lamp with 365 nm wavelength. After all the applied spray processes, the transformation of anatase phase from the initial powders to rutile phase in the coatings occurred. In spite of this transformation, all the coatings exhibited certain photocatalytic activity, which correlated well with their band gap energy calculated from reflectivity. All the coatings offer relatively good mechanical properties and can serve as robust photocatalysts.

Keywords: plasma spraying; high velocity oxy–fuel (HVOF) spraying; flame spraying; titanium dioxide (TiO$_2$); photocatalysis; band gap

1 Introduction

Titanium dioxide (TiO$_2$) photocatalyst has emerged as a promising clean advanced oxidation technology, which could address the ever-increasing global concerns for environmental pollution. TiO$_2$ is an important material for ultra-violet (UV) assisted photocatalysis in the state of fine powders [1–3], thin coatings [3,4] as well as thermally sprayed coatings [5–9], and useful for water and air purification in advanced electrochemical applications. The demand for robust photocatalytic layers is initiated with multiple technical problems in handling with ultrafine powders and thin coatings. Here is a challenge for thermal spray techniques.

The thermal spray techniques used for the current work differ markedly in temperature, gases participating in the employed chemical processes, particle velocity and feed rate, etc. All these factors have influence on the resulting coatings [10–13].

The **high velocity oxy–fuel** (HVOF) process utilizes a combination of oxygen with various fuel gases, including hydrogen, propane, propylene, and kerosene. In the combustion chamber, burning products are expanded and expelled outward through an orifice with very high velocities. Powders to be sprayed via HVOF

* Corresponding author.
E-mail: ctibor@ipp.cas.cz

are injected axially into the expanding hot gases where they are propelled forward, heated and accelerated onto a surface to form a coating. Gas velocities are supersonic, whereas the temperature is about 2300 ℃. The coupling of highly plasticized particles can achieve coatings approaching the theoretical density. Disadvantages include relatively low deposition rates. Recent efforts have been focused on applying thick coatings and improvements in the process control.

Powder **flame spraying** is probably the simplest of all the spray processes—powders are fed through the center bore of a nozzle where they melt and are carried by the escaping oxy–fuel gases to the work piece. Unfortunately, this approach yields coatings with high porosity content. The density of the supporting gas influences the feed rate, and there is an optimum amount that can be carried in a gaseous stream for any particular powders. It depends on the velocity and volume of the gases used.

Plasma is an ionized gaseous jet composed of free electrons, positive ions, neutral atoms and molecules. In a commercial technology, plasma is considered a hot stream reaching temperatures greater than 10 000 ℃. The generator is essentially an electric arc working in a constricted space. Two electrodes, front (anode) and rear (cathode), are contained in a chamber, as is the arc through which the operating gas passes. Plasma generators work on the principle that—if sufficient voltage is applied to the two electrodes, separated by a small gap—the gas in the gap is ionized, becoming electrically conductive. Plasma subsequently exits the chamber as a plasma jet. This concept represents a **gas-stabilized plasma** (GSP) torch.

Water-stabilized plasma (WSP) torch has the advantage of combining stabilizing system and cooling system in one. Water is fed into a chamber instead of gases, where it creates a swirl around the walls. Electric arc is burning in the center of the channel formed by the swirl, between the graphite cathode inside the chamber and a rotating copper anode outside. The anode is internally water-cooled. Feedstock powders are introduced into the plasma jet outside the gun using two injectors, which can be positioned in various distances (feeding distance, FD) from the exit nozzle. The oxidizing power of oxygen released from the water is balanced by the reducing power of hydrogen, thus the overall chemical influence of the stabilizing water in the flame is rather neutral. In general, the structures of coatings made by GSP and WSP are relatively similar.

The main goal of the present paper is to investigate spraying TiO_2 by all techniques described above. TiO_2 represents a ceramic material with moderate refractory character, medium hardness and toughness. It is—according to our previous experience (mainly based on WSP spraying)—suitable for thick coatings. From the coating's mechanical integrity point of view, TiO_2 is relatively insensitive to spraying parameters; however, it is sensitive to reduction, undergoes phase transformation at cooling, and its numerous physical properties can be tailored only after studying the phenomena associated with the individual spraying methods.

2 Experiment

2.1 Powders and spraying

TiO_2 powder feedstock for thermal spray (Altair Nanomaterials, Reno, NV, USA) was prepared from nanometric (~ 20 nm) powders by agglomeration. The fraction of 75–106 μm in size was utilized for WSP spraying experiments. Samples were produced using the spraying system WSP 500 (Institute of Plasma Physics, Prague, Czech Republic). Carbon steel as well as stainless steel coupons was used as substrates. Some of the coatings were then stripped off for further characterizations. The spraying parameters of WSP are in Table 1. Argon was used as a powder feeding gas—a pressure of 2.5 bar and a flow rate of 3.25 slpm were employed. Argon was also used for cooling the substrate and the just deposited coating. The cooling tube was installed on a robot, and after each pass, it copied the movement pattern of the spray gun over the substrate. The temperature was monitored by a two-color pyrometer not to exceed 250 ℃.

An overview of the used spraying parameters for all other techniques is also presented in Table 1. Where not indicated, compressed air was used as the powder feeding gas. TopGun Medipar 02–555 (Medicoat, Maegenwil, Switzerland) was used for HVOF, plasma torch Plasmatechnik F4 (Sulzer Metco, Winterthur, Switzerland) was used for GSP spraying, and conventional flame spray equipment was employed. All processes other than WSP were performed at SAM Ltd. (Miletin, Czech Republic).

Table 1   Parameters used for spraying

Process	Feeding distance, FD (mm)	Spray distance, SD (mm)	Torch power (kW)	Gases used	Substrate temperature (℃)	Powder feed rate (kg/h)	Feedstock powder size (μm)	Feedstock anatase content (%)[**]
WSP	120	400	150	H$_2$O vapor, Ar feeding	250	15	75–106	95
GSP	—	130	9	Ar + H$_2$ (80/35[*]), Ar feeding	130	3	20–50	77
HVOF	—	220	—	oxygen, ethylene C$_2$H$_4$; N$_2$ feeding	170	1.2	75–106	95
Flame	—	220	—	oxygen, acetylene C$_2$H$_2$	—	—	20–50	77

* flow rates in standard liters per minute; ** the balance is rutile.

Different size distributions of the powders for the different spraying techniques are necessity. The fact that TiO$_2$ powders with different sizes also have different anatase contents (verified by X-ray diffraction (XRD)) is a producer-governed feature, difficult to avoid if we work with a commercial material.

The parameters for spraying with various techniques were selected having in mind the main target to obtain coatings relatively comparable with the WSP coating from the viewpoint of the oxygen non-stoichiometry. At both GSP and WSP spraying, titania is usually reduced and the coating has a lower content of oxygen compared to the starting powders [14]. In the case of our GSP spraying, a very low power level was selected as a means contributing to the oxygen loss. Low power level represented a low powder temperature. Low feed rate was used at the same time. A small quantity of powder particles in a plasma plume with low temperature was supposed to maximize the influence of plasma-constituent gases and minimize the influence of the ambient air surrounding the plasma plume. The temperature of substrate was kept relatively low (150 ℃) to support cooling. As a result, oxygen-deficient coating was expected.

HVOF parameters were selected also with the aim to support reduction of the powders. Relatively low feed rate was used together with nitrogen as a powder feeding gas, both with the aim to limit the contact of the powders with the ambient air.

For the same reason, a reductive flame with high acetylene-to-oxygen ratio was applied in the case of flame spraying.

2. 2   Characterization techniques

The porosity was studied by optical microscopy on polished cross sections. The micrographs were taken with a CCD (charge-coupled device) camera and processed using an image analysis (IA) software (Lucia G, Laboratory Imaging, Czech Republic). Ten images of microstructures taken from various areas of the cross section for each sample were analyzed. The magnification was 250 in all cases. For a better description of the porosity, certain additional criteria were introduced [16].

The microstructures of selected coatings were also studied by a scanning electron microscope (SEM) EVO® MA 15 (Carl Zeiss SMT Inc., Germany) with point resolution of 2 nm at 30 kV. TiO$_2$ coatings made by GSP were analyzed also by high-resolution transmission electron microscope (HR-TEM) Jeol JEM 3010.

XRD was performed on a 2-theta Bragg–Brentano diffractometer SIEMENS D500, using cobalt Kα radiation to obtain information about the phases present in the feedstock powders and coatings. A "beamknife" attachment improving the peak-to-background ratio was employed.

The microhardness was measured by a Hanemann microhardness head (Carl Zeiss, Germany) mounted on an optical microscope with a fixed load of 1 N and a Vickers indenter. Twenty indentations in various areas of the cross section for each sample were made.

The slurry abrasion response (SAR) of coatings was measured based on an ASTM standard [17] using an in-house designed apparatus. SAR test was based on measuring the mass loss rate of a standard-shaped block (7.5 mm × 12 mm × 25 mm) lapped in slurry. The test was run for 8 h in 2 h increment with mass loss being measured at the end of each increment. The applied force was 22 N per specimen. After each run, the samples were ultrasonically cleaned and weighted. The slurry consisted of 150 g water and 150 g alumina powders having size of 40–50 μm. The accuracy of measurement was approximately ±5%. The parameter "inverse wear rate" was higher for a coating with better wear resistance.

The diffuse reflectance was measured by an ultra-violet/visible/near-infrared (UV/Vis/NIR) scanning spectrophotometer (Shimadzu, Japan) with a multi-purpose large sample compartment. Prior to the actual measurement, the calibration was conducted using the $BaSO_4$ reference mirror in order to minimize the possible error from the experimental environment. The measurement accuracy guaranteed by the equipment manufacturer was as follows: the accuracy of wavelength was less than ±0.3 nm, and the uncertainty of the measurement was less than 0.2%. The wavelength of incident light used for the reflectance measurement was in the range from 250 nm to 2000 nm, and the diameter of the measured area was about 20 mm. The corresponding band gap width was estimated based on the values converted to absorbance and recalculated to band gap energy E_{bg} [18].

The kinetics of photocatalytic degradation of gaseous acetone was measured using a self-constructed stainless steel photoreactor with a fluorescent lamp Narva LT8WT8/073BLB and a black lamp with a wavelength of 365 nm and input power of 8 W (light intensity 6.3 mW/cm^2). The gas concentration was measured with a quadrupole mass spectrometer JEOL JMS-Q100GC and gas chromatograph Agilent 6890N. A high-resolution gas chromatography column (19091P-QO4, J&W Scientific) was used. The sample from the reactor was taken via a sampling valve at a time interval of 2 h. The reactor with a total volume of 3.5 L was filled with oxygen at a flow rate of 1 L/min.

3 Results and discussion

The microstructures of coatings are shown in Figs. 1(a)–1(d). The coatings have different thicknesses—only about 150 μm for HVOF and flame spraying, whereas plasma-sprayed coatings are markedly thicker.

The HVOF coating contains unmelted particles with nano-structure exhibiting several tens of nanometer-large individual particles separated by fine pores (Fig. 2(a)). The structure of this coating is bimodal, similar to another publication [15]—besides the mentioned nanostructural particles, the rest of the coating exhibits a microstructure with splats built from completely resolidified TiO_2. This feature is not observed in any

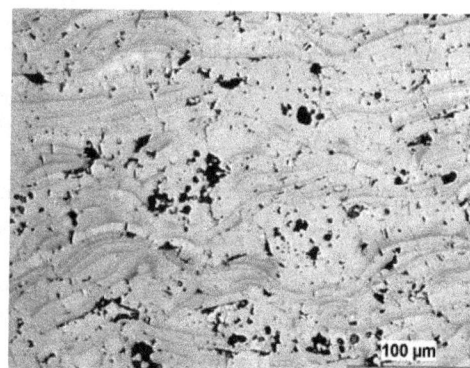

(a)

(b)

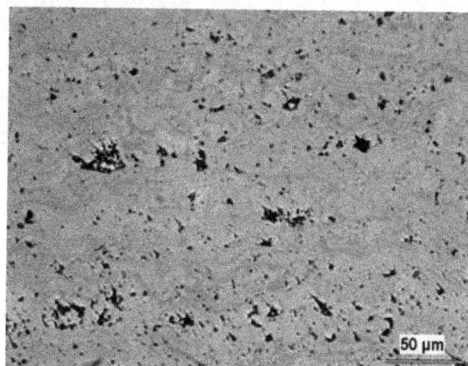

(c)

(d)

Fig. 1 Optical microscopy of the cross-sections of (a) WSP coating, (b) flame-sprayed coating, (c) GSP coating, and (d) HVOF coating.

other coatings in our set of experiments. The reason is that the high velocity of the particles impacting under relatively cold conditions onto the substrate, results in the disintegration of the agglomerated powders.

Certain splat-to-splat variability of the level of grey is visible in the GSP coating (Fig. 2(b)), most probably associated with the subtle variation in stoichiometry. The WSP and flame-sprayed coatings exhibit some globular particles disturbing the general arrangement

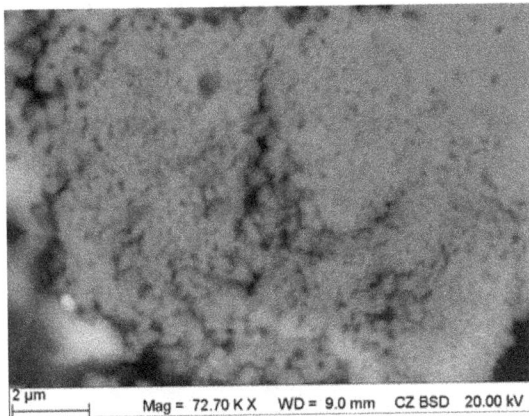

(a)

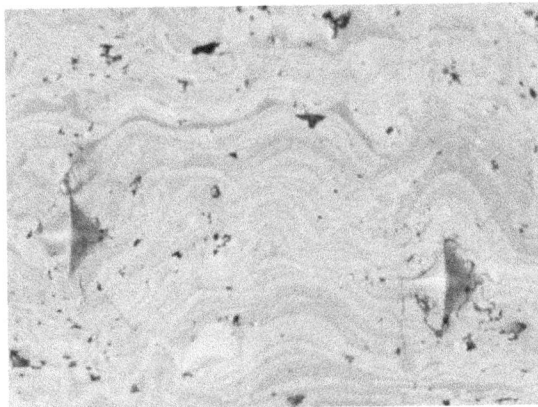

(b)

Fig. 2   (a) Detail of the cross section of HVOF coating; (b) detail of two indents on the cross section of the GSP coating.

of lamellas aligned perpendicularly to the spraying direction. Such particles are not flattened well at impact because of the low impact velocity, but they retain the original size and shape of the feedstock.

The surface roughness is the biggest for WSP coating and similar for all other coatings (smaller than WSP) (Table 2). Finer starting powders lead typically to finer roughness, and comparing the particle impact velocity of WSP and HVOF, the second one is markedly higher leading to finer roughness. The porosity is similar for WSP and flame-sprayed coatings, whereas it is lower for HVOF and even lower for GSP. The mean pore size for HVOF is the lowest due to a large impact velocity resulting in good splat contact, whereas for all other coatings it is similar. The pores are globular particularly for GSP and HVOF, whereas for WSP and flame spray, they are flatter (lower minimal circularity). This is probably due to a low feed rate used at HVOF and GSP spraying enabling more homogeneous melting of the powders, particularly in the case of GSP.

In the case of HVOF, low microhardness is in connection with the bimodal zones depicted in Fig. 2(a). The wear resistance of the HVOF coating is also the worst, being less than one half of the best value obtained with the GSP coating. The inverse wear rate values have the same trend as microhadrdness. In fact, HVOF spraying of ceramics needs relatively sophisticated optimization [15] to get perfect melting of the powders and building of the coating with this low-temperature process leading to good mechanical properties. Plasma spraying offers better melting of the ceramic powder particles (indicated by low roughness), resulting in good inter-splat contact in the coatings. That is why they exhibit higher microhardness than flame and HVOF coatings. Also, the lower standard deviation of the microhardness values obtained for both plasma-sprayed coatings (compared to flame and HVOF) is due to better homogeneity of well bonded

Table 2   Characteristics of the coatings

Parameter	WSP	GSP	HVOF	Flame
Phase according to XRD	Rutile	Rutile	Rutile	Rutile
E_{bg} at indirect transition (eV)	3.12 (2.22)	3.08	2.82 (2.34)	3.25
Surface roughness R_a (μm)	13.4 ± 0.5	4.3 ± 0.14	5.7 ± 0.2	5.9 ± 1.4
Surface roughness $R_{y,max}$ (μm)	97.7 ± 6.4	35.2 ± 3.8	50.6 ± 4.7	48.8 ± 2.4
Porosity (%)	8.5	3.4	4.4	8.3
Mean pore size (μm)	13.6	12.9	7.8	13.9
Pore circularity minimum	0.095	0.144	0.158	0.051
Microhardness at 1 N (GPa)	9.46 ± 1.31	10.40 ± 1.60	4.95 ± 2.11	8.42 ± 2.35
Inverse wear rate (m/mm^3)	29.3	50.1	18.4	21.7

splats.

From the viewpoint of wear resistance, the GSP coating is markedly superior over all other coatings. This is mainly because of its low porosity. The HVOF coating is the worst; its porosity is also relatively low, but its bimodal structure exhibits defects like clusters of pores, particularly near the substrate (Fig. 1(d)).

According to the XRD in all coatings, the only identified phase is rutile (Fig. 3). The difference in the intensities of peaks corresponding to rutile planes (indicated in the top image) is due to the different surface qualities of samples. The absence of peaks of known sub-stoichiometric titanium oxides is due to the fact that an oxygen-deficient structure can be, besides sub-stoichiometric titanium oxides, also formed predominantly by vacancies [19,20] and crystallographic stacking faults [14]. In the HR-TEM micrograph (Fig. 4(a)), a region with point defects is marked by a circle. In Fig. 4(b), a crystallographic stacking fault is aligned in the direction of the long arrow and creates a misalignment between the crystal planes marked by short arrows.

As a result of the photocatalytic test, the relative concentration of the substances $C/C(0)$, where $C(0)$ is the initial concentration, is displayed versus time (Fig. 5). The HVOF and GSP coatings have very similar photocatalytic characteristics.

Under the UV light among all coatings, a 10% degradation of acetone within 16 h is reached by the GSP and HVOF coatings. At the same time, WSP and flame coatings reach 6% and 3% of acetone

decomposition, respectively (Fig. 5(a)). The CO_2 uptake is relatively similar for all coatings with exception of the flame coating (Fig. 5(b)). Also, the CO uptake is relatively similar for HVOF, WSP and GSP coatings, while for the flame coating it is again significantly higher (Fig. 5(c)).

Band gap energies are the lowest for HVOF and GSP, 2.82 eV and 3.08 eV respectively (Fig. 6). The WSP coating is proven as a medium quality photocatalyst among the studied coatings. Also, its band gap is slightly higher, 3.12 eV. Finally, the flame-sprayed coating is the least photoactive one and its band gap energy is the highest, 3.25 eV. For the HVOF and WSP coatings, secondary absorption edges also exist at 2.34 eV and 2.22 eV, respectively. The

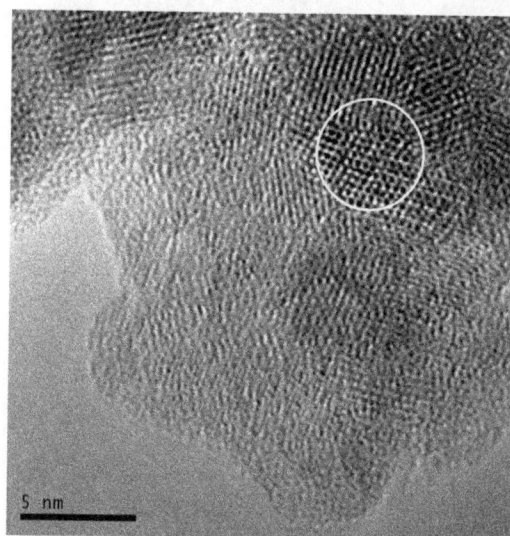

(a)

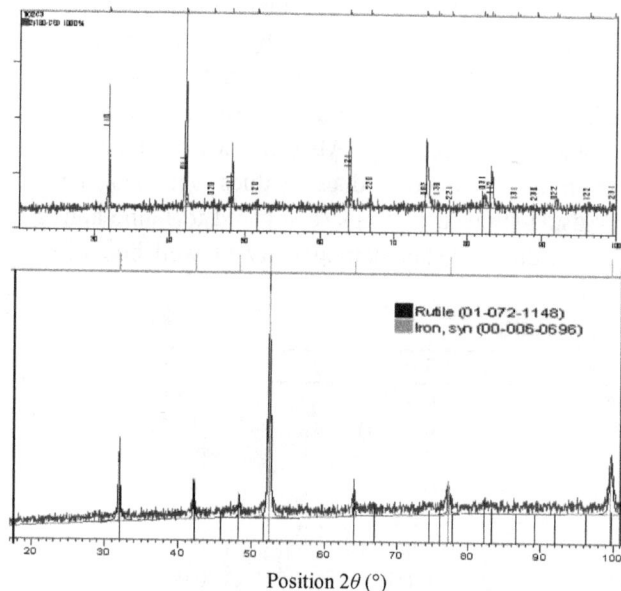

(b)

Fig. 4 HR-TEM micrographs of (a) point defects and (b) stacking fault in GSP TiO$_2$ coating.

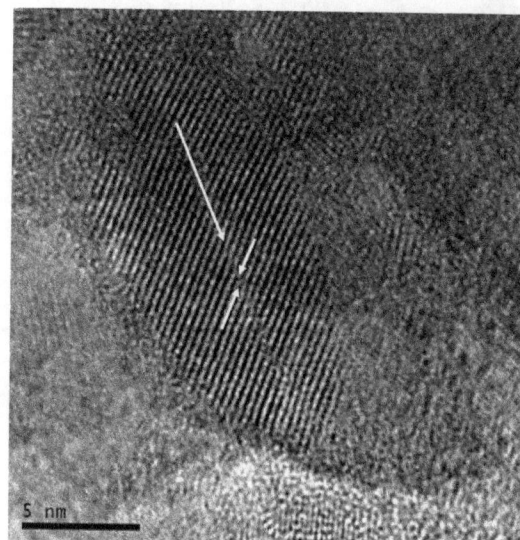

Fig. 3 XRD patterns of flame-sprayed coating (top) and HVOF coating (bottom).

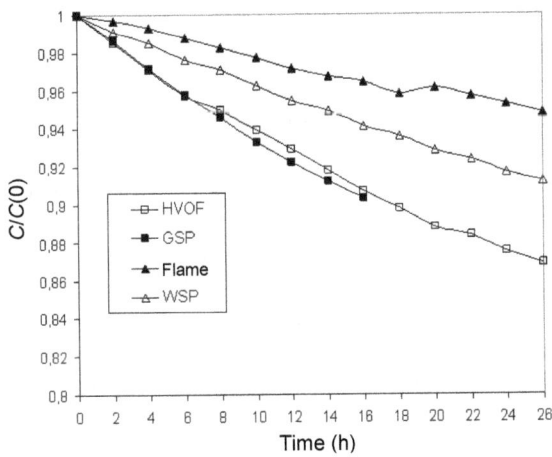

(a)

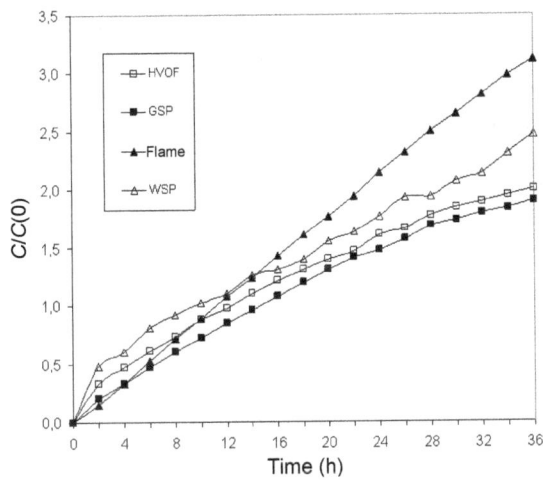

(b)

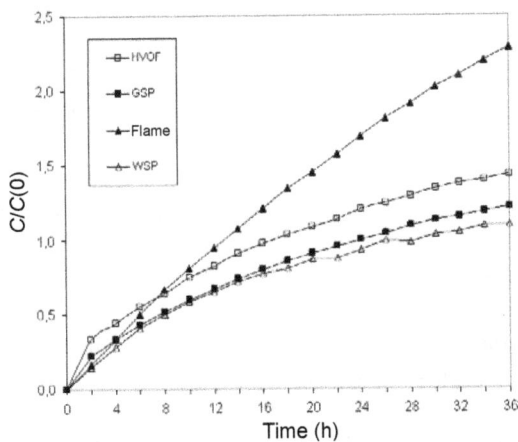

(c)

Fig. 5   Kinetics of (a) acetone concentration decrease, (b) carbon dioxide concentration increase, and (c) carbon monoxide concentration increase.

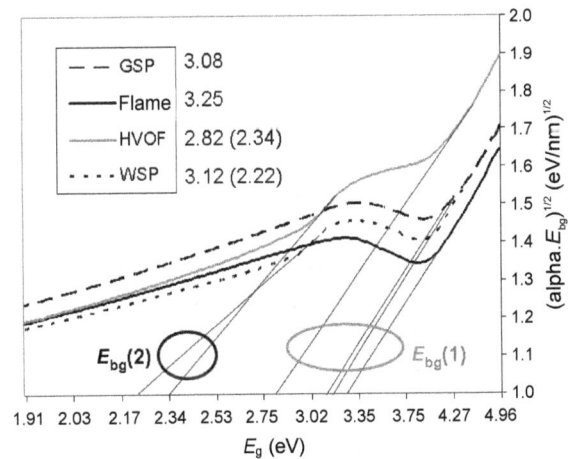

Fig. 6   Band gap estimation for all coatings. The energies of primary (and secondary) absorption edge (eV) are in the legend.

low excitation energy [19,21,22]. The existence of a secondary absorption edge does not seem to correlate tightly with the acetone photo-decomposition.

GSP spraying process due to H_2 and HVOF spraying process due to ethylene both provide more free hydrogen atoms for reactions leading to the reduction of TiO_2. In the case of WSP and flame spraying, the working gas does not provide efficient conditions for reduction. Anatase is more sensitive to reaction with hydrogen from ethylene C_2H_4 than from acetylene C_2H_2 [23,24]. Anatase is present in higher content in the feedstock for HVOF spraying than in the feedstock used for flame spraying. That is why the flame-sprayed coating should have less oxygen-deficiency than the HVOF coating and therefore less photoactivity. Concerning the GSP coating, the high rutile content in the feedstock exhibits the possibility to be highly reduced directly by H_2 plasma gas. Flame coating has the highest band gap, and according to previous considerations, it should be less oxygen-deficient.

4   Conclusions

TiO_2 was sprayed by four different thermal spraying techniques—plasma spraying using a water-stabilized torch, plasma spraying using a gas-stabilized torch, high velocity oxy–fuel gun, and oxy–acetylene flame. Plasma spraying was realized with high coating thickness. The last two techniques—usually considered as non-optimal for spraying ceramics because they exhibited significantly lower temperature accessible with the heating medium—were employed successfully,

main absorption edge at $E_{bg}(1)$ corresponds to non-stoichiometric TiO_2 features in rutile, and the secondary absorption edge at $E_{bg}(2)$ corresponds to an absorption tail of delocalized electronic states with a

too. The target of our work was to obtain all coatings with the main feature typical for plasma spraying of TiO_2, i.e., oxygen-deficient structure. The other feature inherent to all applied spray processes was the transformation of anatase phase in the initial powders to rutile phase in the coatings. In spite of this transformation, all coatings exhibited certain photocatalytic activity, which correlated well with their band gap energy. This feature was present thanks to the oxygen-deficient structure exhibiting vacancies and crystallographic stacking faults. At the same time, all the coatings offer relatively good mechanical properties and can serve as robust photocatalysts.

All techniques used also offered processing parameter ranges suitable for further optimization of titania spraying especially from the viewpoint of the mechanical integrity of the coatings. At GSP, the very low feed rate used brought good photocatalytic performance and good mechanical properties, whereas the HVOF coating, having GSP-similar photoactivity, was mechanically inferior. This fact resulted from the markedly different jet temperatures for these two techniques.

Acknowledgements

This work was supported by the Czech Science Foundation under Project P108/12/1872.

References

[1] Li D, Haneda H, Hishita S, et al. Fluorine-doped TiO_2 powders prepared by spray pyrolysis and their improved photocatalytic activity for decomposition of gas-phase acetaldehyde. *J Fluorine Chem* 2005, **126**: 69–77.

[2] Matsuda S, Hatano H, Tsutsumi A. Ultrafine particle fluidization and its application to photocatalytic NO_x treatment. *Chem Eng J* 2001, **82**: 183–188.

[3] Herrmann J-M. Heterogeneous photocatalysis: Fundamentals and applications to the removal of various types of aqueous pollutants. *Catal Today* 1999, **53**: 115–129.

[4] Dumitriu D, Bally AR, Ballif C, et al. Photocatalytic degradation of phenol by TiO_2 thin films prepared by sputtering. *Appl Catal B: Environ* 2000, **25**: 83–92.

[5] Colmenares-Angulo J, Zhao S, Young C, et al. The effects of thermal spray technique and post-deposition treatment on the photocatalytic activity of TiO_2 coatings. *Surf Coat Technol* 2009, **204**: 423–427.

[6] Ctibor P, Pala Z, Sedláček J, et al. Titanium dioxide coatings sprayed by a water-stabilized plasma gun (WSP) with argon and nitrogen as the powder feeding gas: Differences in structural, mechanical and photocatalytic behavior. *J Therm Spray Techn* 2012, **21**: 425–434.

[7] Ctibor P, Štengl V, Píš I, et al. Plasma sprayed TiO_2: The influence of power of an electric supply on relations among stoichiometry, surface state and photocatalytic decomposition of acetone. *Ceram Int* 2012, **38**: 3453–3458.

[8] Ctibor P, Stengl V, Zahalka F, et al. Microstructure and performance of titanium oxide coatings sprayed by oxygen–acetylene flame. *Photochem Photobiol Sci* 2011, **10**: 403–407.

[9] Lima RS, Marple BR. Process–property–performance relationships for titanium dioxide coatings engineered from nanostructured and conventional powders. *Mater Design* 2008, **29**: 1845–1855.

[10] Fauchais P, Vardelle A, Dussoubs B. Quo vadis thermal spray? *J Therm Spray Techn* 2001, **10**: 44–66.

[11] Bozorgtabar M, Rahimipour M, Salehi M. Novel photocatalytic TiO_2 coatings produced by HVOF thermal spraying process. *Mater Lett* 2010, **64**: 1173–1175.

[12] Jeffery B, Peppler M, Lima RS, et al. Bactericidal effects of HVOF-sprayed nanostructured TiO_2 on pseudomonas aeruginosa. *J Therm Spray Techn* 2010, **19**: 344–349.

[13] George N, Mahon M, McDonald A. Bactericidal performance of flame-sprayed nanostructured titania–copper composite coatings. *J Therm Spray Techn* 2010, **19**: 1042–1053.

[14] Skopp A, Kelling N, Woydt M, et al. Thermally sprayed titanium suboxide coatings for piston ring/cylinder liners under mixed lubrication and dry-running conditions. *Wear* 2007, **262**: 1061–1070.

[15] Leivo J, Varis TE, Turunen E, et al. Influence of the elementary mixing scale on HVOF-sprayed coatings derived from nanostructured aluminosilicate/mullite feedstock. *Surf Coat Technol* 2008, **203**: 335–344.

[16] Ctibor P, Neufuss K, Chraska P. Microstructure and

abrasion resistance of plasma sprayed titania coatings. *J Therm Spray Techn* 2006, **15**: 689–694.

[17] ASTM International. ASTM G75-95. Standard Test Method for Determination of Slurry Abrasivity (Miller Number) and Slurry Abrasion Response of Materials (SAR Number). 1995.

[18] Reddy KM, Manorama SV, Reddy AR. Bandgap studies on anatase titanium dioxide nanoparticles. *Mater Chem Phys* 2003, **78**: 239–245.

[19] Moustakas NG, Kontos AG, Likodimos V, *et al.* Inorganic–organic core–shell titania nanoparticles for efficient visible light activated photocatalysis. *Appl Catal B: Environ* 2013, **130–131**: 14–24.

[20] Chen X, Liu L, Yu PY, *et al.* Increasing solar absorption for photocatalysis with black hydrogenated titanium dioxide nanocrystals. *Science* 2011, **331**: 746–750.

[21] Tsuyumoto I, Uchikawa H. Nonstoichiometric orthorhombic titanium oxide, $TiO_{2-\delta}$ and its thermochromic properties. *Mater Res Bull* 2004, **39**: 1737–1744.

[22] Barajas-Ledesma E, García-Benjume ML, Espitia-Cabrera I, *et al.* Determination of the band gap of TiO_2–Al_2O_3 films as a function of processing parameters. *Mat Sci Eng B* 2010, **174**: 71–73.

[23] Panpranot J, Kontapakdee K, Praserthdam P. Selective hydrogenation of acetylene in excess ethylene on micron-sized and nanocrystalline TiO_2 supported Pd catalysts. *Appl Catal A: Gen* 2006, **314**: 128–133.

[24] Tieng S, Kanaev A, Chhor K. New homogeneously doped Fe(III)–TiO_2 photocatalyst for gaseous pollutant degradation. *Appl Catal A: Gen* 2011, **399**: 191–197.

5

Blister defect formation within partially stabilized zirconia film during constrained sintering

Kais HBAIEB[a,b,*]

[a]*Strategic Technology Unit, First Floor, Room G-08, Taibah University, P.O. Box 344, Al-Madinah Al-Munawwara, Kingdom of Saudi Arabia*
[b]*Mechanical Department, College of Engineering, Taibah University, P.O. Box 344, Al-Madinah Al-Munawwara, Kingdom of Saudi Arabia*

Abstract: Yttria stabilized zirconia (YSZ) film has been screen printed and sintered on a rigid substrate. The constrained sintering caused the formation of multiple microcracks and most critically large "blister" defects. The morphology of such defects has been characterized by scanning electron microscopy (SEM). It was revealed that the film surface exhibits noticeable roughness. Microhardness testing revealed little variation in green density distribution. Rheological measurement, however, showed that some agglomerations are present in the YSZ ink. The existence of agglomerations in the screen printing ink in combination with debonding at the film/substrate interface is potentially the cause for the formation of blister defects.

Keywords: constrained sintering; defects; yttria stabilized zirconia (YSZ); desintering; debonding

1 Introduction

Sintering of films is a necessary processing stage in many industrial applications. For example, in fabricating solid oxide fuel cell (SOFC) layer, anode, electrolyte and other layers have to be sintered together. Many technical difficulties may be encountered during sintering of these layers. In anode supported fuel cell, anode and electrolyte are co-sintered with the potential result of warpage if sintering is not well controlled [1,2]. In contrary, in substrate supported fuel cell, warpage is not possible as the substrate is usually thick and pre-sintered. However, constrained sintering of the electrolyte may result in defect formation and evolution that cause loss of leak tightness against gas flow through the electrolyte, a required condition for SOFC functionality.

Constrained sintering is a process in which the sintering of a ceramic body is slowed because of some constraint. Lange [3,4] reported that a phenomenon of desintering occurs concurrently with densification of powder compact when two particles forming necks undergo coarsening, constrained from sintering as a result of inclusions or pulled apart due to tension applied to the sintering body. When the green compact is not uniformly packed, the dense regions may act as hard inclusions in hindering the sintering of the surrounding matrix and desintering can also result. Although this phenomenon is observed in every sintered body, it is mostly obvious in layered structures when multiple layers are deposited onto a rigid substrate. Stresses arise either during sintering or upon cooling of the sintered structure. Because of sharp

* Corresponding author.
E-mail: hbaiebkais@gmail.com

transition of material properties, stress concentration can develop at the interface of adjoining layers. While sintering stress, which is the driving force for sintering, tends to cause the film to shrink, the substrate prevents the deformation; the densification completely stops parallel to the layers due to the bonding of layers but is free to take place normal to the layers (no constraint normal to the layers). Although the tensile stresses are not high in magnitude, they can trigger formation of fracture origins in the film as they develop in the very early stages where the particles have not bonded together, i.e., the film is weak [5–9]. Moreover, it is favorable for cracks to propagate further through the entire thickness of a constrained sintering film due to the high tangential sliding constraint that prevents particle rearrangement and induces large near crack tip stresses [10].

Anisotropic shrinkage in constrained sintering film has been observed in many studies [10–18]. Although this phenomenon is obvious in the presence of inclusions and inhomogeneous green density distribution within the pre-sintered film, it also arises when the film is uniformly deposited onto the substrate. Guillon et al. [13] showed anisotropic sintering develops in alumina film grown on a rigid substrate of the same material. Microstructure analysis of polished cross-sectional image in the thickness direction revealed preferential orientation of pores along the thickness direction. This anisotropy developed further as film density increased. Calata et al. [14] reported coarser pore distribution near the interface of a cordierite glass film with a rigid substrate as compared to the rest of the film. This was attributed to the constraint and poor wetting conditions at the film/substrate interface. Mohanram et al. [15] noticed higher pore density near the interface compared with the rest of the film material as the in-plane tensile stresses arisen from constrained sintering are higher at the film/substrate interface. Lu and Xiao [16] showed that relative density gradually increases as distance from interface increases to nearly reach a plateau far from the interface. In a recent paper, Martin and Bordia [17] used discrete element model to study the microstructure evolution of sintering film on a rigid substrate. They showed that constrained sintering at the interface causes large number of pores preferentially orienting themselves perpendicular to the substrate/film planes and confirmed the observation of Lu and Xiao [16]. Martin and Bordia [17] also reported that the substrate does not negate the sintering dynamics of

the particles but slows down the mobility of particles near the interface with some particle contacts growing at the expense of others.

In this paper, we discuss the effect of constrained sintering on the development of sintering defects within partially stabilized zirconia material screen printed and sintered on a rigid substrate.

2 Experimental procedure

Commercial 3 mol% yttria partially stabilized zirconia (YSZ, MEL chemicals, Manchester, UK) has been used for this study. The powder was mixed with terpinol based binder to produce a screen printing ink. The substrate used is pre-sintered porous magnesia-rich spinel substrate (Advanced Ceramics Ltd., UK). Prior to YSZ deposition, a porous layer of the same material (YSZ) with ~30 μm thickness was applied to the substrate to reduce roughness. In some cases, additional anode layer is printed to mimic the SOFC layer sequence. The ink was screen printed on a 40 mm × 40 mm printing area using screen printer (DEK Printing Machines Ltd., UK). In the printing process, a square blade was applied with a load of 12 kg on a stainless steel stencil with 165 mesh, with printing speed of 60 mm/s and 3 mm gap. For each sample, two consecutive printings (back and forth) were conducted. Once the printing was completed, the sample was dried at 120 ℃. The firing was conducted in a high-temperature furnace (Carbolite, UK) with a ramp rate of 2 ℃/min to 300 ℃, dwelled of 1 h and further heated to 1350 ℃ with a ramp rate of 3 ℃/min. The sintering temperature was hold for 1 h before cooling down to room temperature at a rate of 3 ℃/min. Selected samples were polished and gold coated for scanning electron microscopy (SEM) characterization.

Characterization for the samples for roughness and non-uniform green density distribution within the YSZ layer was also considered. Roughness of the zirconia film prior to sintering was measured using classical stylus profilometer. The density variation was studied by measuring microhardness across the electrolyte surface and relating hardness to density. The likelihood of density variation was thought to be due to either screen printing or presence of agglomerates within the YSZ ink. The latter hypothesis was also checked by further mixing the ink using ultrasonic processor. Viscosity and shear stress at various shearing rates of a stored ink and a fresh (ultrasonically) mixed ink were measured using a rheometer (LAUDA Brinkmann, USA).

3   Results and discussion

As the YSZ film undergoes substantial shrinkage during sintering on a prefired substrate, residual tensile stresses arise and trigger formation of desintering defects. These typical defects seen in constrained sintering film have been reported in several studies and are shown in Fig. 1. It is believed that these defects are formed due to the tensile stresses developed during sintering due to the constraint applied by the substrate. As shown in Fig. 1, these defects prevent the electrolyte layer from reaching full density, but as long as they do not interlink or form a percolating network, the film may exhibit leak tightness against gas flow. A more critical type of defects has the form of a blister or volcano shape with large voids (several microns) that are very detrimental as they are permeable for gas flow. This was initially attributed to high internal pressure potentially due to the confinement of organic substances within the film; gases with increasing gas pressure (due to increase of temperature) eventually punch through the film causing such volcano type defect. However, this hypothesis is refuted for two reasons. First, organic binder shall be totally burned out at ~500 ℃ where the film is still porous and gas flow channels are abundant. Moreover, the film surface morphology for a film fired at 500 ℃ for 1 h has been recorded under SEM and is shown in Fig. 2(a). The absence of such defects is obvious; however, the micrograph reveals some unevenness of the surface profile. This has triggered the investigation of surface roughness evolution during printing and sintering process.

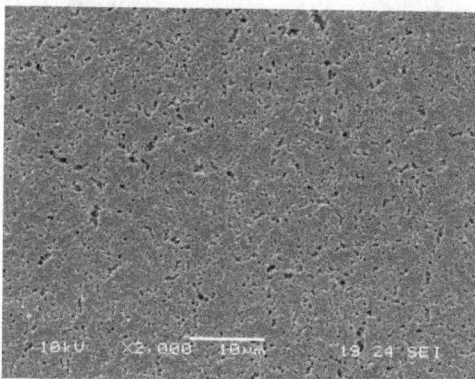

Fig. 1 Microcracks within electrolyte film due to constrained sintering.

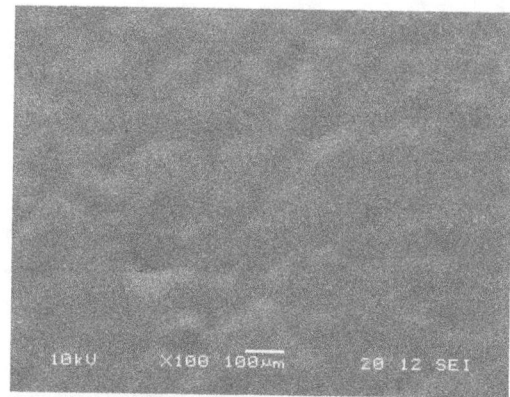

(a)

(b)

Fig. 2 Surface morphology of the electrolyte film: (a) fired at 500 ℃ for 1 h; (b) after complete sintering.

Roughness measurement of the substrate was conducted using classical stylus profilometer. Three measurements were conducted in extrusion and transversal directions. The roughness in the extrusion direction is usually higher than that in the transversal direction. The average roughness is ~4–4.5 μm. Three samples were used for these measurements. Porous zirconia thick film and anode were subsequently printed and co-sintered on the substrates (film thickness > 40 μm). Roughness was again measured and the average roughness was ~3 μm. The YSZ electrolyte was then printed and the sample was fired at 1000 ℃ for 1 h (thickness ≈ 10 μm). Roughness was measured to be ~1.5 μm. Obviously the roughness reduces progressively as more films are deposited. The reduction in roughness is expected and easily conceptualized. Since roughness is an absolute quantity and sintering induces a thickness reduction irrespective of the original thickness, it is expected that the unevenness is reduced by the same shrinkage strain. That is, if the material is shrinking by 30% in the thickness direction, also shall the roughness reduce by the same amount. Note that a roughness of 1.5 μm is considerable and may affect the sintering evolution within the film. It is interesting to note that roughness is quite noticeable after sintering as illustrated in Fig. 2(b). This could be the consequence of the original roughness but also potentially due to shrinkage differences that have taken place in the electrolyte film. That is, some regions within the film are shrinking by different amounts as compared to others. This non-uniformity of shrinking may be attributed to inhomogeneous green density distributions across the film.

The measurement of density variation across the electrolyte surface prior to sintering was more involved. It consists of hardness measurements not only at various locations on the electrolyte surface, but also on thin powder compacts made of electrolyte powder prepared using different compaction pressures to ensure different densities of the electrolyte powder compact. The hardness measurements of the thin powder compacts are necessary to correlate hardness to density. Weight and geometric dimensions were measured and density of the compacts was inferred. Hardness of powder compacts fired at 1000 °C for 1 h was measured at three locations (two opposite locations near the edge of the sample and one location at the center of the powder compacts). Dimensions before and after firing at 1000 °C were measured but shrinkage was infinitesimally small. The average of the hardness values for the different compacts was determined, and in combination of the density values a calibration curve was constructed that relates hardness to density (Fig. 3). In total five samples were prepared and we have used compaction pressure ranging from 1 ton to 5 tons with a 10 mm die. Higher pressure was not used to avoid cracking of the pellets. However, the calibration curve was fitted to a polynomial curve to extrapolate the densities beyond the measuring range when necessary. The hardness measurements were conducted using microhardness tester, where loads of 10 gf and 25 gf were applied. The samples were gold coated prior to hardness measurements for better clarity of the indentation shape.

Microhardness across the electrolyte film surface fired at 1000 °C was measured at various locations spaced by 4 mm in the x- and y-directions and distributed throughout the entire surface. The load used for these measurements was the lowest, i.e., 10 gf (Fig. 4(a)). The density variation was then determined

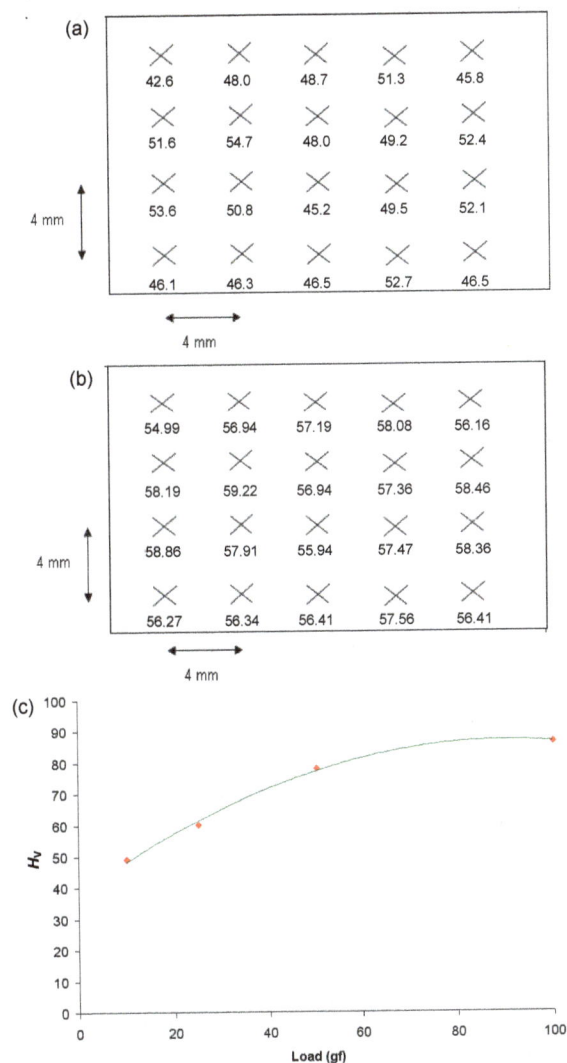

Fig. 4 (a) Microhardness results recorded at different locations across the x- and y-axes and using indentor load of 10 gf. (b) Density variation across the electrolyte surface determined using the measured microhardness results and the calibration curve. (c) Microhardness variation with the applied load. A plateau is reached at high indentor load which suggests that the hardness at high load is indeed the substrate hardness.

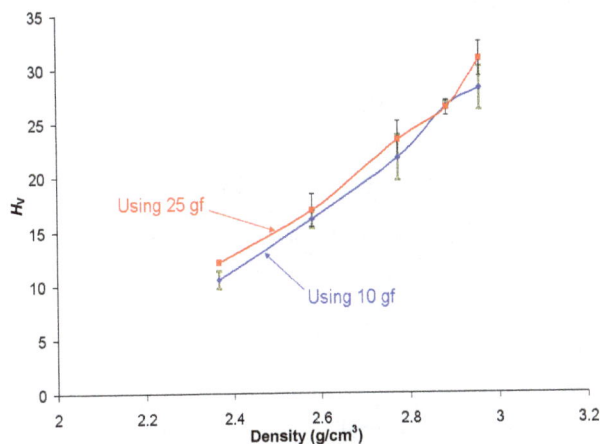

Fig. 3 Calibration curves relating hardness to density of powder compacts.

and given in Fig. 4(b). The geometrical in-plane dimensions measured were ~20 μm (dimensions of the indent imprint), which corresponds to a depth of ~8 μm. Obviously, this depth value is quite substantial relative to the electrolyte film thickness (~10 μm), and consequently the substrate underneath the electrolyte film must have influenced the measurements. This assertion was verified by conducting more hardness tests using higher loads and we have realized that the hardness increases to reach a plateau at high loads, up to 100 gf (Fig. 4(c)). It is believed that the hardness corresponding to high loads is no longer the hardness of the electrolyte but rather of the substrate. Using nanoindentation was not successful possibly due to the relatively large roughness (micron range). Despite the lengthy experimental duration, most of the measurements at different locations were not recorded. Referring to the most meaningful results using the lowest load of 10 gf, it seems that the density variation across the film is not large. Note that such analysis is not conclusive, not only because the substrate has affected the quantities measured, but also because the hardness is measured over a large film area (several microns) and hence only averaging values are determined.

The shelf life of a standard screen printing ink shall commonly extend for months. Ultrasonic processor was used for further mixing the electrolyte YSZ screen printing ink. Rheological characterization was conducted by measuring shear stress and viscosity versus shear rate. The results are shown in Fig. 5. As shown, both viscosity and shear stress have been reduced after ultrasonic processing which indicates the potential presence of agglomerations within the ink.

Figures 6(a), 6(b) and 6(c) illustrate three dimensional micrographs, showing several surface blister cracks, close-up of one defect and its cross-

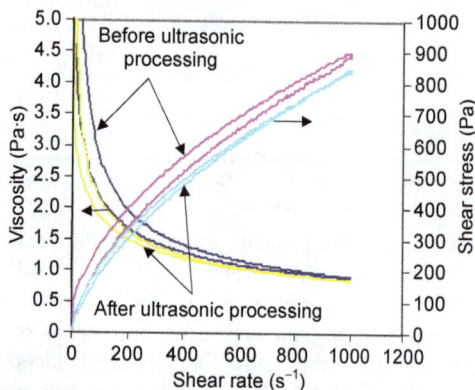
(a)

(b)

(c)

(d)

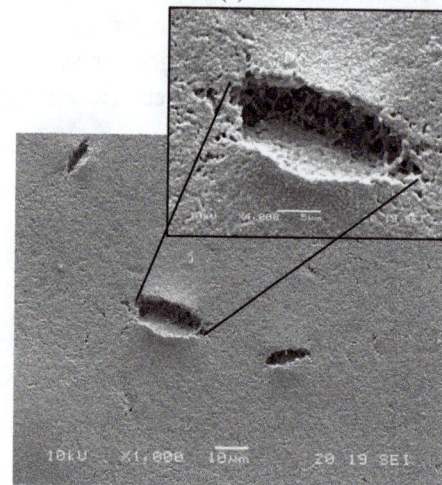

Fig. 5 Viscosity and shear stress versus shear rate before and after ultrasonic processing.

Fig. 6 (a) Several surface blister cracks, (b) close-up of one defect, (c) cross-sectional view of a blister defect and (d) one blister defect and its magnified shape extending much larger than 10 μm.

sectional view, respectively. Figure 6(d) shows one defect and its magnified shape extending much larger than 10 μm. Two smaller defects are randomly located in the proximity of the large crack. In general, these defects are randomly distributed and do not follow a defined pattern which contests the possibility of originating from screen printing mesh markings and reinforces the hypothesis that such defects are originated from low density regions that fail at the constrained sintering stress due to their poor initial packing. However, the blistering effect must be caused by a different phenomenon. Martin and Bordia [17] claimed that the dynamic of the sintering particles has an important effect on the microstructural evolution. The high viscous drag imposed by the substrate causes lack of mobility of particles at the film/substrate interface. Particles near the substrate do not sinter as much as those far from the substrate. The first few particle layers exhibit large porous microstructure with pores preferentially oriented along the thickness direction. The geometric and viscous drag imposed by the substrate causes several particle contacts to disappear for others to grow. When defects due to loss of particle contacts exceed critical size, such defects will grow further. This is manifested in local debonding of the film with the substrate. Henrich *et al.* [10] reported that once a defect is formed within a constrained sintering film it is readily to propagate through the thickness due to the presence of large near crack tip tensile stresses. The combination of surface cracking and debonding at the interface will create a free surface across the entire thickness. Free sintering in the near surface region will develop including curling that usually takes place in free standing film due to the in-plane sintering strains. The resulting shape resembles that of a blister or volcano shape protruding beyond the film surface level. We note that the porous structure is mostly developed at the substrate interface due to the high constraint applied by the substrate. Such constraint reduces as we go away from the interface. We speculate that the high sintering activity of the deposited film will accentuate such phenomenon at the film/substrate interface with the debonding and porosity density increasing. This is because the high sintering will intensify not only the densification process but also the growth of the large pores. In this particular study the YSZ film is very sintering active; dilatometry study showed that a powder compact sintered at 3 ℃/min to 1350 ℃ and hold for 1 h at the sintering temperature reaches full density with total engineering sintering strain of 27%. Such engineering strain corresponds to true strain of 32%. Tillman *et al.* [19] also reported that the sintering stress will gradually decrease as we go away from the interface to nearly disappear on the top surface. Hence, when a thick film is growing it is possible that the particles near the top surface will not feel any constraint from the substrate and behave as free sintering layer. Therefore, we recommend that graded layers shall be printed on the substrate with the low sintering active layer at the very bottom followed with gradual increasing sintering active layers. As such both debonding and surface crack may be eliminated or considerably reduced.

4 Conclusions

Yttria partially stabilized zirconia film has been screen printed and sintered on a rigid pre-sintered thick substrate. The constraint applied by the substrate resulted in formation of blister defects. The formation of such defects has been attributed to existence of agglomerations that have acted as fracture origins and to the high densification rate of the YSZ powders resulting in debonding at the interface.

References

[1] Li W, Hasinska K, Seabaugh M, *et al.* Curvature in solid oxide fuel cells. *J Power Sources* 2004, **138**: 145–155.

[2] Guo H, Iqbal G, Kang BS. Development of an *in situ* surface deformation and temperature measurement technique for a solid oxide fuel cell button cell. *Int J Appl Ceram Tec* 2010, **7**: 55–62.

[3] Lange FF. Densification of powder compacts: An unfinished story. *J Eur Ceram Soc* 2008, **28**: 1509–1516.

[4] Lange FF. De-sintering, a phenomenon concurrent with densification within powder compacts: A review. In *Sintering Technology*. German RM, Messing GL, Cornwall RG, Eds. New York: Marcel Dekker Inc., 1996: 1–12.

[5] Bordia RK, Raj R. Sintering behavior of ceramic films constrained by a rigid substrate. *J Am Ceram Soc* 1985, **68**: 287–292.

[6] Cheng T, Raj R. Flaw generation during constrained sintering of metal–ceramic and metal–glass multilayer

films. *J Am Ceram Soc* 1989, **72**: 1649–1655.

[7] Bordia RK, Jagota A. Crack growth and damage in constrained sintering films. *J Am Ceram Soc* 1993, **76**: 2475–2485.

[8] Wang X, Kim J-S, Atkinson A. Constrained sintering of 8 mol% Y_2O_3 stabilised zirconia films. *J Eur Ceram Soc* 2012, **32**: 4121–4128.

[9] Wang X, Atkinson A. Microstructure evolution in thin zirconia films: Experimental observation and modelling. *Acta Mater* 2011, **59**: 2514–2525.

[10] Henrich B, Wonisch A, Kraft T, *et al.* Simulations of the influence of rearrangement during sintering. *Acta Mater* 2007, **55**: 753–762.

[11] Fu Z, Dellert A, Lenhart M, *et al.* Effect of pore orientation on anisotropic shrinkage in tape-cast products. *J Eur Ceram Soc* 2014, **34**: 2483–2495.

[12] Heunisch A, Dellert A, Roosen A. Effect of powder, binder and process parameters on anisotropic shrinkage in tape cast ceramic products. *J Eur Ceram Soc* 2010, **30**: 3397–3406.

[13] Guillon O, Weiler L, Rödel J. Anisotropic microstructural development during the constrained sintering of dip-coated alumina thin films. *J Am Ceram Soc* 2007, **90**: 1394–1400.

[14] Calata JN, Matthys A, Lu G-Q. Constrained-film sintering of cordierite glass–ceramic on silicon substrate. *J Mater Res* 1998, **13**: 2334–2341.

[15] Mohanram A, Lee S-H, Messing GL, *et al.* Constrained sintering of low-temperature co-fired ceramics. *J Am Ceram Soc* 2006, **89**: 1923–1929.

[16] Lu X-J, Xiao P. Constrained sintering of YSZ/Al_2O_3 composite coatings on metal substrates produced from eletrophoretic deposition. *J Eur Ceram Soc* 2007, **27**: 2613–2621.

[17] Martin CL, Bordia RK. The effect of a substrate on the sintering of constrained films. *Acta Mater* 2009, **57**: 549–558.

[18] Green DJ, Guillon O, Rödel J. Constrained sintering: A delicate balance of scales. *J Eur Ceram Soc* 2008, **28**: 1451–1466.

[19] Tillman M, Yeomans JA, Dorey RA. The effect of a constraint on the sintering and stress development in alumina thick films. *Ceram Int* 2014, **40**: 9715–9721.

Effect of grain size on dielectric and ferroelectric properties of nanostructured Ba$_{0.8}$Sr$_{0.2}$TiO$_3$ ceramics

Venkata Ramana MUDINEPALLI[a,*], Leng FENG[b], Wen-Chin LIN[a], B. S. MURTY[c]

[a]*Department of Physics, "National Taiwan Normal University", Taipei 11677, Taiwan, China*
[b]*Department of Chemistry, Shenzhen Graduate School, Peking University, Shenzhen 518055, China*
[c]*Nanotechnology Laboratory, Department of Metallurgical and Materials Engineering,
Indian Institute of Technology-Madras, Chennai 600036, India*

Abstract: Barium strontium titanate (Ba$_{0.8}$Sr$_{0.2}$TiO$_3$, BST) nanocrystalline ceramics have been synthesized by high energy ball milling. As the sintering temperature increases from 1200 ℃ to 1350 ℃, the average grain size of BST ceramics increases from 86 nm to 123 nm. The X-ray diffraction (XRD) studies show that these ceramics are tetragonal. The phase and grain size of the sintered pellets have been estimated from the XRD patterns, scanning electron microscopy (SEM) and transmission electron microscopy (TEM) images. The effect of grain size on dielectric and ferroelectric properties is studied. The dielectric and piezoelectric parameters are greatly improved at room temperature with increase in grain size. The Curie transition temperature is found to shift slightly towards higher temperatures as the grain increases from 86 nm to 123 nm. The coercive field decreases and the remnant polarization and spontaneous polarization increase as the grain size of BST nano ceramics increases. These ceramics are promising materials for tunable capacitor device applications.

Keywords: nanocrystalline ceramics; lead free ceramics; crystallization; electron microscopy; dielectric properties; hysteresis

1 Introduction

Piezoelectric and ferroelectric ceramic materials are mature and ubiquitous materials for advanced technology. These ceramic materials are active elements in a range of piezoelectric devices and perform functions such as sensing and actuation. Ferroelectric materials are characterized by a switchable macroscopic polarization, which have drawn extensive attention due to their applications in nonvolatile memories, micro-electromechanical systems, nonlinear optics and sensors. However, at nano-scale, the ferroelectric structure will exhibit quite different properties from the bulk materials. For example, the ferroelectric properties, including the Curie temperature, mean polarization, area of hysteresis loop, coercive electric field, piezoelectric strain and remnant polarization, will become size dependent.

Barium strontium titanate (($\text{Ba}_{1-x}\text{Sr}_x$)TiO$_3$, BST) is

* Corresponding author.
E-mail: materialphysics.ramana@gmail.com

the most common ferroelectric oxide in the perovskite ABO_3 structure. Insulating BST is widely used as dielectrics in capacitors because of its high dielectric constant. The high dielectric constant value makes BST one of the promising candidates for high speed random access memories (FRAMs), dynamic random access memories (DRAMs), multilayer ceramic capacitors, piezoelectric transducers and wireless communication devices, and pyroelectric elements have also increased in recent years. The tendency of the electronic industry towards miniaturization and the need to achieve higher performances in smaller structures lead to high interest in understanding the changes of properties while passing from bulk to the nanosized systems, and determining the ultimate structure (particle or grain size (GS) whilst still preserving ferroelectric properties) as well [1,2].

It is well known that ferroelectric materials, such as $BaTiO_3$ and $PbTiO_3$, show "size effect". The size effect means that the properties of ferroelectrics are dependent on the size of materials and show different behavior from a single crystal [3]. However, there are many causes for size effect in ferroelectrics and it is often difficult to separate true size effect from other factors that change with the size of the system. For instance, the results have been reported in the literature about the size effect in ferroelectric particles associated with the variety of synthesis routes and processing techniques adopted [4–8].

In the present study, the dependence of the dielectric properties on the average grain size down to tens of nanometers range in dense BST ceramics has been investigated. In particular, we explored the possibility of achieving a system, which is expected to realize temperature dependent transformation from one phase to another phase and elevate Curie transition temperature T_C. For that purpose, the nanostructure powders prepared from a blend of elemental oxides by high energy ball milling were compacted by means of conventional sintering technique. The grain size of these nano ferroelectric materials has been modified by varying the sintering temperature from 1200 ℃ to 1350 ℃. The microstructure results obtained by X-ray diffraction (XRD), scanning electron microscopy (SEM) and transmission electron microscopy (TEM) were correlated with measurements of relative permittivity, ferroelectric loop and piezoelectric coefficient d_{33} calculations. The phase analysis and dielectric property measurements in these specimens with different grain size revealed several novel features of these materials.

Recently, a mechanical alloying technique (i.e., high energy ball milling) has successfully been used to synthesize the nanocrystalline ferroelectric materials [9,10]. Indeed, mechanical alloying is found superior to the high temperature solid state reaction method/wet chemical process because it lowers the calcination and sintering temperature due to the nanocrystalline nature of the resultant powders. The properties of nanocrystalline materials are found superior to those of conventional polycrystalline coarse-grained materials.

2 Experimental conditions

2. 1 Sample preparation

The nano ceramic $Ba_{0.8}Sr_{0.2}TiO_3$ samples were prepared through high energy ball milling, i.e., mechanical alloying. Stoichiometric amounts of $BaCO_3$, $SrCO_3$ and TiO_2 were weighed with "Sr" content. The milling was carried out in a Fritsch Pulverisette P5 planetary ball mill at room temperature for different milling time 0, 5 h, 10 h, 15 h and 20 h, in tungsten carbide (WC) milling media at a speed of 300 rpm and a ball-to-powder weight ratio of 10:1. The milling was suspended for 30 min after every half an hour of milling to cool down the system. The powders thus obtained were compacted at 100 MPa to form the pellets with size 10 mm in diameter and 2 mm in thickness by using microprocessor controlled uniaxial compaction unit. The pellets were sintered at different sintering temperatures (1200–1350 ℃) for 2 h in air. The densities were determined by Archimedes method, and this densification process then enabled better electric characterization of the ceramics. The measured density of the sintered BST pellets was found to be within 95%–99% of their theoretical value.

2. 2 Microstructural investigations

For XRD measurements, the sample surfaces were ground and polished to remove the sintering layer. The phase identification was performed with an X-ray diffractometer (Philips, PW-1710 XRD) with Cu Kα radiation, $\lambda = 0.15418$ nm. Typical 2θ angular scans ranging from 20° to 80° in a 0.02° step were used in these experiments. The lattice parameters were also calculated. The average crystalline size was calculated from X-ray peak broadening using the Scherer formula after eliminating the instrumental broadening [11] and strain contribution which was confirmed by TEM and SEM. Microstructure characterization measurements were performed on polished and chemically etched samples by means of SEM. For etching, a solution of 5 mL HCl and 95 mL distilled water, containing a few drops of HF acid, was used. 10 s exposure gave clear images of the grains. TEM was used to characterize the present sample products by using a Philips CM 20 microscope.

2. 3 Electrical measurements

The flat polished surfaces of the sintered pellets were electroded with high-purity silver paste and then dried at 150 ℃ before making electrical measurements. The capacitance measurements were made on polled pellets by using a HIOKI-3535-50 LCR Hi-Tester under an electric field (with the maximum magnitude 1 V) in the temperature range of 25–150 ℃ at different frequencies varying from 100 kHz to 100 MHz at a heating rate of 5 ℃/min. The dielectric constant can be calculated by the equation $\varepsilon_r = Cd/(\varepsilon_0 A)$, where C is the capacitance (F); ε_0 is the free space dielectric constant value (8.854×10^{-12} F/m); A is the capacitor area (m^2); and d is the thickness (m) of the ferroelectric material.

Electroded pellets were poled in silicon oil at 150 ℃ by applying a DC field of 10 kV/cm for 2 h. This gave balanced poling processes to enable observations of any changes in piezoelectric properties independent of poling levels. The piezoelectric coefficient (d_{33}) was measured using a Piezo d_{33} meter. Ferroelectric hysteresis measurements were done at room temperature by using an automatic PE loop tracer based on Radiant Technologies ferroelectric test system with virtual ground mode at 1 kHz.

3 Results and discussion

3. 1 Phase, structural and crystallite size analysis

XRD analysis was used to determine lattice parameters and degree of tetragonality of each ceramic specimen. Figure 1(a) compares the XRD patterns of starting

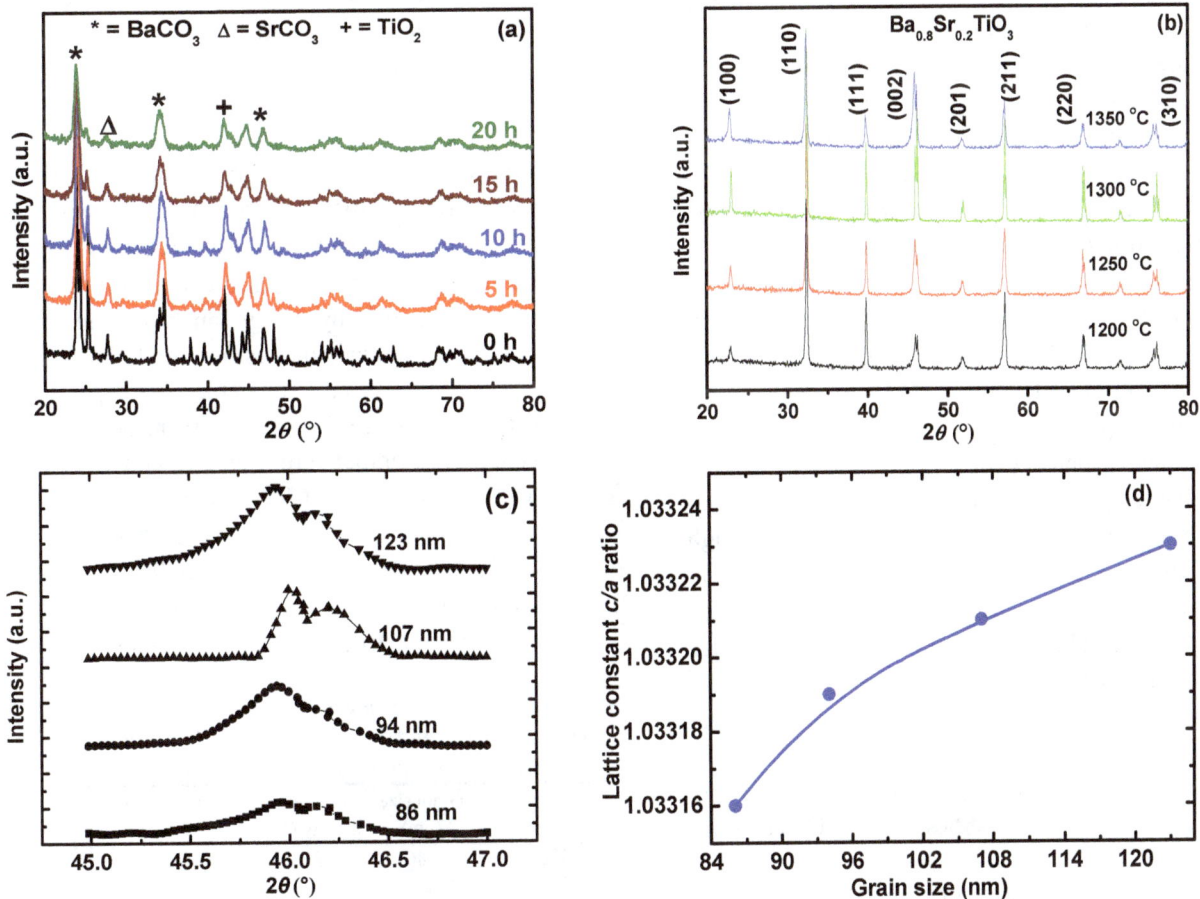

Fig. 1 (a) XRD patterns of mechanically activated BST ceramics at different ball milling time; (b) XRD of BST composition at different temperatures; (c) XRD profile around 45° to 47°; (d) lattice constants of the samples dependent on the mean grain size.

powder mixtures of BST subjected to different durations of milling. After 20 h of milling, alloying starts and some of the XRD peaks of the starting oxides have vanished. The peaks become increasingly broader with milling time and a number of new broadened peaks appear. The new peak patterns have tetragonal structure suggesting the formation of a new phase during mechanical alloying of mixtures of oxides. However, to form single phase tetragonal BST, it requires sintering. Figure 1(b) compares the XRD patterns of mechanically alloyed BST pellets sintered at different temperatures (1200–1350 ℃). The indexing of XRD peaks and the determination of lattice parameters of BST were carried out using a software package POWD. All these peak patterns depict the formation of a single phase with a tetragonal perovskite structure. The results also clearly show that there are no diffraction peaks from impurity phases after sintering. As the sintering temperature is increased, the peaks are sharpened suggesting the increase in the crystallite size. The grain growth of BST nano ceramics as a function of sintering temperature is shown in Table 1. As the sintering temperature increases from 1200 ℃ to 1350 ℃, the crystallite size also increases from ~80 nm to ~120 nm. In order to realize the sub-micron microstructure in the sintered ceramics, it is essential to have the nanocrystalline powders as starting milled powders. Figure 1(c) shows the XRD patterns of the (002) peak as a function of grain size. The diffraction peaks are obtained in the range of $2\theta = 45°$–$47°$ determined by analyzing the individual peak using Lorentzian curve fit function. As expected from the shape of the XRD spectra, the fit procedure yields a significant increase of the tetragonal phase content with increase in grain size. As the temperature decreases, the grain size is too small for domain twinning to occur; resulting high internal stress is expected to reduce the tetragonal distortion. From the XRD patterns, the lattice distortion is calculated as a function of grain size shown in Fig. 1(d). As the grain size increases, the c/a ratio also increases. These results are in agreement with the previous observations of PMN-PT relaxor ferroelectric ceramics and the grain and particle size effects on microwave properties of BaTiO$_3$ ceramics [4,12].

Average grain sizes of the Ba$_{0.8}$Sr$_{0.2}$TiO$_3$ ferroelectric ceramics are estimated using SEM and high resolution TEM. Figure 2(i) shows the typical

Table 1 Lattice parameters of BST nanocrystalline ferroelectric ceramics

Composition	Preparation method	Parameter	Sintering temperature			
			1200 ℃	1250 ℃	1300 ℃	1350 ℃
Ba$_{0.8}$Sr$_{0.2}$TiO$_3$	Ball milled for 20 h	a (Å)	3.9982	3.9981	3.9982	3.9982
		c (Å)	4.1308	4.1307	4.1310	4.1311
		c/a	1.03316	1.03319	1.03321	1.03323
		Crystallite size from XRD (nm)	86	94	107	123

SEM micrographs of nanocrystalline BST ferroelectric ceramics sintered at different temperatures. The ceramics are very dense, smooth and crack free. The average grain sizes of the present ceramics are estimated in the range of 150 nm to 300 nm after sintering, suggesting that the nanocrystalline nature is retained even after sintering at high temperature. Figure 2(ii) shows the typical TEM micrograph of the Ba$_{0.8}$Sr$_{0.2}$TiO$_3$ ferroelectric ceramics; it reveals that, the grain sizes in the range of 20–50 nm are much smaller than those reported elsewhere [3,4,12]. Such fine-grained microstructure is produced for the first time in BST ceramics and the size of these particles is comparable to crystallite sizes calculated by the XRD peak broadening. The results of the SEM and TEM analyses reveal fine morphology of the nanocrystalline BST ferroelectric ceramics.

The relative density and grain size of the nanocrystalline BST ferroelectric ceramics densified at different temperatures in the range of 1200–1350 ℃ are shown in Fig. 3. The values are determined on polished cross sections by image analysis and are averaged over all chemical compositions and plotted in dependence of the sintering temperature. As expected, the density of sintered samples increases with increasing sintering temperature (Fig. 3(a) and Table 2). The values range from about 95% theoretical density for lower sintering temperature to about 99% theoretical density after sintering at 1350 ℃. Figure 3(b) displays that, the grain size increases from 86 nm to 123 nm as the sintering temperature increases from

Table 2 Variations in grain size, density, dielectric constant at T_R and T_C and piezoelectric coefficient d_{33} with different sintering temperatures

Sintering temperature (℃)	Grain size (nm)	Relative density (%)	ε' at T_R	ε' at T_C	d_{33} (pC/N)
1200	86	95	731	1520	160
1250	94	97	1028	2071	182
1300	107	98	1123	2504	198
1350	123	99	1530	3062	212

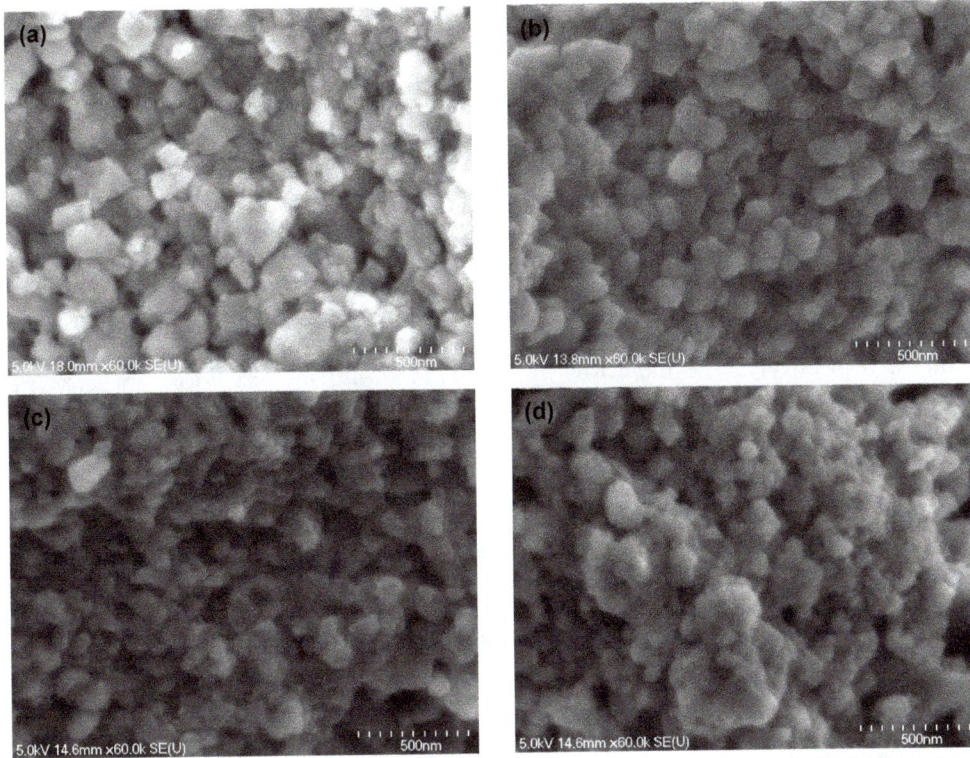

Fig. 2(i) SEM of BST ceramics ball milled for 20 h and sintered at (a) 1200 ℃, (b) 1250 ℃, (c) 1300 ℃ and (d) 1350 ℃ for 2 h.

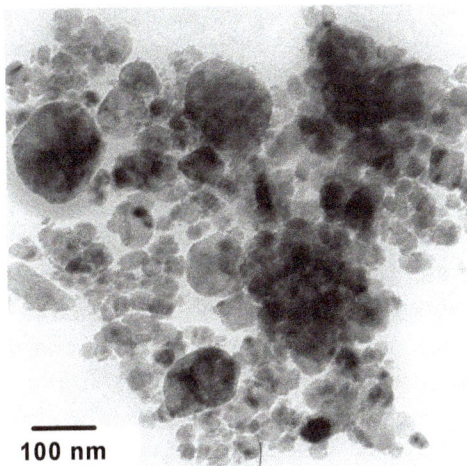

Fig. 2(ii) TEM of BST ceramics ball milled for 20 h and sintered at 1350 ℃ for 2 h.

1200 ℃ to 1350 ℃. The reduced grain sizes of the specimens in the present work can be related to the fine morphology of the starting powders and a good grain connection (Fig. 2(i)) for each constituent phase in the ceramic specimens.

3. 2 Dielectric constant

The grain size dependence of dielectric constant of nanocrystalline BST compositions at 100 kHz is shown

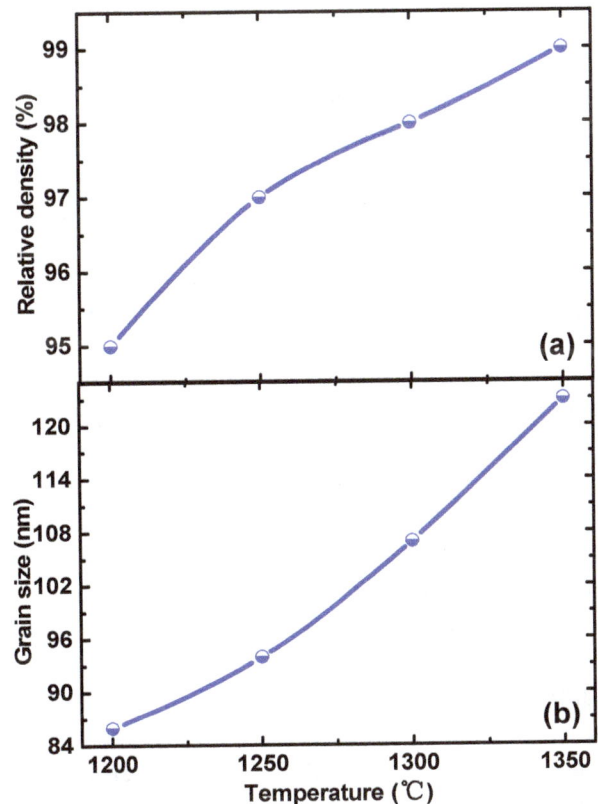

Fig. 3 (a) Relative density and (b) grain size as a function of sintering temperature.

in Fig. 4(a) as a function of temperature. The dielectric constant increases with increasing temperature for all the grain sizes. As expected, it reaches a maximum value at transition temperature (T_C) and then decreases, with further increase in temperature. Dielectric constant is observed to increase from 731 to 1530 as the sintering temperature increases; this is due to the increase in the density of the samples. Chen *et al.* [3] and Xu *et al.* [13] also observed similar variation in dielectric constant in these compositions prepared by thin films and ceramics, respectively. The dielectric anomaly is clearly observable for all the samples, indicating a ferroelectric behavior. The hopping of charge carriers is thermally activated with temperature rise; hence, the dielectric polarization increase causes an increase in ε_r. The dielectric constants at T_R and T_C are also shown in Fig. 4(b). From Fig. 4(b), it is described that the dielectric constants at T_R and T_C increase with the grain size increasing. It is found that at T_C the dielectric constant increases as 1520, 2071, 2504 and 3062 with grain size of 86 nm, 94 nm, 107 nm and 123 nm, respectively (Table 2). The dielectric constant of the nanocrystalline samples is remarkably less sensitive to temperature in contrast to the coarse ceramics. As is well known, the dielectric anomaly can be ascribed to a ferroelectric phase (tetragonal)–paraelectric phase (cubic) transition at T_C and it is also noticed that the Curie transition temperature shifts from 70 ℃ to 80 ℃ as the grain size increases.

There are two possibilities to gain the high dielectric properties of BST nano ceramics. The first possibility is the crystal structure of the phase. The smaller c/a means the smaller tetragonality, which also means that the crystal structure is closer to the cubic structure. This leads to an easier transformation from tetragonal to cubic phase, which explains why the dielectric constant at room temperature decreases with decreasing c/a. The second possibility is the grain size which also has a significant effect on the dielectric properties. The dielectric properties of BST ceramics are affected by both crystalline structure of phase and microstructure. Grain size also has significant effect on the dielectric constant. A ferroelectric single crystal often has multiple ferroelectric domains separated by interfaces called domain walls, which are transition between different dipole alignment. The dielectric constant is dependent on the population of domains and mobility of the domain walls [14]. It is also believed that when the grain size is uniform, the

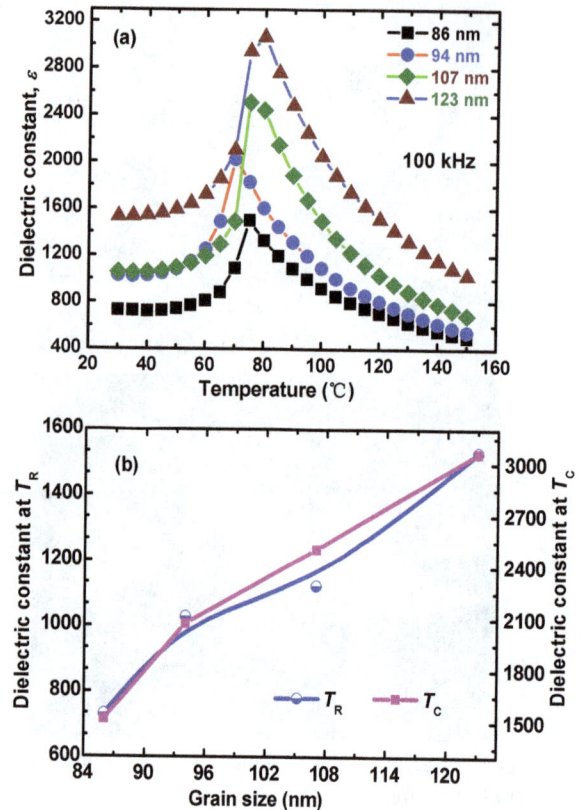

Fig. 4　(a) Temperature dependence of the dielectric constant at 100 kHz with different grain sizes; (b) grain size dependence of the dielectric constant at T_R and T_C.

domain wall movement is relatively easy and regular, and dielectric constant increases. Internal stress arises in the material at Curie transition temperature due to the mechanical deformation of the unit cells that comes from the phase transformation. It is understandable that this internal stress is affected by the grain size and uniformity. Uniform grain size generates less internal stress, resulting in easier domain wall motion.

3.3　Piezoelectric coefficient

The nanocrystalline BST ferroelectric ceramics were poled in silicon oil at 150 ℃ by applying a DC field of 10 kV/cm for 2 h. This gives balanced poling processes to enable observations of any changes in the piezoelectric independent of poling levels. In general, the piezoelectric properties at room temperature decrease consistently with grain size reduction [15–17]. The grain size dependence of piezoelectric coefficient (d_{33}) is shown in Fig. 5. It can be noticed that the piezoelectric constant of our experiment can be decreased from 212 pC/N for 123 nm to 160 pC/N for 86 nm (Table 2). Generally the piezoelectric constant is

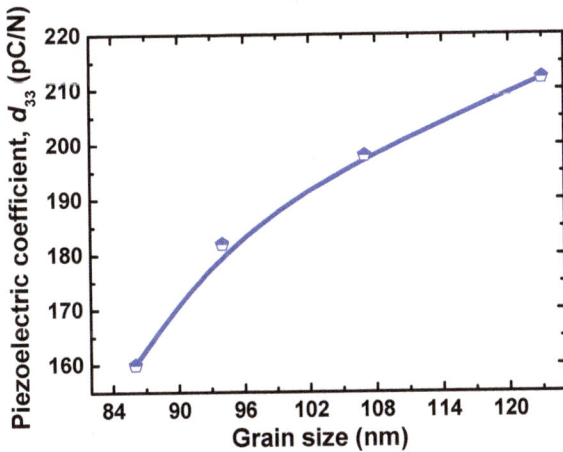

Fig. 5 Grain size dependence of the piezoelectric constant d_{33} of BST ceramics.

determined by crystal structure, relative density as well as grain size. Since there is no apparent difference in crystal structure and relative density, the grain size should be a main factor such that the piezoelectric constant increases apparently with increasing grain size. However, a study by Shao *et al.* [18] indicated that for the conventional sintered BaTiO$_3$ ceramics with larger grain size, the piezoelectric constant decreases from 419 pC/N to 185 pC/N when the average grain size increases from 7 μm to 19 μm. Hence, there could be a critical grain size with which the piezoelectric constant of BST has a maximum value.

3. 4 *P–E* hysteresis loop behavior

Figure 6(a) shows the polarization versus electric field (*P–E*) hysteresis loops of the nanocrystalline BST ferroelectric ceramics with different grain sizes at room temperature. It is found that the hysteresis loops change significantly with grain size. The grain size dependence of the remanent polarization (P_r) and the coercive electric field (E_c) of BST nano ceramics are shown in Fig. 6(b). It is found that the coercive electric field (E_c) decreases as grain size increases. For the small grain (86 nm) $E_c = 3.2$ kV/cm, and for the largest grain (123 nm) $E_c = 2.04$ kV/cm. Energy barrier for switching ferroelectric domain must be broken through and energy barrier decreases as grain size increases. So reversal polarization process of a ferroelectric domain is much easier inside a large grain than in a small grain [19]. Moreover, it is also found that the remanent polarization (P_r) increases with the increase of grain size of BST ceramics (Table 3). Effect of grain

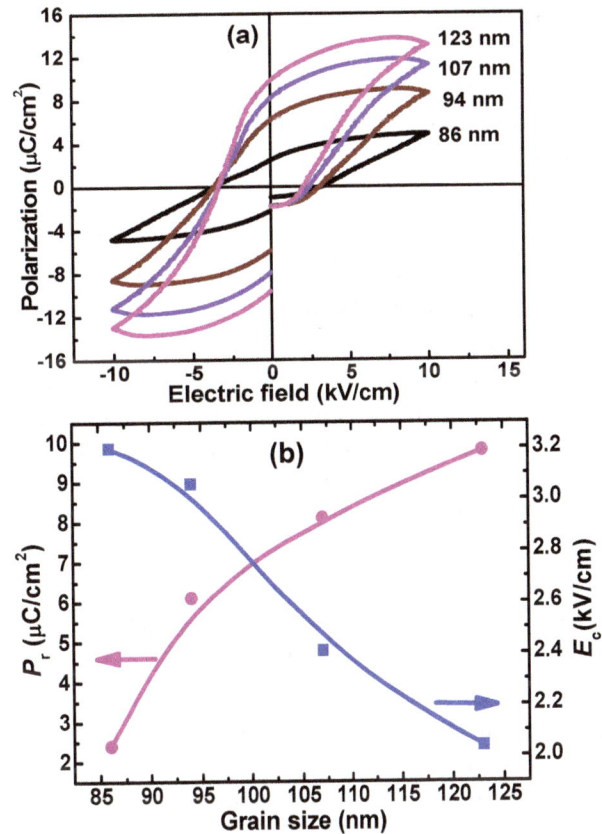

Fig. 6 (a) Ferroelectric hysteresis loops at room temperature and 1 kHz with different grain sizes; (b) grain size dependence of the remanent polarization (P_r) and the coercive field (E_c) of BST ceramics.

Table 3 Values of E_c, P_{max} and P_r obtained from the hysteresis loops of BST ceramics with different grain sizes

Grain size (nm)	E_c (kV/cm)	P_r (μC/cm^2)	P_{max} (μC/cm^2)
86	3.20	2.4	4.82
94	3.06	6.1	8.56
107	2.41	8.1	11.24
123	2.04	9.8	13.06

boundary on polarization includes two facets. On one hand, grain boundary is a low-permittivity region. That means the grain boundary has poor ferroelectricity. Therefore, polarization of grain boundary may be little, and even none. On the other hand, space charges in grain boundary exclude polarization charge on grain surface, and depletion layer on grain surface can be formed. That results in polarization discontinuity on grain surface to form depolarization field, and polarization decreases. The number of grain boundary increases as grain size decreases. Consequently, the remanent polarization increases as the grain size increases.

4 Conclusions

In this study, nanocrystalline barium strontium titanate ferroelectric ceramics have been synthesized from the fine constituent powders by high energy ball milling at sintering temperatures of 1200–1350 ℃. From the results of XRD, it is evident that BST is of perovskite crystal structure. As the sintering temperature increased, the average grain size of BST ceramics increased from 86 nm to 123 nm. As the grain size increased, the dielectric constant both at room temperature and the Curie transition temperature increased. It can be noticed that the piezoelectric constant of our experiment can be increased with grain size. It is found that the ferroelectric hysteresis loops change significantly with grain size, and the coercive electric field (E_c) decreases as grain size increases. For the small grain (86 nm) $E_c = 3.2$ kV/cm, and for the largest grain (123 nm) $E_c = 2.04$ kV/cm. In addition, it is also found that the remnant polarization (P_r) increases with the increase of grain size of BST ceramics. So, BST systems are very attractive as an alternative promising lead free material for piezoelectric and tunable capacitor device applications.

References

[1] Akdogan EK, Leonard MR, Safari A. Chapter 2. Size effects in ferroelectric ceramics. In *Handbook of Low and High Dielectric Constant Materials and Their Applications*, *Vol. 2*. Nalwa HS, Ed. London: Acadamic Press, 1999.

[2] Ahn CH, Rabe KM, Triscone J-M. Ferroelectricity at the nanoscale: Local polarization in oxide thin films and heterostructures. *Science* 2004, **303**: 488–491.

[3] Chen H, Yang C, Fu C, *et al.* The size effect of $Ba_{0.6}Sr_{0.4}TiO_3$ thin films on the ferroelectric properties. *Appl Surf Sci* 2006, **252**: 4171–4177.

[4] McNeal MP, Jang S-J, Newnham RE. The effect of grain and particle size on the microwave properties of barium titanate ($BaTiO_3$). *J Appl Phys* 1998, **83**: 3288.

[5] Tang X-G, Chen HL-W. Effect of grain size on the electrical properties of $(Ba,Ca)(Zr,Ti)O_3$ relaxor ferroelectric ceramics. *J Appl Phys* 2005, **97**: 034109.

[6] Tang XG, Wang J, Wang XX, *et al.* Effects of grain size on the dielectric properties and tunabilities of sol–gel derived $Ba(Zr_{0.2}Ti_{0.8})O_3$ ceramics. *Solid State Commun* 2004, **131**: 163–168.

[7] Hu Z, Wang G, Huang Z, *et al.* Effects of thickness on the infrared optical properties of $Ba_{0.9}Sr_{0.1}TiO_3$ ferroelectric thin films. *Appl Phys A* 2004, **78**: 757–760.

[8] Jin BM, Kim J, Kim SC. Effects of grain size on the electrical properties of $PbZr_{0.52}Ti_{0.48}O_3$ ceramics. *Appl Phys A* 1997, **65**: 53–56.

[9] Parashar SKS, Choudhary RNP, Murty BS. Ferroelectric phase transition in $Pb_{0.92}Gd_{0.08}(Zr_{0.53}Ti_{0.47})_{0.98}O_3$ nanoceramic synthesized by high-energy ball milling. *J Appl Phys* 2003, **94**: 6091.

[10] Parashar SKS, Choudhary RNP, Murty BS. Size effect of $Pb_{0.92}Nd_{0.08}(Zr_{0.53}Ti_{0.47})_{0.98}O_3$ nanoceramic synthesized by high-energy ball milling. *J Appl Phys* 2005, **98**: 104305.

[11] Wu YJ, Uekewa N, Kakegawa K, *et al.* Compositional fluctuation and dielectric properties of $Pb(Zr_{0.3}Ti_{0.7})O_3$ ceramics prepared by spark plasma sintering. *Mater Lett* 2002, **57**: 771–775.

[12] Wagner S, Kahraman D, Kungl H, *et al.* Effect of temperature on grain size, phase composition, and electrical properties in the relaxor-ferroelectric-system $Pb(Ni_{1/3}Nb_{2/3})O_3$–$Pb(Zr,Ti)O_3$. *J Appl Phys* 2005, **98**: 024102.

[13] Xu Q, Zhang X-F, Liu H-X, *et al.* Effect of sintering temperature on dielectric properties of $Ba_{0.6}Sr_{0.4}TiO_3$–MgO composite ceramics prepared from fine constituent powders. *Mater Design* 2011, **32**: 1200–1204.

[14] Shaw TM, Trolier-McKinstry S, McIntrye PC. The properties of ferroelectric films at small dimensions. *Annu Rev Mater Sci* 2000, **30**: 263–298.

[15] Kamel TM, de With G. Grain size effect on the poling of soft $Pb(Zr,Ti)O_3$ ferroelectric ceramics. *J Eur Ceram Soc* 2008, **28**: 851–861.

[16] Randall CA, Kim N, Kucera JP, *et al.* Intrinsic and extrinsic size effects in fine-grained morphotropic-phase-boundary lead zirconate titanate ceramic. *J Am Ceram Soc* 1998, **81**: 677–688.

[17] Shen Z-Y, Li J-F. Enhancement of piezoelectric constant d_{33} in $BaTiO_3$ ceramics due to nano-domain structure. *J Ceram Soc Jpn* 2010, **118**: 940–943.

[18] Shao S, Zhang J, Zhang Z, *et al.* High piezoelectric properties and domain configuration in $BaTiO_3$ ceramics obtained through the solid-state reaction route. *J Phys D: Appl Phys* 2008, **41**: 125408.

[19] Leu C-C, Chen C-Y, Chien C-H, *et al.* Domain structure study of $SrBi_2Ta_2O_9$ ferroelectric thin films by scanning capacitance microscopy. *Appl Phys Lett* 2003, **82**: 3493.

Development and characterization of 3CaO·P₂O₅–SiO₂–MgO glass-ceramics with different crystallization degree

Juliana Kelmy M. F. DAGUANO[a], Paulo A. SUZUKI[a], Kurt STRECKER[b],
José Martinho Marques de OLIVEIRA[c], Maria Helena Figueira Vaz FERNANDES[d],
Claudinei SANTOS[a,e,*]

[a]Universidade de São Paulo - Escola de Engenharia de Lorena, USP-EEL - Pólo Urbo-Industrial, s/n, Gleba AI-6, Lorena-SP, CEP 12600-000, Brazil
[b]Universidade Federal de São João del-Rei, – UFSJ-CENEN, Campus S° Antônio - Praça Frei Orlando 170 – Centro, S. J. del-Rei-MG. CEP 36307-352, Brazil
[c]Escola Superior Aveiro Norte, Edifício Rainha, 3720-232 O. Azeméis, Portugal
[d]Universidade de Aveiro, Campus Universitário de Santiago 3810-193 Aveiro, Portugal
[e]Universidade do Estado do Rio de Janeiro – Faculdade de Tecnologia de Resende – UERJ-FAT – Rod. Presidente Dutra, km, 298, Resende-RJ, CEP 27537-000, Brazil

Abstract: The CaO–P₂O₅–SiO₂–MgO system presents several compounds used as biomaterials such as hydroxyapatite (HA), tricalcium phosphate (TCP) and TCP with magnesium substituting partial calcium (TCMP). The β-TCMP phase with whitlockite structure has interesting biological features and mechanical properties, meeting the requirements of a bioactive material for bone restoration. In this work, the production of Mg-doped TCP, β-TCMP, has been investigated by crystallization from a glass composed of 52.75 wt% 3CaO·P₂O₅, 30 wt% SiO₂ and 17.25 wt% MgO (i.e., 31.7 mol% CaO, 10.6 mol% P₂O₅, 26.6 mol% MgO and 31.1 mol% SiO₂) using heat treatments between 775 ℃ and 1100 ℃ for up to 8 h. The devitrification process of the glass has been accompanied by differential scanning calorimetry (DSC), high-resolution X-ray diffraction (HRXRD), relative density and bending strength measurements. The characterization by HRXRD and DSC revealed the occurrence of whitlockite soon after the bulk glass preparation, a transient non-cataloged silicate between 800 ℃ and 1100 ℃, and the formation of diopside in samples treated at 1100 ℃ as crystalline phases. The overall crystalline fraction varied from 26% to 70% depending on the heat treatments. Furthermore, contraction of the a-axis lattice parameter and expansion of the c-axis lattice parameter of the whitlockite structure have been observed during the heat treatments, which were attributed to the β-TCMP formation with the partial substitution of Ca²⁺ by Mg²⁺. Relative densities near 99% and 97% for the glass and glass-ceramics respectively indicated a discrete reduction as a function of the devitrification treatment. Bending strengths of 70 MPa and 120 MPa were determined for the glass and glass-ceramic material crystallized at 975 ℃ for 4 h, respectively.

Keywords: glass-ceramics; heat treatment; high-resolution X-ray diffraction (HRXRD); bending strength

* Corresponding author.
E-mail: claudinei@demar.eel.usp.br

1   Introduction

Calcium phosphate ceramics are used as artificial bone material, in the form of solid pieces, filling powders and porous films. Currently, calcium phosphate ceramics are widely used in orthopedic and dental applications as repair for bone defects, increasing and maintaining of alveolar bone crests, reallocation of tooth root, auricular implants, consolidation of spine, and coating of metal implants [1]. The advantage of these ceramic materials consists in their chemistry based on calcium and phosphor, which contribute to the ionic equilibrium between the biological liquids and the ceramic [2]. The use of glass and glass-ceramics as bone implants may be limited by inadequate mechanical properties. Therefore, fracture strength becomes a relevant factor to be evaluated for their use as structural implants, which require the ability to withstand stress without failure.

Calcium phosphate-based ceramics are only available as blocks or granules used in small implants and in non-load bearing areas, or as coatings for metals, because there is a potential risk of failure due to their very slow resorption kinetics and poor mechanical properties. Of all calcium phosphates, tricalcium phosphate (TCP) and hydroxyapatite (HA) have been studied most intensively because of their use in biological applications for their high compatibility with natural bone. HA is considered as the most similar artificial and bioactive material to bone and teeth [3]. However, the natural bone mineral component is mainly non-stoichiometric HA (i.e., Ca:P molar ratio is other than 1.67 and differs in crystallinity and specific surface area), making it more reactive in a biological environment. On the other hand, TCP is known for its resorption by body, because it has a Ca:P molar ratio of lower than 1.67 and therefore greater solubility and resorption capacity *in vivo*, ensuring its bone-conductivity and bioactivity [4].

The TCP with composition of $Ca_3(PO_4)_2$ exhibits a Ca:P molar ratio of 1.5. The TCP exits in four crystal modifications depending on temperature and pressure. β-TCP is stable until 1125 ℃, transforming into α-TCP at higher temperatures. At 1430 ℃, the transformation into γ-TCP occurs. γ-TCP and a high-pressure form are difficult to obtain [5]. Of these four polymorphs, β-TCP phase is the most promising for bioceramic implant materials because of its low ratio of dissolution, chemical stability and mechanical strength,

attending the requirements for a material in regenerative medicine [6].

A β-TCP variation is the β-TCMP phase with whitlockite-like structure, corresponding to a solid solution with composition of $Ca_{2.589}Mg_{0.411}(PO_4)_2$. β-TCMP is formed by the substitution of Ca^{2+} by Mg^{2+} ions, preferentially on Ca(5) sites in the β-TCP structure [7]. Hence, the preparation of whitlockite is interesting for applications in both bone and tooth replacements. TCP can be obtained either by solid-state reaction or by sintering calcium-deficient apatites. However, the limits of solid-state reaction are inhomogeneity, non-uniformity in the particle size distribution, and a higher chance of impurities in the final product as well [8,9]. On the other hand, β-TCMP was present in glass-ceramics based on $3CaO \cdot P_2O_5$–SiO_2–MgO system via crystallization glass bulk [10].

In recent years, the chemical modification of β-TCP through ionic substitutions has received much attention, considering that major components of biological tissues are composed of a calcium phosphate mineral phase containing a variety of other elements [11]. Mg^{2+} is considered the most important ion used in calcium substitution in the crystal structure of β-TCP, leading to a change in the biological and chemical behavior of these materials. Incorporation of Mg^{2+} ions suppresses the β→α TCP phase transition and the bone metabolism improves, reducing cardiovascular diseases by promoting catalytic reactions and controlling biological functions [8]. Banerjee *et al.* [12] demonstrated better cell attachment and proliferation for doped β-TCP in an *in vitro* cell–material interaction study and in an *in vivo* study as well. The authors reported that bone forms more quickly in doped samples than in control samples.

As reported by Schroeder *et al.* [13], β-TCP crystallizes in the rhombohedral space group *R3c* (161) with unit cell parameters of $a = b = 10.439$ Å and $c = 37.375$ Å (hexagonal setting). Whitlockite (β-TCMP) is very similar to the β-TCP structure with unit cell parameters of $a = b = 10.337$ Å and $c = 37.068$ Å. Enderle *et al.* [14] examined β-TCP powders with Mg additions obtained by a solid-state reaction, using X-ray powder diffraction combined with the Rietveld method. A maximum of about 16 mol% Mg^{2+} substitution on Ca(4) and Ca(5) sites in the β-TCP structure was found. Exceeding 15 mol% Mg^{2+} will result in either un-reacted magnesium chloride (Mg adsorbs on the calcium phosphate surface) or even the precipitation of Mg-rich phases, such as stanfieldite.

The present work investigated an alternative route to the conventional powder processing method to prepare β-TCMP by crystallization from a bulk glass based on the $3CaO \cdot P_2O_5$–SiO_2–MgO system. Furthermore, the resulting mechanical properties such as hardness and fracture toughness were related to the degree of crystallization.

2   Experimental procedure

2.1   Synthesis of glasses and glass-ceramics

High-purity powders of $CaCO_3$ (SYNTH), MgO (SYNTH), SiO_2 (Fluka) and $Ca(H_2PO_4)_2 \cdot H_2O$ (SYNTH) were used as starting materials, and the glass with composition of 52.75 wt% $3CaO \cdot P_2O_5$, 30 wt% SiO_2 and 17.25 wt% MgO was prepared. This glass composition has already been studied and combines good bioactivity with considerable mechanical strength [15,16]. The powders were mixed by ball milling for 4 h using isopropilic alcohol as vehicle. After milling, the powder mixture was dried at 90 ℃ for 24 h and passed through a sieve 230 Mesh with openings of 64 μm for deagglomeration.

The melting of the glass was done in a platinum crucible at 1600 ℃ for 4 h. The liquid glass was poured into a metallic mould, casting bars of 15 mm × 15 mm × 50 mm. Immediately after casting, the specimens were annealed at 700 ℃ for 2 h and then cooled down to room temperature at a rate of 3 ℃/min.

Samples were further treated at different temperatures for 0.25 h, 2 h, 4 h and 8 h, and also cooled down at a rate of 3 ℃/min in order to induce crystallization of the bioglass. Sample designation of the heat-treated materials is summarized in Table 1.

Table 1   Sample designation of glass-ceramics obtained by heat treatment

Heat treatment	Time (h)			
(℃)	0.25	2	4	8
775	V775-1	—	V775-4	V775-8
800	V800-1	V800-2	V800-4	V800-8
850	V850-1	V850-2	V850-4	V850-8
900	V900-1	V900-2	V900-4	V900-8
975	V975-1	V975-2	V975-4	V975-8
1000	—	—	V1000-4	—
1050	—	—	V1050-4	—
1100	—	—	V1100-4	—

Additional heat treatments were made at 700 ℃ for 4 h (V700-4) and 950 ℃ for 4 h (V950-4).

2.2   Characterizations

The non-isothermal crystallization kinetics was studied using differential scanning calorimetry (DSC, Model 404, NETZSCH, Germany). Glass powders of different particle size and monolithic pieces were treated in a platinum crucible at 10 ℃/min from room temperature until 1200 ℃.

The densities of the V700-4, V775-4, V900-4, V975-4 and V1100-4 bulk samples were determined by the immersion method proposed by Archimedes. Real densities of the samples were evaluated by a helium picnometer model AccuPyc 1330-Micrometrics using crushed samples with particle size smaller than 63 μm.

2.3   Phase analysis

The heat-treated samples were analyzed by high resolution X-ray diffraction (HRXRD), using a diffractometer with multiple axes (Huber, Germany). The samples were crushed and sieved until particle size smaller than 32 μm. The measurements were realized in a setup of two coupled concentric circles (ω–2θ), with a monochromatic X-ray beam of 10 keV ($\lambda = 1.2398$ Å). The powders were put in a cylindrical support of 10 mm in diameter and 2 mm in depth and maintained rotating in order to promote randomness of orientation of the crystallographic planes. The diffracted beam was collected by a germanium crystal (200) and a scintillation detector. The powders were analyzed under diffraction angles ranging from 7° to 50°, with a step width of 0.01° and 1 s exposure time per position.

The amount of the crystalline phase contained in the glass-ceramic samples was determined by an adoption of the methods used by Krimm and Tobolsky [17], where the percentage of crystallinity I_c is calculated by the ratio of the crystalline area A_c present in the diffractogram of the glass-ceramics to the total area A_t (A_t = amorphous + crystalline) present in these diffractograms, using the following equation:

$$I_c = (A_c / A_t) \times 100\% \qquad (1)$$

After determining the degree of crystallinity, the amount of the whitlockite phase has been estimated, using an internal standard. This technique consists of relating the peak intensities of a crystal phase with the peaks of the internal standard material, added in a known proportion. For this analysis, the sample materials and metallic tin as internal standard were mixed in certain proportions,   as shown in Table 2, and analyzed by X-ray diffraction using Cu Kα radiation, in the 2θ range of 8° to 62°, with a step width of 0.05° and 3 s exposure time per position.

Table 2   Sample to metallic tin mass ratio (internal standard)

Sample	m_{glass} (g)	m_{tin} (g)	m_{glass}/m_{tin}
V775-4	0.875	1.595	0.55
V850-4	0.446	0.849	0.53
V950-4	0.349	0.701	0.50
V975-4	0.704	1.320	0.53
V1100-4	0.704	1.329	0.53

The relationship of the peak intensity of the whitlockite phase and the internal standard was determined by the area of the (200) plane of whitlockite and the area of the (211) plane of tin (Fig. 1).

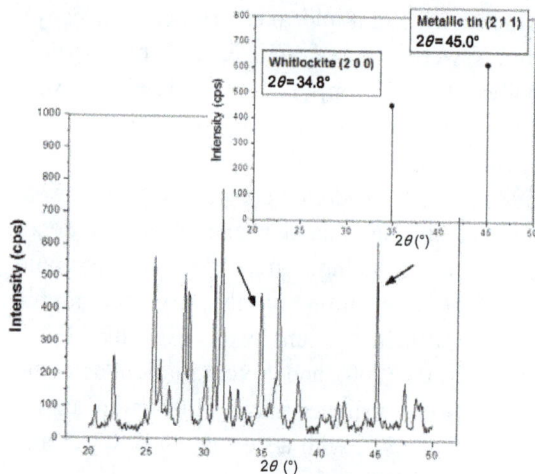

Fig. 1   Quantification of the whitlockite phase in the glass-ceramic materials using Sn as internal standard.

The crystal structure of the samples was analyzed by X-ray diffraction and refined by the Rietveld analysis. This analysis permits to evaluate the influence of the thermal treatment on the structure of whitlockite by the variations of the lattice parameters and the volume of the unit cell in functions of the treatment temperature.

2. 4   Mechanical properties

The strength of the glass and glass-ceramics heat treated at 700 ℃, 775 ℃, 900 ℃, 975 ℃ and 1100 ℃ for 4 h was determined by four-point bending tests, using a MTS 810 (50 N) universal testing machine. Batches of six samples were grinded and polished, obtaining bars of 2.0 mm × 1.5 mm × 25.0 mm

according to ASTM C 1161-02c. The tests were conducted using a four-point bending device with outer and inner spans of 20 mm and 10 mm, respectively, and a velocity of the crosshead displacement of 0.2 mm/min. The bending strength of the samples was calculated by

$$\sigma_f = \frac{3}{2} F_A \frac{I_1 - I_2}{bh^2} \tag{2}$$

where σ_f is the bending strength (MPa); F_A the rupture load (N); b the width of the sample (mm); h the height of the sample (mm); I_1 the outer span distance (mm) and I_2 the inner span distance (mm).

3   Results and discussion

3. 1   Thermal analysis

The results of DSC (Fig. 2) for a monolithic sample and powder samples of size ranges of 125–75 μm and 38–22 μm respectively indicate that the heat-treated glass exhibits a step near to 715 ℃, which is attributed to the glass transition temperature T_g. Two exothermal peaks are also detected at $T_{p1} = 830$ ℃ and at T_{p2} varying between 915 ℃ and 975 ℃ depending on the particle size.

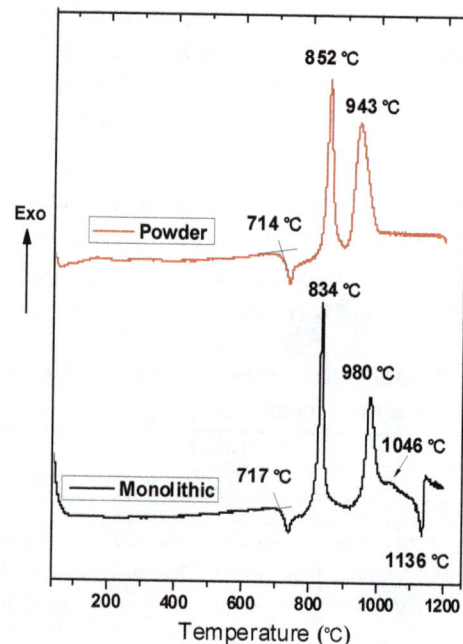

Fig. 2   Thermal analysis of $3CaO \cdot P_2O_5 - SiO_2 - MgO$ glass.

Another exothermal reaction at 1040 ℃ has been observed in the bulk sample, but of small intensity, suggesting a phase transformation of the material. This phase transformation may be confirmed later by the analysis of the crystal phases present. At last during cooling of the glass an exothermal peak at $T \approx 1115$ ℃ has been observed.

Previous studies [15,16] suggested that the first peak corresponds to the formation of the TCP with partial substitution of Mg by Ca (whitlockite) of composition $3(Ca,Mg)O \cdot P_2O_5$, and the second corresponds to the precipitation of enstatite with partial substitution of Ca by Mg, $(Mg,Ca)O \cdot SiO_2$. Later results of the phase analysis by X-ray diffraction (Fig. 3) confirm the crystallization of the primary phase whitlockite, but enstatite has not been found. It is only known that the second exothermal peak is due to the formation of a metastable silicate that recrystallizes at temperatures above 1050 ℃, as indicated by the third exothermal peak, into the stable phase diopside, $CaMgSi_2O_6$.

Furthermore, it is possible to affirm that the whitlockite phase presents internal crystallization, because no variation of the transformation temperature with the particle size of the materials occurs. On the other hand, the crystallization of the silicate, indicated by the second exothermal peak, varies with the particle size of the material. With increasing surface (decreasing particle size), the crystallization temperature decreases, indicating that the crystallization of the diopside phase occurs preferentially at the surface. This crystallization behavior has been studied by Ray et al. [18] in the thermal analysis of glasses.

3.2 Crystallization behavior

The crystallization process of the materials can be accompanied by analyzing Fig. 3, showing a decrease of the glassy phase and the formation of the whitlockite phase, the transient silicate and the diopside.

Sample V700-4 exhibits an amorphous, glassy structure, because no diffraction peak is observed but only a diffuse high background characteristic of materials without long-range atomic order. In sample V775-4, the formation of whitlockite can be verified besides the amorphous phase. The samples V800-4, V900-4 and V975-4 exhibit quite similar diffractogramms, with whitlockite and the transient silicate as crystal phases, only differing in their crystallization degree which tends to increase with increasing heat-treatment temperature. In sample V1100-4, diopside instead of the transient silicate has been detected, suggesting a phase transformation has occurred.

All glass-ceramics studied in this work present the whitlockite phase (PDF#87-1582) with Ca substituting Mg, $(Ca,Mg)O \cdot P_2O_5$. It is believed that the formation of whitlockite is induced by the phase separation

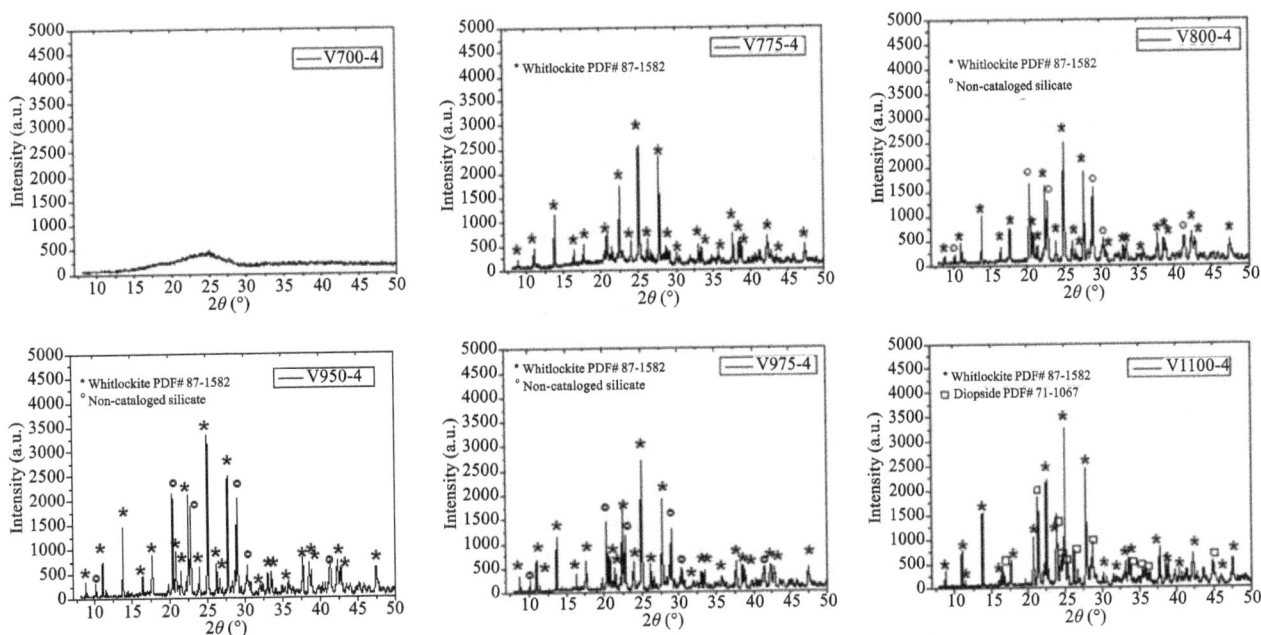

Fig. 3　Crystal phases present in glass-ceramics of the system $3CaO \cdot P_2O_5$–SiO_2–MgO after heat treatments at different temperatures.

already existent in the glass in form of silica droplets [19]. This phase is TCP with partial substitution of Ca by Mg with composition of $Ca_{2.5725}Mg_{0.4275}(PO_4)_2$, also known as β-TCMP.

A second phase starts to crystallize at 800 ℃. In a previous study [20], it has been suggested that this phase is enstatite, $(Mg,Ca)O·SiO_2$, but it has been verified that some of the main peaks do not correspond to this phase. Furthermore, after carefully comparing the X-ray diffraction results with the JCPDS data base, no compatible phase has been identified. Therefore, the authors believed that the "transient silicate" is a metastable phase, originated from the residual glassy matrix after crystallization of whitlockite. This phase disappears in samples treated at 1100 ℃ (Fig. 3). Metastable phases are quite commonly observed during the crystallization of a glass by thermal treatment [21].

A third crystalline phase, diopside (PDF#71-1067), $CaMgSi_2O_6$, has been found only in the sample V1100-4. This phase is considered to be the result of a transformation of the transient silicate. When glass-ceramics with metastable phases are heated up to temperatures high enough that solid-state reactions may occur, transformations into stable phases take place, in this case, diopside. The results of the DSC analysis enforce this supposition as indicated by the third exothermal peak at 1046 ℃, suggesting a phase transformation characterized by smaller enthalpy.

For a comparative analysis in regard to the crystallization temperatures of the phases, the diffractograms are plotted in the same figure (Fig. 4). For this a 2θ range between 17° and 27° is chosen, because the most intense peak referring to whitlockite

is observed at approximately 25°.

In general, the main crystallographic changes of the structure of the material occur between 700 ℃ and 775 ℃ as shown in Figs. 3 and 4, and between 975 ℃ and 1100 ℃, due to the start of the crystallization process and the transformation into diopside, respectively.

As the phase transformations and the increasing crystallinity interfere directly in the bioactivity and mechanical properties, an estimate of the amount of whitlockite present in the glass-ceramics has been made using an internal standard. In the quantitative X-ray diffraction analysis using an internal standard, the peak intensities are related to the volumetric proportion of the phases present. In the glass-ceramics analyzed in this work, an amorphous phase, whitlockite, a transient silicate (between 800 ℃ and 975 ℃) and diopside (above 1050 ℃), have been identified. However, the peak intensities of whitlockite and tin (the internal reference material) cannot be related directly with their respective volume amounts, because the X-ray absorption of each phase is different. To solve this problem, it would be necessary to analyze mixtures containing known proportions of whitlockite and tin, which is not possible. On the other hand, the ratio of the peak intensities of each phase is associated to the volume proportion of whitlockite and tin in each sample, which makes it possible to determine the relative variation of whitlockite by comparison of different samples.

For this comparison, equal proportions between the mass of the sample and tin are necessary. A correction factor, CF, has been calculated in order that the ratio between the mass of whitlockite and tin always equals 1:1 (arbitrary value) and this is multiplied with the peak intensity of tin. In this way, it may be concluded, for example, that an increasing proportion of whitlockite in the samples indicates a decrease of the amount of glass or the silicate phase and vice versa.

The integrated peak intensities of whitlockite and tin, as well as the correction factor CF used are listed in Table 3. The term x_w relates the peak intensities of whitlockite and tin with the sample mass and the internal standard using the correction factor, and represents a numerical value associated to the whitlockite volumetric fraction of different samples. The analysis of this numerical value permits to establish the interaction of crystal phases in function of the temperature of the heat treatment.

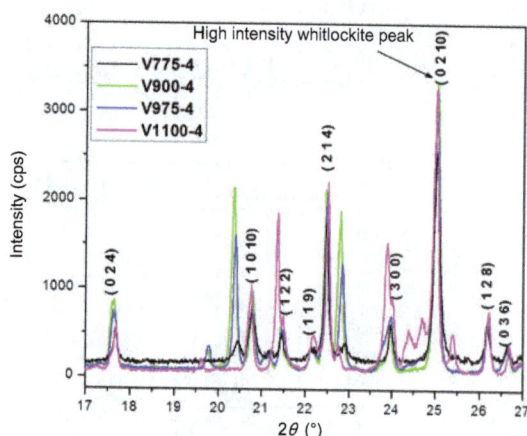

Fig. 4 Comparison of the diffractograms obtained after different heat treatments.

Table 3   Estimation of the amount of whitlockite in some glass-ceramics using Sn as internal standard

Sample	Integrated intensity whitlockite I_w	Integrated intensity tin I_E	$CF = m_{sample}/m_{tin}$	$x_w = I_w/(CF \times I_E)$
V775-4	41.7	36.9	0.55	2.06
V850-4	181.4	54.3	0.53	6.30
V950-4	176.3	71.4	0.50	4.94
V975-4	133.3	102.6	0.53	2.45
V1100-4	84.1	62.8	0.53	2.53

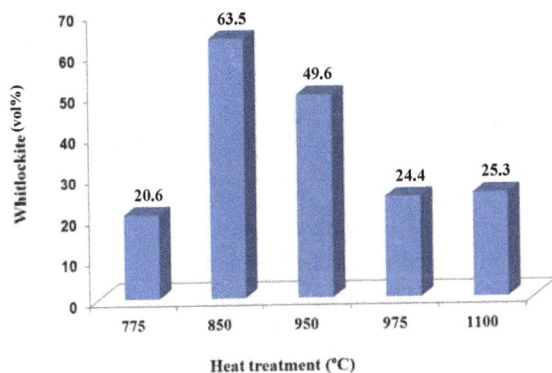

Fig. 5   Relative volume percentage of whitlockite in the glass-ceramics treated at different temperatures.

Figure 5 shows the volume proportions of the whitlockite phase in the samples V775-4, V850-4, V950-4, V975-4 and V1100-4.

Based on the X-ray diffractograms shown in Figs. 3 and 4, the results of the quantitative analysis of whitlockite listed in Table 3 and illustrated in Fig. 5, it can be concluded that in the sample V775-4, whitlockite formation has started but still in a low proportion in regard to the glass phase. In the sample V850-4, the amount of whitlockite has increased to approximately 63 vol%. The heat treatments at 950 ℃ and 975 ℃ result in a decrease of the whitlockite fraction of the crystallized portion, probably due to the formation of the transient silicate. In the sample V1100-4, the proportion of whitlockite remains almost the same as in the sample V975-4.

The X-ray diffractograms of glass-ceramic powders obtained by heat treatments of 0.25 h, 2 h, 4 h and 8 h show that time does not significantly influence the formation of new phases during crystallization, except for samples treated at 800 ℃, where it has been possible to observe the crystallization of the silicate increases with the time of heat treatment (Fig. 6).

In order to evaluate the effect of the heat-treatment time on the crystallization, the crystallinity has been

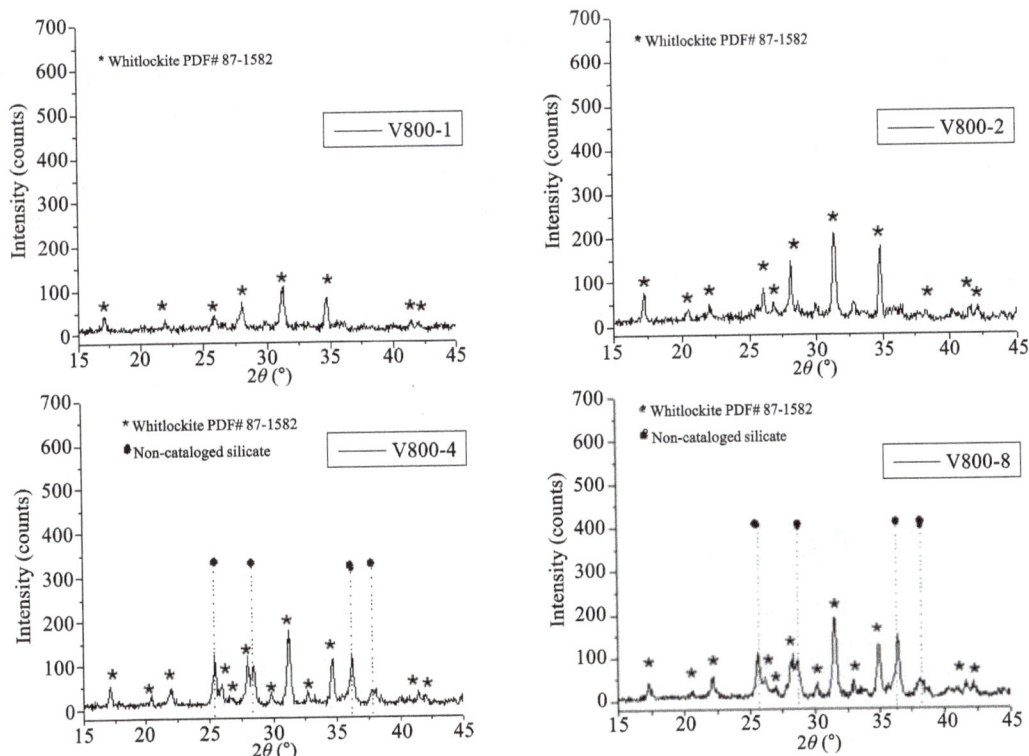

Fig. 6   Effect of the time of heat treatment on the formation of the "transient silicate" at 800 ℃ for 0.25 h (V800-1), 2 h (V800-2), 4 h (V800-4) and 8 h (V800-8).

estimated and compared. Therefore, the method used by Krimm and Tobolsky [17] has been taken as basis (Eq. (1)), and the percentages of crystal phases obtained during the treatments at varying temperatures are shown in Fig. 7.

In general, crystallinity increases with increasing duration of the heat treatment. Treatments of 8 h results in the highest crystallinity, because more time is given for that the structural rearrangement of atoms may occurr. The highest gain in crystallinity can be observed in samples treated at 775 ℃ for 4 h and 8 h with an increase of 150%. For samples treated at 800 ℃, only small variations for treatments between 0.25 h and 2 h and between 4 h and 8 h can be noted, but crystallinity changes 50% in the intervall between 2 h and 4 h. In this case, the increase is attributed to the formation of the transient silicate.

3.3 Structural analysis of whitlockite

The variations of the lattice parameters of the whitlockite structure in functions of the heat-treatment temperature are presented in Fig. 8.

In functions of the heat-treatment temperature, it has been possible to calculate the volume change of the unit cell of the whitlockite structure, related to the incorporation of Mg^{2+} ions into the structure. As reported by Schroeder et al. [13], the β-TCP phase has rhomboedric geometry, space group $R3c$ (161), and lattice parameters of $a = b = 10.439$ Å and $c = 37.375$ Å. The structure of whitlockite, β-TCMP, is quite similar to the β-TCP structure, but of smaller lattice parameters of $a = b = 10.337$ Å and $c = 37.068$ Å, in agreement with the results of the glass-ceramics prepared in this work. This result is the consequence of the substitution of Ca^{2+} ions by smaller Mg^{2+} ions ($r_{Mg^{2+}} = 0.75$ Å and $r_{Ca^{2+}} = 1.05$ Å), reducing the

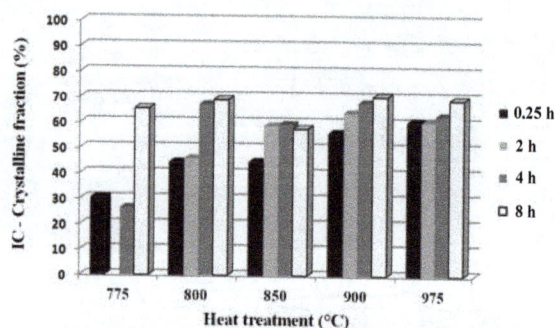

Fig. 8   Variations of the lattice parameters a and c of whitlockite due to the heat treatments.

interatomic spacings.

Due to the substitution of Ca^{2+} by Mg^{2+} ions in the structure of β-TCP, the lattice parameter a diminishes from 10.349 Å to 10.332 Å. Studies conducted by Enderle et al. [14] and Kannan et al. [8] related an expansion of the parameter c of the β-TCP phase for substitutions of higher than 10 mol% Ca. They attributed this behavior to the observation that until 10 mol%, the Mg^{2+} ions are allocated in Ca(5) sites resulting in shrinkage along the c axis, and after this site has been completely occupied, the Ca(4) sites start to be occupied and the lattice parameter c increases up to the maximum substitution. The solubility limit is 14.25 mol% of Mg^{2+} ions in the Ca(4) and Ca(5) sites of the β-TCP structure. In this way, it is believed that

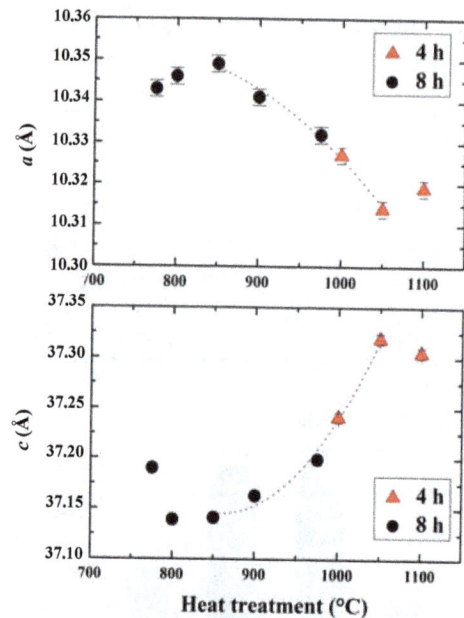

Fig. 7   Effect of the duration of heat treatments at different temperatures on the crystallinity of the materials.

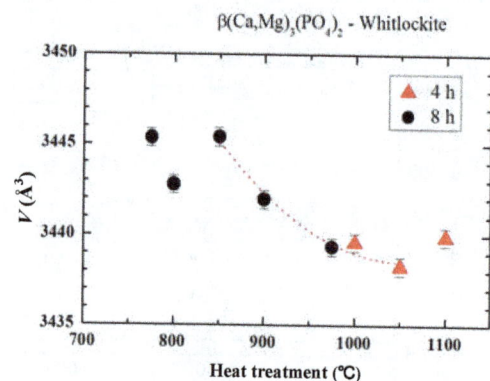

Fig. 9   Variation of the unit cell volume of whitlockite in function of the heat-treatment temperature.

the percentage of Mg^{2+} ions present in the structure of whitlockite or β-TCMP of the glass-ceramics varies between 10 mol% and 14.25 mol%.

The contraction of the whitlockite structure may be better understood by the analysis of the unit cell volume, as shown in Fig. 9. Initially, the sample V775-8 presents a unit cell volume of about 3445 $Å^3$, and with increasing temperature of the heat treatment, the volume decreases to about 3438 $Å^3$ for the sample V1050-4, corresponding to a volume reduction of less than 1%. Whitlockite (PDF#87-1582) presents a unit cell volume of 3439.05 $Å^3$, in agreement with the standard used in this work, indicating that the measurements and the refinements done using the high-definition X-ray diffractograms are in good agreement with the literature [14].

The results of density, relative density and bending strength are presented in Table 4.

Table 4 Density, relative density and bending strength of the samples

Sample	ρ_{real} (g/cm^3)	$\rho_{apparent}$ (g/cm^3)	Relative density (%)	Bending strength (MPa)
V700-4	2.933 ± 0.003	2.87 ± 0.01	98.8 ± 0.3	71 ± 4
V775-4	2.913 ± 0.003	2.86 ± 0.09	98.3 ± 0.3	105 ± 7
V900-4	3.012 ± 0.002	2.94 ± 0.01	97.6 ± 0.3	116 ± 10
V975-4	3.043 ± 0.002	2.96 ± 0.01	97.3 ± 0.3	119 ± 12
V1100-4	3.079 ± 0.002	2.97 ± 0.12	96.3 ± 2.5	69 ± 4

The real density of the samples increases as a function of the heat-treatment temperature because of the crystallization of phases. The sample V700-4 (nucleated glass) exhibits a density of 2.933 g/cm^3, while sample V975-4 with some residual glass phase, whitlockite and the "transient silicate" as crystal phases shows an increase of 3.5% in real density. The V1100-4 sample, with residual glass phase, whitlockite and diopside, shows a further density increase of almost 5%. The crystalline phases, whitlockite and diopside, with theoretical densities of 3.7 g/cm^3 and 3.28 g/cm^3, respectively, cooperate directly with the increased density of the samples.

This behavior is well accepted, because with increasing crystallinity of the material, a volume decrease is expected due to the restructuring approach of atoms from the amorphous glass matrix. However, the relative density of these samples decreases with increasing heat-treatment temperature. This behavior can be related to the formation of voids in the structure due to the rearrangement of atoms by the crystallization process. Recent studies [22,23] showed the formation of internal porosity in the diopside–albanite glass-ceramic system, associated to diopside crystallization, called "induced crystallization porosity". Porosity is generally of the closed and intragranular type, and found in the center of the samples.

The bending strength values range between 70 MPa and 120 MPa for the glass and glass-ceramics, respectively. It is possible to note an increase of about 70% in bending strength of the samples thermally treated at 975 ℃, compared with the untreated glass sample. This improvement in strength is due to the higher atomic ordering found in materials with a high crystallinity (crystalline phase ≈ 65%), which lowers the defect energy and thus decreases the tendency to fracture. Possibly, the "silicate transient" phase may be contributing to the improvement of the flexural strength of the material, since a higher content of this phase is found in the sample V975-4. In contrast, at temperature of 1100 ℃, a lower modulus of rupture is found, which can be related to the lower relative density of 98.5%.

4 Conclusions

Time and temperature of the thermal treatments influence the crystallization behavior of whitlockite in the glass-ceramics system of 3CaO·P_2O_5–SiO_2–MgO. Increasing the duration of the heat treatment results in higher degrees of crystallinity, while higher temperatures result in different crystal phases. For all glass-ceramics studied, the major crystal phase formed has been whitlockite or β-TCMP, 3(Ca,Mg)·P_2O_5. Furthermore, a metastable silicate phase has been found in samples treated in the temperature range between 800 ℃ and 1000 ℃. Close to 1050 ℃, this phase transforms into the stable phase diopside, $CaMgSi_2O_6$. By variation of the heat treatments, it has been possible to obtain glass-ceramics with a degree of crystallinity superior to 20% and a maximum volume percentage of 60% of whitlockite in the crystallized part of the material, after a treatment at 850 ℃ for 4 h. The substitution of Ca^{2+} by Mg^{2+} ions in the whitlockite structure has been confirmed by Rietveld analysis with the variations of the lattice parameters of the β-TCP structure, suggesting the incorporation of up to 14 mol% of Mg^{2+} ions. Furthermore, the crystallization of phases from the glass leads to a decrease of the relative density. However, a high whitlockite-phase content in the glass-ceramics leads to increasing bending strength, approximately 120 MPa for samples heat treated at 975 ℃ for 4 h.

Acknowledgements

The authors would like to thank LNLS - Laboratório Nacional de Luz Síncrotron for technical support, and FAPESP for financial support, under grant No. 07/50510-4. We also acknowledge Prof. E. D. Zanotto and the LaMaV, UFSCar, for melting of glass and DSC analysis.

References

[1] LeGeros RZ. Properties of osteoconductive biomaterials: Calcium phosphates. *Clin Orthop Relat Res* 2002, **395**: 81–98.

[2] Park JB, Lakes RS. Ceramic implant materials. In *Biomaterials: An Introduction*, 3rd edn. New York: Springer, 2007: 139–171.

[3] Rodrigues CVM, Serricella P, Linhares ABR, *et al.* Characterization of a bovine collagen–hydroxyapatite composite scaffold for bone tissue engineering. *Biomaterials* 2003, **24**: 4987–4997.

[4] Billotte WG. Ceramic biomaterials. In *Biomaterials: Principles and Applications*. Park JB, Bronzino JD, Eds. Boca Raton: CRC Press, 2003: 21–54.

[5] Elliot JC. *Structure and Chemistry of the Apatites and Other Calcium Orthophosphates*, 2nd end. *Vol.18 Studies in Inorganic Chemistry*. Elsevier Science & Technology, 1994.

[6] Ribeiro GBM, Trommer RM, dos Santos LA, *et al.* Novel method to produce β-TCP scaffolds. *Mater Lett* 2011, **65**: 275–277.

[7] Bigi A, Cojazzi G, Panzavolta S, *et al.* Chemical and structural characterization of the mineral phase from cortical and trabecular bone. *J Inorg Biochem* 1997, **68**: 45–51.

[8] Kannan S, Ventura JM, Ferreira JMF. Aqueous precipitation method for the formation of Mg-stabilized β-tricalcium phosphate: An X-ray diffraction study. *Ceram Int* 2007, **33**: 637–641.

[9] Araújo JC, Sader MS, Moreira EL, *et al.* Maximum substitution of magnesium for calcium sites in Mg–β-TCP structure determined by X-ray powder diffraction with the Rietveld refinement. *Mater Chem Phys* 2009, **118**: 337–340.

[10] Daguano JKMF, Santos C, Suzuki PA, *et al.* Improvement of the mechanical properties of glasses based on the $3CaO \cdot P_2O_5$–SiO_2–MgO system after heat-treatment. *Mater Sci Forum* 2010, **636–637**: 41–46.

[11] Bose S, Tarafder S, Banerjee SS, *et al.* Understanding *in vivo* response and mechanical property variation in MgO, SrO and SiO_2 doped β-TCP. *Bone* 2011, **48**: 1282–1290.

[12] Banerjee SS, Tarafder S, Davies NM, *et al.* Understanding the influence of MgO and SrO binary doping on the mechanical and biological properties of β-TCP ceramics. *Acta Biomater* 2010, **6**: 4167–4174.

[13] Schroeder LW, Dickens B, Brown WE. Crystallographic studies of the role of Mg as a stabilizing impurity in β-$Ca_3(PO_4)_2$. II. Refinement of Mg-containing β-$Ca_3(PO_4)_2$. *J Solid State Chem* 1977, **22**: 253–262.

[14] Enderle R, Götz-Neurnhoeffer F, Göbbels M, *et al.* Influence of magnesium doping on the phase transformation temperature of β-TCP ceramics examined by Rietveld refinement. *Biomaterials* 2005, **26**: 3379–3384.

[15] Oliveira AL, Oliveira JM, Correia RN, *et al.* Crystallization of Whitlockite from a glass in the system $CaOP_2O_5SiO_2MgO$. *J Am Ceram Soc* 1998, **81**: 3270–3276.

[16] Oliveira JM, Correia RN, Fernandes MH. Surface modifications of a glass and a glass-ceramic of the MgO–$3CaO \cdot P_2O_5$–SiO_2 system in a simulated body fluid. *Biomaterials* 1995, **16**: 849–854.

[17] Krimm S, Tobolsky AV. Quantitative X-ray studies of order in amorphous and crystalline polymers. Quantitative X-ray determination of crystallinity in polyethylene. *J Polym Sci* 1951, **7**: 57–76.

[18] Ray CS, Yang Q, Huang W, *et al.* Surface and internal crystallization in glasses as determined by differential thermal analysis. *J Am Ceram Soc* 1996, **79**: 3155–3160.

[19] Oliveira AL, Oliveira JM, Correia RN, *et al*. Phase separation and crystallization in 3CaO·P$_2$O$_5$–SiO$_2$–MgO glasscs. In *Proceedings of the 5th International Otto Schott Colloquium. Vol.67 Glass Science and Technology*. Dt. Glastechn. Ges., 1994: 367–370.

[20] Queiroz CMGA. Cristalização de biomateriais vitrocerâmicos e mineralização em meio fisiológico simulado. Doctoral Thesis. Portugal: Universidade de Aveiro, 2005.

[21] Holand W, Beall GH. *Glass-Ceramic Technology*. New York: Wiley-Blackwell, 2002.

[22] Karamanov A, Pelino M. Induced crystallization porosity and properties of sintereds diopside and wollastonite glass-ceramics. *J Eur Ceram Soc* 2008, **28**: 555–562.

[23] Karamanov A, Pelino M. Sinter-crystallisation in the diopside–albite system: Part I. Formation of induced crystallisation porosity. *J Eur Ceram Soc* 2006, **26**: 2511–2517.

Influence of sintering temperature on electrical properties of $(K_{0.4425}Na_{0.52}Li_{0.0375})(Nb_{0.8825}Sb_{0.07}Ta_{0.0475})O_3$ ceramics without phase transition induced by sintering temperature

Shaohua QIAN, Kongjun ZHU[*], Xuming PANG,
Jing WANG, Jinsong LIU, Jinhao QIU

*State Key Laboratory of Mechanics and Control of Mechanical Structures,
Nanjing University of Aeronautics and Astronautics, Nanjing 210016, China*

Abstract: Lead-free $(K_{0.4425}Na_{0.52}Li_{0.0375})(Nb_{0.8825}Sb_{0.07}Ta_{0.0475})O_3$ (KNLNST) piezoelectric ceramics are synthesized by the conventional solid-state reaction method. The sintering temperature and poling temperature dependence of ceramic properties are investigated. Previous studies have shown that variation of sintering temperature can cause phase transition, similar to the morphotropic phase boundary (MPB) behavior induced by composition changes in $Pb(Zr,Ti)O_3$ (PZT). And the best piezoelectric performance can be obtained near the phase-transition sintering temperature. In this research, phase transition induced by sintering temperature cannot be detected and excellent piezoelectric properties can still be obtained. The sintering temperature of the largest piezoelectric coefficient of such composition is lower than that of the highest density, which is considered in composition segregation as a result of intensified volatilization of alkali metal oxides. Combined with the effect of poling temperature, the peak values of the piezoelectric properties are $d_{33} = 313$ pC/N, $k_p = 47\%$, $\varepsilon_r = 1825$, $\tan\delta = 0.024$, $T_{o-t} = 88$ ℃, and $T_C = 274$ ℃.

Keywords: tetragonal; piezoelectric properties; sintering temperature; doped; poling temperature

1 Introduction

The speedy development of piezoelectric devices urgently calls for environment-friendly materials as substitutes for the widely-used lead zirconate titanate $(Pb(Zr,Ti)O_3$, PZT). Among the various lead-free piezoelectric materials, alkali niobate $(K,Na)NbO_3$ (KNN) is considered one of the most promising candidates for lead-free piezoelectric ceramics due to its high Curie temperature (about 420 ℃) and strong ferroelectricity [1–5]. However, the high volatility of alkaline elements at high temperature makes it difficult to achieve the densification and obtain well-sintered KNN ceramics from pure KNN ceramics using ordinary sintering process, which has hindered the research for a long time [2,6]. Thus far, pure KNN ceramics prepared by conventional solid-state reaction method have poor d_{33} values of 80 pC/N and low densities of 4.2 g/cm^3 [7]. Various fabricating techniques, such as hot-pressed sintering [8], spark–plasma sintering [9], microwave sintering [10],

* Corresponding author.
E-mail: kjzhu@nuaa.edu.cn

two-step sintering [11] and hydrothermal synthesis [12], have been utilized to improve the properties of lead-free piezoelectric ceramics. However, all of these techniques are unsuitable in industrial applications.

The addition of a small amount of sintering aid (CuO [13], MnO_2 [14], $AgTaO_3$ [15], SnO_2 [16], ZnO [17], and $BiMnO_3$ [18]) is an effective method to reduce the sintering temperature and enhance the density. However, the piezoelectric properties of the ceramics could also obviously decrease. Recent studies have shown that Li/Ta/Sb doping can increase the sintering density of KNN ceramics and significantly improve its piezoelectric properties. Saeri et al. [19] investigated the effects of Li doping on the property of KNN lead-free ceramics, which exhibit d_{33} = 215 pC/N and $\varepsilon_r = 560$. Su et al. [20] reported that $(K_{0.7}Na_{0.3})(Nb_{0.95}Sb_{0.05})O_3$ lead-free piezoelectric powders are successfully synthesized by hydrothermal treatment. Saito et al. [21] used reactive template growth method to investigate $(Na,K)NbO_3$–$Li(Ta,Sb)O_3$. The as-obtained ceramics exhibited piezoelectric properties as high as 416 pC/N, which is comparable to that of PZT ceramics. Partial substitutions of Li at the A-site and Sb and/or Ta at the B-site could induce a sharp change in the lattice parameters of KNN and hence lead to phase transition. However, Pang et al. [22] reported a similar phase transition of (K,Na,Li)(Nb,Sb,Ta) ceramics by varying the sintering temperature from 1110 ℃ to 1140 ℃. The peak value of the piezoelectric coefficient (d_{33}) can be obtained while the sintering temperature is close to the phase-transition sintering temperature.

The superior properties of KNN-based ceramics are generally attributed to the presence of phase structure similar to the morphotropic phase boundary (MPB). Accordingly, it is significant to investigate whether such a similar MPB behavior induced by different sintering temperatures could impact the properties of KNN-based ceramics.

In this paper, the ceramic $(K_{0.4425}Na_{0.52}Li_{0.0375})$ $(Nb_{0.8825}Sb_{0.07}Ta_{0.0475})O_3$ (KNLNST) was chosen to investigate the effects of the variation of sintering temperature from 1110 ℃ to 1150 ℃ on the ceramic phase structure and piezoelectric properties.

2   Experimental procedure

K_2CO_3 (99%), Na_2CO_3 (99.8%), Li_2CO_3 (98%), Nb_2O_5 (99.5%), Ta_2O_5 (99.99%), and Sb_2O_3 (99.5%) were used as raw materials to prepare KNLNST ceramics by conventional mixed-oxide method. The stoichiometric powders were mixed in ethanol by ball-milling for 12 h, and then dried and calcined at 900 ℃ for 5 h. The calcined powders were then mixed with 3 wt% polyvinyl alcohol (PVA) solution and uniaxially pressed into pellets with a diameter of 1.5 cm under 300 MPa pressure. After burning out the PVA, the green disks were sintered in air at selected temperatures (1110–1150 ℃) for 3 h. The microstructure was observed by a scanning electron microscope (SEM, JSM-5610LV/Noran-Vantage). Powder X-ray diffraction (XRD, D8 Advance) was utilized to identify the crystal structures and phases. Silver paste electrodes were formed at the two circular surfaces of the disk-shaped specimens after firing at 700 ℃ for 10 min to measure the dielectric and piezoelectric properties. The piezoelectric constant d_{33} was measured using a static piezoelectric constant testing meter (ZJ-3A, Institute of Acoustics, Chinese Academy of Science, Beijing, China). Dielectric properties as functions of temperature and frequency were measured by an impedance analyzer (HP4294A). Polarization versus electric field hysteresis loops were measured using a ferroelectric tester (TF Analyzer 2000). The measurement of piezoelectric and electromechanical properties was only carried out 24 h after a poling process.

3   Results and discussion

Figure 1(a) shows the XRD patterns of the samples sintered from 1110 ℃ to 1150 ℃. The enlarged XRD patterns of the ceramics in the range of 2θ from 44° to 48° are shown in Fig. 1(b). All the ceramics show a tetragonal phase, and the phase transition does not occur as the sintering temperature increases. A secondary phase is detected in all doped samples when 2θ is approximately 28.5°. The secondary phase could be assigned to the tetragonal tungsten–bronze (TTB) type structure phase, which does not disappear as the composition changes. The occurrence of the TTB secondary phase is attributed to the volatilization and segregation of the alkali elements during the sintering process for Li/Ta-modified KNN material [23], which induces B-site ion excess that is accommodated through TTB phase formation. With the increase in temperature, the (002) and (200) peaks shift toward lower angles, which is attributed to the easy volatilization of sodium and potassium during high-temperature sintering. Sodium volatilizes faster than potassium, resulting in the presence of a niobium-rich phase [24]. The lattice parameters increase (Fig. 2) because the radius of K^+ (1.38 Å) is larger than that of Na^+ (1.02 Å), which could lead to the gradual increase in space distance. According to Bragg's equation, $2d\sin\theta = \lambda$, $\lambda = 1.5416$ Å, increasing d leads to a decrease in θ.

Figure 3 depicts the micrographs of the KNLNST ceramics sintered at different temperatures. All the ceramics show a bimodal distribution with many fine grains located at the boundaries of the coarse grains. Most figures demonstrate apparent square grains,

except for Fig. 3(a). The boundaries of the grains are ambiguous, and the shape of the grains is not structured according to the KNLNST ceramics sintered at 1110 ℃. Figure 3(a) illustrates that sintering temperature of 1110 ℃ is unsuitable to obtain square grains. When the sintering temperature is increased from 1120 ℃ to 1150 ℃, the shape of matrix grains exhibits more tacticity, but no significant change in the size of grain occurs. A comparison among Figs. 3(b) to 3(e) shows that the shape and tetragonality of the grains are the best in Figs. 3(d) and 3(e). The grains in Fig. 3(e) are more uniform than the grains in Fig. 3(d).

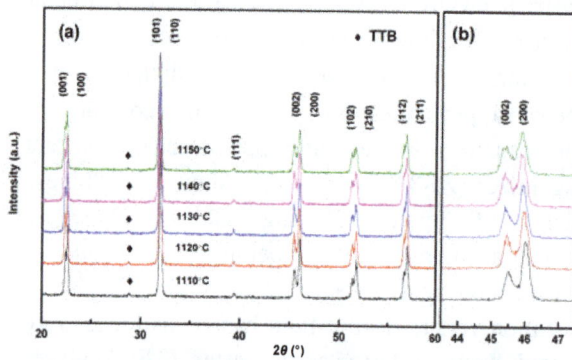

Fig. 1   XRD patterns of the KNLNST ceramics sintered at different temperatures in the range of 2θ: (a) from 20° to 60°; (b) from 44° to 48°.

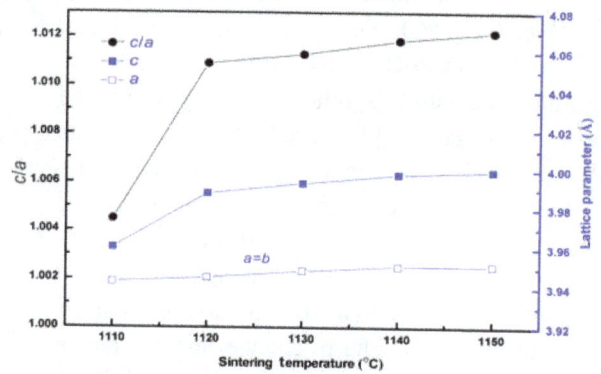

Fig. 2   Lattice parameter evolutions as functions of sintering temperature.

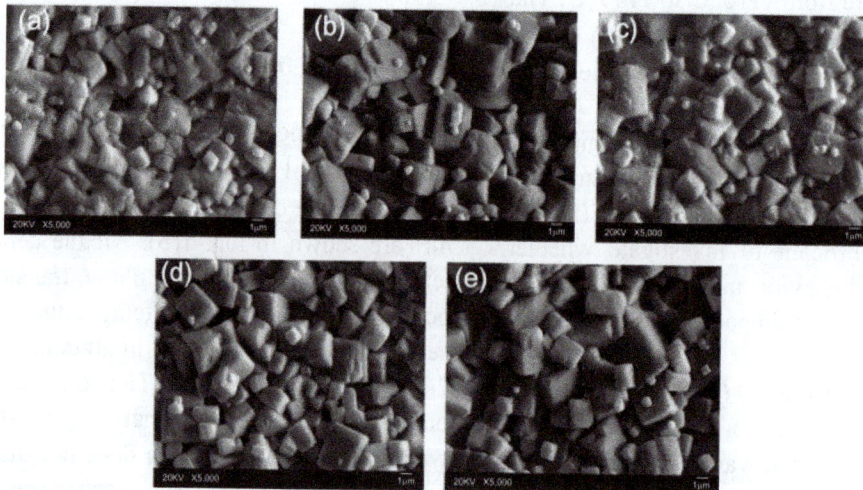

Fig. 3   SEM images of the KNLNST ceramics sintered at different temperatures: (a) 1110 ℃, (b) 1120 ℃, (c) 1130 ℃, (d) 1140 ℃, and (e) 1150 ℃.

Figure 4 displays the measured density of KNLNST as a function of sintering temperature. The measured density of the KNLNST samples increases from 4.33 g/cm^3 to 4.52 g/cm^3, while the sintering temperature increases by a scope of 30 ℃ from 1110 ℃ to 1140 ℃. The highest density is obtained for the KNLNST sample sintered at 1140 ℃, and then tends to decrease as the temperature exceeds 1140 ℃.

Influence of sintering temperature on electrical properties of (K0.4425Na0.52Li0.0375)...

61

A relatively higher temperature is helpful in achieving higher density, but may also lead to higher volatilization of K and Na for KNN-based ceramics. So when the effect of temperature on density increase cannot compensate the volatilization of alkali metal, the density decreases.

Figures 5(a) and 5(b) show the piezoelectric property (d_{33}) and electromechanical coupling factor (k_p). As the sintering temperature increases, both d_{33} and k_p initially increase rapidly from 230 pC/N to 313 pC/N and from 0.32 to 0.47, respectively. The further increase of the sintering temperature leads to obvious decreases in d_{33} and k_p which are partly caused by the volatilization of Na and K during high-temperature sintering. The peak value of d_{33} appears when the sintering temperature is 1130 ℃. Although phase transition does not occur in the sintering temperature range of 1110 ℃ to 1150 ℃, excellent piezoelectric properties like those mentioned in previous reports can still be obtained, indicating that the excellent piezoelectric property does not result from phase transition caused by the variation in sintering temperature. Mechanical quality factor (Q_m) (seen in Fig. 5(c)) reveals an opposite trend to d_{33}, showing an obvious "valley" region within the sintering temperature range of 1120 ℃ to 1140 ℃. The lowest value of Q_m (29) appears at 1130 ℃ corresponding to the highest value of d_{33}. The relative permittivity (ε_r) value variation of KNLNST ceramics can be seen in Fig. 5(d). It shows that the influence of sintering temperature on ε_r is similar to d_{33}, increasing with increasing sintering temperature, reaching the peak value of 1825 at 1130 ℃. Over the peak value, ε_r value decreases sharply to 1567 at 1140 ℃. While the

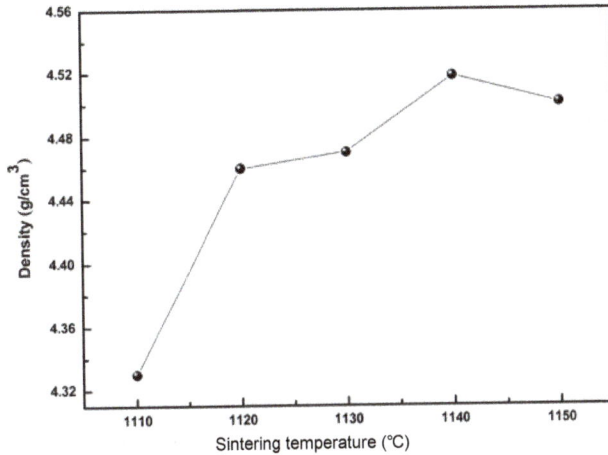

Fig. 4 Dependence of measured density as a function of sintering temperature.

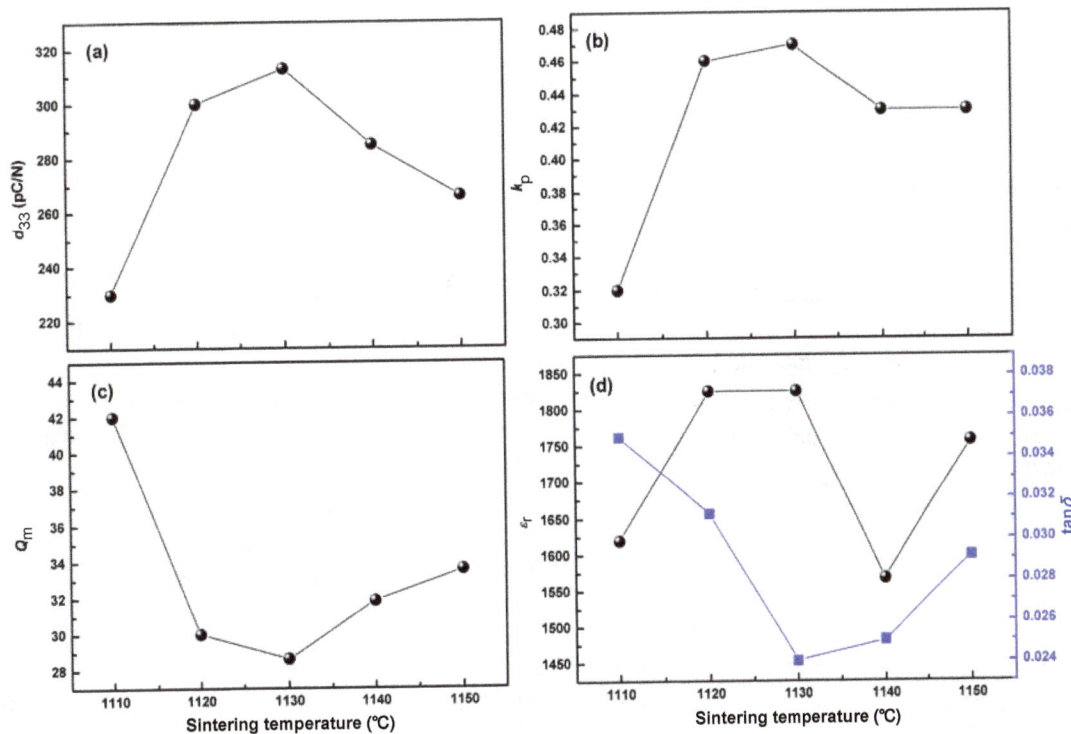

Fig. 5 Dependence of electrical properties of the KNLNST ceramics sintered at different temperatures: (a) piezoelectric coefficient (d_{33}), (b) planar mode electromechanical coupling coefficient (k_p), (c) mechanical quality factor (Q_m), and (d) relative permittivity (ε_r) and dielectric loss (tanδ).

sintering temperature reaches 1150 ℃, ε_r value again exhibits a significant increase from 1567 to 1757. This abnormal variation is not well-understood at present, and needs to be studied further. Dielectric loss (tanδ) exhibits normal variation process, decreasing incipiently and then gradually increasing with the sintering temperature exceeding 1130 ℃. The minimal tanδ appears at 1130 ℃, indicating that the ceramic owns fewer defects. Therefore, the optimum sintering temperature for KNLNST ceramics is 1130 ℃, and the properties are $d_{33} = 313$ pC/N, $\varepsilon_r = 1825$, $k_p = 0.47$, $Q_m = 29$, and tan$\delta = 0.024$.

Figure 6(a) shows the P–E hysteresis loops of the KNLNST ceramics as functions of the sintering temperature from 1110 ℃ to 1150 ℃. It shows that all the P–E loops are well saturated. Remanent polarization P_r produces slight fluctuation as the sintering temperature varies, and the value of P_r settles into the 15.5 μC/cm^2 to 16.3 μC/cm^2 range (as shown in Fig. 6(b)). The varying trend of coercive field E_c (Fig. 6(c)) is similar to that of P_r, exhibiting inactivity as the sintering temperature changes. The sample sintered at 1150 ℃ shows the lowest E_c, which is distinct from the E_c of other sintering temperatures.

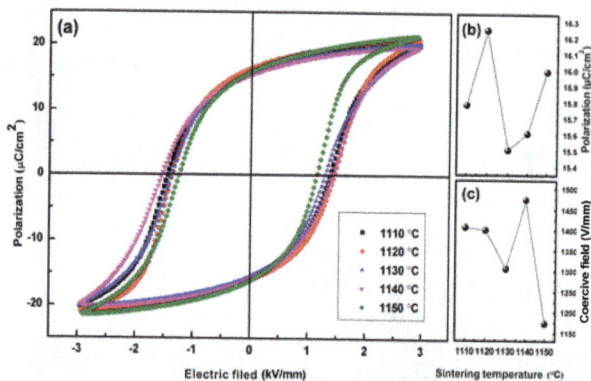

Fig. 6 (a) Polarization–electric field hysteresis loops of the KNLNST ceramics sintered at different temperatures; (b) the variation of polarization (P_r) value as a function of sintering temperature; and (c) the coercive field (E_c) value of KNLNST ceramics sintered at different temperatures.

Figure 7(a) shows the temperature dependence of the dielectric constant (measured at 10 kHz) for the KNLNST ceramics with different sintering temperatures. The two peaks can be detected in the temperature ranges of 40 ℃ to 110 ℃ and 250 ℃ to 300 ℃, respectively. These peaks correspond to the transition from the orthorhombic phase to the

tetragonal phase (T_{o-t}), and from the tetragonal phase to the cubic phase (T_C), respectively. T_{o-t} fluctuates with the increase of the sintering temperature, shifting to higher temperature. T_{o-t} for all samples synthesized at 1110 ℃, 1130 ℃, and 1150 ℃ are 47 ℃, 88 ℃, and 101 ℃, respectively (as shown in Fig. 7(b)). T_C shows the same variation trend, shifting to higher temperature but changing slightly. The maximum dielectric constant at T_C can be obtained at 1130 ℃. The values of T_C are 271 ℃, 274 ℃, and 287 ℃, respectively (seen in Fig. 7(c)). Referring to a previous study, T_C decreases as a result of Ta/Sb doping. However, Choi et al. [25] pointed out that the T_C value of ceramics can be improved by enhancing the tetragonality of the ceramics, which is decided by the lattice ratio of c/a. The illustration of the lattice parameters of the different sintering temperatures shows that the value of c/a increases with temperature. Therefore, the variation trend of T_C is more complicated, which may be decided by two factors, namely, Ta content and tetragonality. For the stability of the Ta content, the change in the tendency of T_C is decided by the tetragonality of the ceramics. In addition, no apparent broadening of the dielectric constant can be observed in the figure, indicating that the ceramics do not possess relaxation properties.

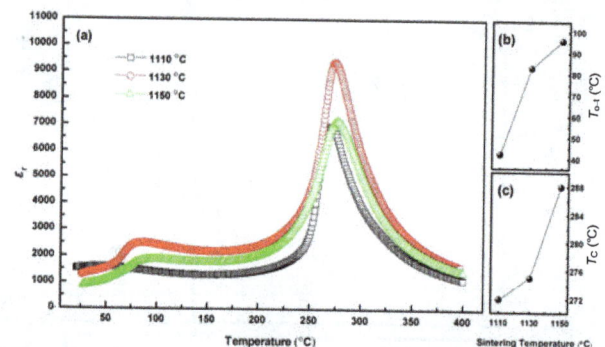

Fig. 7 (a) Temperature dependence of the dielectric constant (ε_r) for the KNLNST ceramics sintered at 1110 ℃, 1130 ℃, 1150 ℃ for 3 h measured at 10 kHz; (b) the phase transition temperature (T_{o-t}) of KNLNST ceramics sintered at 1110 ℃, 1130 ℃, 1150 ℃; and (c) the Curie temperature (T_C) of KNLNST ceramics as a function of sintering temperature.

The effect of polarization temperature ranging from 30 ℃ to 150 ℃ on the value of piezoelectric coefficient (d_{33}) is also investigated (Fig. 8). The optimal d_{33} can be obtained when poling temperature is

70 ℃, which approaches T_{o-t}. This finding is attributed to the metastability of the phase structure near the T_{o-t}, where domains possess higher activity and are more likely to reverse. Moreover, d_{33} decreases monotonously from 313 pC/N to 276 pC/N as the polarization temperature continuously increases to 150 ℃.

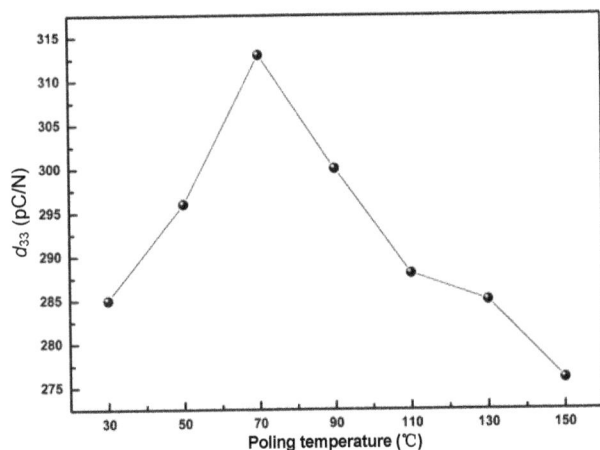

Fig. 8   Poling temperature dependence of the piezoelectric coefficient (d_{33}) for the KNLNST ceramics sintered at 1130 ℃ for 3 h.

4   Conclusions

The effects of sintering temperature on the microstructure, dielectric, piezoelectric, and ferroelectric properties of KNLNST ceramics were studied. No obvious phase transition behavior for the KNLNST ceramics with Ta = 4.75 mol% was observed with changes in sintering temperature ranging from 1110 ℃ to 1150 ℃. However, excellent piezoelectric properties can still be obtained. Accordingly, the excellent piezoelectric property did not result from such similar MPB behavior caused by the variation in sintering temperature. Moreover, by combining the effect of sintering temperature and poling temperature, the KNLNST ceramics exhibited optimum electrical properties as follows: d_{33} = 313 pC/N, k_p = 0.47, ε_r = 1825, tanδ = 0.024, T_C = 274 ℃, T_{o-t} = 88 ℃, P_r = 15.5 μC/cm^2, and E_c = 1316 V/mm.

Acknowledgements

This study was supported by the National Natural Science Foundation of China (No. 51172108), and the project funded by the Priority Academic Program Development of Jiangsu Higher Education Institutions.

References

[1] Haertling GH. Properties of hot-pressed ferroelectric alkali niobate ceramics. *J Am Ceram Soc* 1967, **50**: 229–330.

[2] Jaeger RE, Egerton L. Hot pressing of potassium–sodium niobates. *J Am Ceram Soc* 1962, **45**: 209–213.

[3] Hu W, Tan X, Rajan K. Combinatorial processing libraries for bulk BiFeO$_3$–PbTiO$_3$ piezoelectric ceramics. *Appl Phys A* 2010, **99**: 427–431.

[4] Kosec M, Kolar D. On activated sintering and electrical properties of NaKNbO$_3$. *Mater Res Bull* 1975, **10**: 335–339.

[5] Narayana Murty S, Ramana Murty KV, Umakantham K, *et al.* Modified (NaK)NbO$_3$ ceramics for transducer applications. *Ferroelectrics* 1990, **102**: 243–247.

[6] Cross E. Materials science: Lead-free at last. *Nature* 2004, **432**: 24–25.

[7] Egerton L, Dillon DM. Piezoelectric and dielectric properties of ceramics in the system potassium–sodium niobate. *J Am Ceram Soc* 1959, **42**: 438–442.

[8] Li K, Li FL, Wang Y, *et al.* Hot-pressed K$_{0.48}$Na$_{0.52}$Nb$_{1-x}$Bi$_x$O$_3$ (x = 0.05–0.15) lead-free ceramics for electro-optic applications. *Mater Chem Phys* 2011, **131**: 320–324.

[9] Zhang B-P, Li J-F, Wang K, *et al.* Compositional dependence of piezoelectric properties in Na$_x$K$_{1-x}$NbO$_3$ lead-free ceramics prepared by spark plasma sintering. *J Am Ceram Soc* 2006, **89**: 1605–1609.

[10] Xie Z, Gui Z, Li L, *et al.* Microwave sintering of lead-based relaxor ferroelectric ceramics. *Mater Lett* 1998, **36**: 191–194.

[11] Pang X, Qiu J, Zhu K, *et al.* (K, Na)NbO$_3$-based lead-free piezoelectric ceramics manufactured by two-step sintering. *Ceram Int* 2012, **38**: 2521–2527.

[12] Zhou Y, Guo M, Zhang C, *et al.* Hydrothermal synthesis and piezoelectric property of Ta-doping K$_{0.5}$Na$_{0.5}$NbO$_3$ lead-free piezoelectric ceramic. *Ceram Int* 2009, **35**: 3253–3258.

[13] Shao B, Qiu J, Zhu K, *et al.* Influence of sintering

temperature on microstructure and electric properties of CuO doped alkaline niobate-based lead-free ceramics. *J Mater Sci: Mater El* 2012, **23**: 1455–1461.

[14] Lin D, Zheng Q, Kwok KW, *et al.* Dielectric and piezoelectric properties of MnO_2-doped $K_{0.5}Na_{0.5}Nb_{0.92}Sb_{0.08}O_3$ lead-free ceramics. *J Mater Sci: Mater El* 2010, **21**: 649–655.

[15] Wang Y, Liu Q, Zhao F. Phase transition behavior and electrical properties of $[(K_{0.50}Na_{0.50})_{1-x}Ag_x](Nb_{1-x}Ta_x)O_3$ lead-free ceramics. *J Alloys Compd* 2010, **489**: 175–178.

[16] Li Z, Xu G, Li Y, *et al.* Dielectric and piezoelectric properties of ZnO and SnO_2 co-doping $K_{0.5}Na_{0.5}NbO_3$ ceramics. *Physica B* 2010, **405**: 296–299.

[17] Rubio-Marcos F, Romero JJ, Navarro-Rojero MG, *et al.* Effect of ZnO on the structure, microstructure and electrical properties of KNN-modified piezoceramics. *J Eur Ceram Soc* 2009, **29**: 3045–3052.

[18] Jiang M, Liu X, Chen G, *et al.* Dielectric and piezoelectric properties of $BiMnO_3$ doped $0.95Na_{0.5}K_{0.5}NbO_3$–$0.05LiSbO_3$ ceramics. *J Mater Sci: Mater El* 2011, **22**: 876–881.

[19] Saeri MR, Barzegar A, Ahmadi Moghadam H. Investigation of nano particle additives on lithium doped KNN lead free piezoelectric ceramics. *Ceram Int* 2011, **37**: 3083–3087.

[20] Su L, Zhu K, Bai L, *et al.* Effects of Sb-doping on the formation of (K, Na)(Nb, Sb)O_3 solid solution under hydrothermal conditions. *J Alloys Compd* 2010, **493**: 186–191.

[21] Saito Y, Takao H, Tani T, *et al.* Lead-free piezoceramics. *Nature* 2004, **432**: 84–87.

[22] Pang X, Qiu J, Zhu K, *et al.* Influence of sintering temperature on piezoelectric properties of $(K_{0.4425}Na_{0.52}Li_{0.0375})(Nb_{0.8925}Sb_{0.07}Ta_{0.0375})O_3$ lead-free piezoelectric ceramics. *J Mater Sci: Mater El* 2011, **22**: 1783–1787.

[23] Wang Y, Damjanovic D, Klein N, *et al.* High-temperature instability of Li- and Ta-modified (K,Na)NbO_3 piezoceramics. *J Am Ceram Soc* 2008, **91**: 1962–1970.

[24] Jenko D, Bencan A, Malic B, *et al.* Electron microscopy studies of potassium sodium niobate ceramics. *Microsc Microanal* 2005, **11**: 572–580.

[25] Choi SW, Shrout TR, Jang SJ, *et al.* Morphotropic phase boundary in Pb(Mg$_{1/3}$Nb$_{2/3}$)O_3–PbTiO$_3$ system. *Mater Lett* 1989, **8**: 253–255.

9

Fusion bonding and microstructure formation in TiB$_2$-based ceramic/metal composite materials fabricated by combustion synthesis under high gravity

Xuegang HUANG[a,*], Jie HUANG[a], Zhongmin ZHAO[b], Long ZHANG[b], Junyan WU[b]

[a]*Hypervelocity Aerodynamics Institute, China Aerodynamics Research and Development Center, Mianyang 621000, China*
[b]*Department of Vehicle and Electrical Engineering, Mechanical Engineering College, Shijiazhuang 050003, China*

Abstract: The novel ceramic/metal composite materials were successfully fabricated by combustion synthesis in high gravity field. In this paper, the Ti–B$_4$C was selected as the main combustion reaction system to obtain TiB$_2$–TiC ceramic substrate, and the 1Cr18Ni9Ti stainless steel was selected as the metal substrate. It was found that the TiB$_2$–TiC/1Cr18Ni9Ti composite materials exhibited continuously graded composition and hybrid microstructure. The TiC$_{1-x}$ carbides and TiB$_2$ platelets decreased gradually in size and volume fraction from the ceramic to stainless steel. Due to the rapid action of thermal explosion as well as the dissolution of the molten stainless steel into TiB$_2$–TiC liquid, the diffusion-controlled concentration gradient from the ceramic liquid to the alloy liquid was observed. Finally, as a result of the rapid sequent solidification of the ceramic liquid and the melt alloy surface, the laminated composite materials were achieved in multilevel, scale-span hybrid microstructure.

Keywords: combustion synthesis; high gravity field; TiB$_2$–TiC/1Cr18Ni9Ti; hybrid microstructure

1 Introduction

Because of its high hardness, high corrosion resistance, good thermal shock resistance, and high temperature stability, TiB$_2$-based ceramic composites attract a lot of attention as the potential materials in engineering applications [1,2]. The novel TiB$_2$–TiC ceramic possesses more advantages than traditional metals, especially in high temperature techniques [3–5]. However, ceramics generally have low toughness and are hard to be used to manufacture complex parts. So it is necessary to manufacture these complex parts by joining ceramics and metals to meet the special requirements [6]. Over the past 50 years, various ceramic–metal joining techniques have been developed and improved, such as mechanical joining, adhesive joining, friction joining, high energy beam welding, microwave joining, ultrasonic welding, explosive welding, reaction joining, combustion reaction joining, field-assisted bonding, diffusion bonding, transient liquid phase bonding, partial transient liquid phase bonding, and so on [6]. However, most of these methods mentioned above need complicate equipment, extra energy supply, and long preparation time, which lead to the high production costs finally [7–9]. It is

* Corresponding author.
E-mail: emei-126@126.com

generally known that a key problem in joining ceramic and metal is the thermal expansion mismatch between them, which brings about high residual stress at the interface of the ceramic and the metal, thereby deteriorating mechanical properties of the ceramic–metal composite. Bever and Duwez [10] have pointed out that the functionally graded material (FGM) of the ceramic/metal can effectively mitigate the residual stress arising at ceramic–metal interface.

Combustion synthesis which was also called self-propagating high-temperature synthesis (SHS) is being used to produce ceramics, intermetallics, and composite materials [11–14]. Based on this technology, joining a ductile metal with a brittle ceramic composite matrix has considerable potential for substantial improvements in fracture toughness [15]. Recently, a rapid and economical processing named combustion synthesis in high gravity field has been applied to prepare bulk solidified TiB_2–TiC ceramics, and the high performance of TiB_2–TiC composites is presented due to the achievement in fine-grained and ultrafine-grained microstructures [5]. Subsequently, the laminated composite materials of TiB_2–TiC ceramic and Ti alloy in continuously graded microstructures were fabricated by employing combustion synthesis in high gravity field to accomplish fusion bonding of liquid ceramic to Ti alloy substrate [16]. Hence, based on the achievement in combustion synthesis in ultrahigh gravity field, another economical material of the laminated ceramic–steel composite develops. In the thesis, applying combustion synthesis in ultrahigh gravity field to achieve laminated composite materials of the ceramic on stainless steel substrate, and the microstructure and formation mechanism of the laminated composite of the ceramic on stainless steel are discussed.

2 Experimental

Commercial powders of Ti (99.5% purity, <34 μm), B_4C (97% purity, <3 μm), CrO_3 (99.9% purity, <60 μm), and Al (96% purity, <50 μm) were used as raw materials. The Ti to B_4C molar ratio of 3:1 was chosen as the starting composition based on Eq. (1), so the composition of the solidified TiB_2–TiC composite was determined as TiC:$TiB_2 = 1$:2. In order to ensure full-liquid products after the combustion reaction, the 30 wt% (CrO_3+Al) subsystem was added as the activator for increasing the combustion temperature

according to Eq. (2).

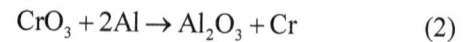

$$3Ti + B_4C \rightarrow TiC + 2TiB_2 \tag{1}$$

$$CrO_3 + 2Al \rightarrow Al_2O_3 + Cr \tag{2}$$

The above powder blends were mechanically homogenized by ball milling for 1 h. The 1Cr18Ni9Ti stainless steel plate with thickness of 10 mm was placed at the bottom of the cylindrical graphite crucible, and then the crucible was filled with the mechanically activated raw blends (500 g) under uniaxially cold-press of about 200 MPa. After that, the prepared graphite crucible was fixed on the centrifugal machine. The combustion reaction was triggered with the electrically heated W wire (diameter of 0.5 mm), while the centrifugal machine provided a high gravity acceleration of 2000g (where g means the gravitational constant, $g = 9.8$ m/s^2). When the combustion reaction was over, the centrifugal machine continued to run for 30 s. After the crucible was cooled to ambient temperature, the laminated composites of the ceramic and stainless steel were obtained.

The whole process of high gravity combustion synthesis and melt-infiltration can be described as the following three steps. First, the TiB_2–TiC ceramic melt and Al_2O_3 melt were obtained by the aluminothermic reactions in Eq. (1) and Eq. (2), respectively, and then the Al_2O_3 melt was separated from the ceramic due to the differences in density and wetting ability, and finally, TiB_2–TiC ceramic melt infiltrated into the fused surface of stainless steel substrate. After that, the Al_2O_3 oxide slag at the top of the sample should be removed by grinding. Figure 1 shows the cross-sectional image of TiB_2–TiC/1Cr18Ni9Ti composite materials without Al_2O_3 oxide slag. It also can be found that the surface of the stainless steel substrate is partially molten.

The phase composition was identified by X-ray diffraction (XRD; D/max-PA 2550PC, Rigaku, Japan) with a step of 0.02° and a scanning rate of 2 (°)/min. The microstructure and fracture morphology were

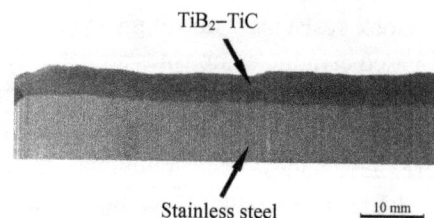

Fig. 1 Cross-sectional image of TiB_2–TiC/1Cr18Ni9Ti composite materials after the Al_2O_3 oxide slag has been removed.

examined by field emission scanning electron microscopy (FESEM; JSM-7001F, JEOL, Japan). Electron probe microanalysis (EPMA) was conducted by energy dispersive spectrometry (EDS).

3   Results and discussion

The results of XRD and FESEM show that the TiB$_2$–TiC ceramic is composed of fine TiB$_2$ platelets as the primary phase, irregular TiC grains as the secondary phase, and a few of Cr-based intermetallic binder as intercrystalline phase, as shown in Fig. 2 and Fig. 3. The density, hardness, and fracture toughness of the TiB$_2$–TiC ceramic are tested as 4.3 g/cm^3, 18±1.5 GPa, and 13.5±1.8 MPa·m$^{1/2}$, respectively. The combustion synthesis process in high gravity field for preparing TiB$_2$–TiC ceramic can be described as follows.

Based on Eq. (1) and Eq. (2), there are three liquid equilibrium products of TiC, TiB$_2$, Cr, and Al$_2$O$_3$ from combustion synthesis in terms of the thermodynamics of chemical reaction among the elements of Al, O, Ti,

C, and B. According to the equilibrium diagram of the TiB$_2$–TiC system [1], the melting temperature of 66.7%TiB$_2$–TiC ceramic is around 2900 ℃, which is much smaller than the designed combustion temperature in current experiments, thus the full-liquid products of TiC, TiB$_2$, Cr, and Al$_2$O$_3$ can be achieved in combustion reaction. Because of the immiscibility and density difference between liquid TiB$_2$–TiC–Cr and Al$_2$O$_3$, the liquid-phase separation characterized by float-up of liquid Al$_2$O$_3$ and settle-down of liquid TiB$_2$–TiC–Cr rapidly takes place in high gravity field, resulting in the presence of the layered melt in the crucible. Finally, as the melt is cooled to ambient temperature, the layered sample is formed to present the solidified TiB$_2$–TiC ceramic with Cr metallic binder covered by the flux layer of Al$_2$O$_3$. At the same time, the surface of steel substrate is also melt due to the extremely high heat releasing of combustion reaction. FESEM images of Fig. 4 show that there is an irregular intermediate layer between the ceramic and the stainless steel, and a few randomly distributed Al$_2$O$_3$ inclusions are also observed in the region between the stainless steel substrate and the TiB$_2$–TiC ceramic, as shown by the arrows in Fig. 4. These isolated Al$_2$O$_3$ inclusions tend to distribute in the intermediate area near the stainless steel substrate. However, it is hard to find the Al$_2$O$_3$ inclusions in the intermediate area near the ceramic matrix. The above phenomenon derives from the good heat conductivity of the stainless steel, which results in the extremely high solidification rate of the TiB$_2$–TiC ceramic melt near the stainless steel substrate. Since some Al$_2$O$_3$ drops cannot be completely expelled out of the ceramic melt timely, these randomly distributed Al$_2$O$_3$ inclusions, then, remain in the intermediate layer.

According to the results of FESEM and EDS in Fig. 5 and Table 1, it can be concluded that the intermediate layer within the network structure is mainly composed of TiB$_2$, TiC, Cr–Fe alloy, and solidified stainless steel. Meanwhile, with the distance increasing from the stainless steel substrate to the TiB$_2$-based ceramic, the volume fraction of Cr–Fe alloy phases drops sharply, whereas those of TiB$_2$ and TiC phases rise rapidly as shown by the EDS map scanning results of Fig. 6. So it can be inferred that with the increased distance in the intermediate from the stainless steel substrate, TiB$_2$ and TiC crystals become the main phases in the intermediate layer. Moreover, the ultrafine-grained microstructure of the

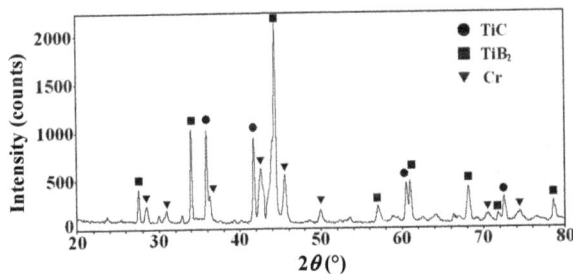

Fig. 2   XRD pattern of the TiB$_2$–TiC ceramic.

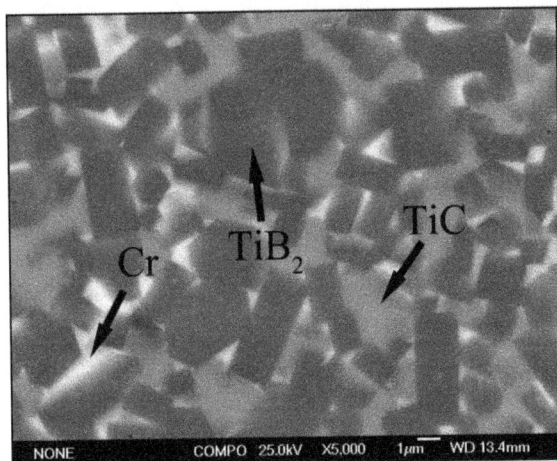

Fig. 3   FESEM microstructure of the TiB$_2$–TiC ceramic.

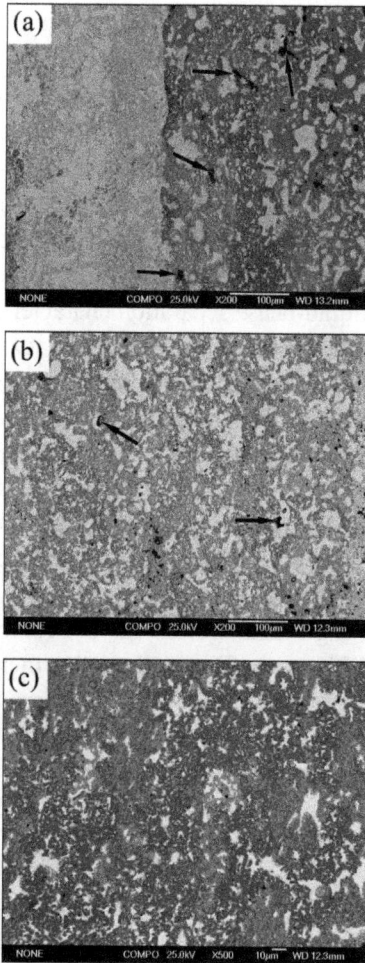

Fig. 4 FESEM images of the joint area of stainless steel and the ceramic: (a) interfacial area between solidified stainless steel and the intermediate layer; (b) the intermediate layer; (c) the intermediate area nearby the ceramic matrix.

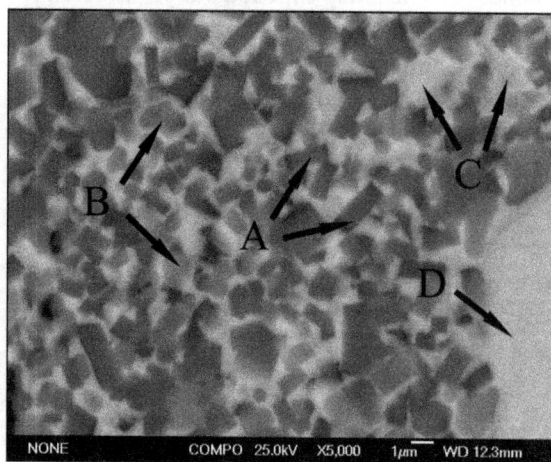

Fig. 5 FESEM image of ultrafine-grained microstructure at the intermediate area between the ceramic and the stainless steel.

Table 1 Composition of the intermediate layer between the ceramic and the stainless steel in Fig. 5 (unit: at%)

Position	Ti	C	B	Cr	Fe	Ni	Determined phase
A	31.88	—	62.91	1.08	1.16	2.97	TiB_2
B	47.02	42.79	—	4.46	2.46	3.27	TiC
C	13.01	—	—	57.10	27.11	2.78	Cr–Fe alloy
D	10.26	—	—	15.74	66.77	7.23	Solidified stainless steel

TiB_2 platelets with thickness less than 1 μm, is clearly observed in the intermediate layer, as shown in Fig. 5.

As discussed in early reports [17], taking ball milling into mechanically activated reactive blends results in a reduced ignition temperature (by over 200 ℃) and an increased combustion velocity. Moreover, the introduction of ultrahigh gravity field causes the full-liquid products to rapidly deposit in the clearance underneath the unreacted blends, resulting in the dramatically increased combustion velocity by enhancing the transfer in heat, mass, and momentum from the liquid products to the reactants [5]. Therefore, by combining mechanical activation with ultrahigh gravity field, the combustion mode has transferred thermal explosion from steady SHS. The excessive chemical energy of combustion reaction can be almost completely used to heat full-liquid products due to the high heat accumulation from thermal explosion. Therefore, the actual combustion temperature in high gravity field is close to the adiabatic temperature of chemical reaction. As a result, following the combustion reaction completion in the crucible, the superfluous reaction heat also makes the surface of the stainless steel underneath the liquid ceramic partially molten. Because of the complete dissolvability among B, C, Ti, Fe, Cr, and Ni elements in liquid, liquid TiB_2–TiC ceramic and the molten stainless steel happen to dissolve into each other, initiating the strong interdiffusion under the concentration-induced chemical potential. In this infiltration process, the atoms of Ti, B, and C in liquid ceramic diffuse toward the molten stainless steel surface, while the atoms of Fe, Cr, and Ni in the molten stainless steel diffuse toward the liquid ceramic. Subsequently, the diffusion-controlled concentration gradient of each element arises between the liquid ceramic and the molten stainless steel, followed by the nucleation and the growth of the phases out of the melt.

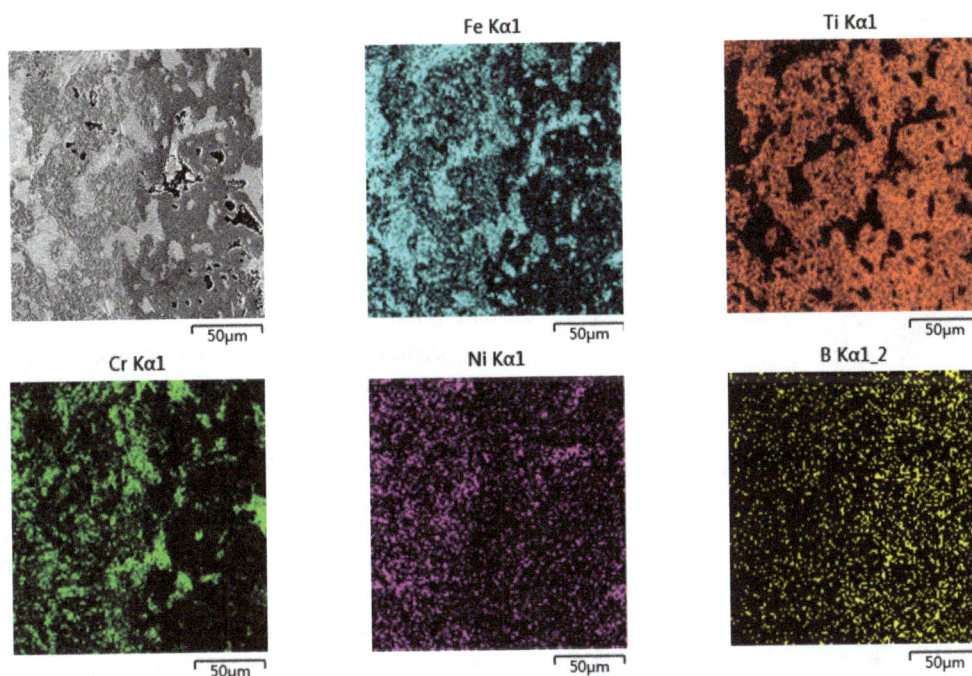

Fig. 6 EDS map scanning images of the joint area of the stainless steel and the ceramic.

In the experiment demonstrated above, the solidified TiB_2–TiC ceramic proves to be $66.7mol\%TiB_2$–TiC based on Eq. (1), which belongs to the hypereutectic group according to the binary equilibrium diagram of TiB_2–TiC system [3,4]. The TiB_2 is considered as the leading phase in nucleation process due to the higher concentration of B atoms than that of C atoms in liquid ceramic. Because crystal growth is usually associated with its melting entropy, and the crystal develops in facet growth mode if Jackson factor α $(=\Delta Sf/R)$ is larger than 2 [18]. The melting entropy of TiB_2 and TiC is calculated to be 3.78 and 2.60, respectively. TiB_2 primary phases are considered to be more inclined to grow in the facet mode, thus the high degree of supercooling is required in the low-velocity growth of TiB_2 crystals. At the same time, TiB_2 has a hexagonal AlB_2-type structure of $P6/mmm$ space group [18], and the hexagonal symmetry of TiB_2 enhances the anisotropy in crystallography. The high melting entropy of TiB_2 also enlarges the difference of different-index crystal planes in growth velocity, so TiB_2 primary phases ultimately develop the morphologies of the hexagonal platelets, which mostly seem to be small-bar grains due to the error of the detecting angle by FESEM, as shown in Fig. 3 and Fig. 5. In contrast, the face-centered structure of TiC strengthens the isotropy in crystallography, which makes it more inclined to develop the high-velocity non-facet growth in despite of its melting entropy

larger than 2. Consequently, the TiB_2 platelets are hard to develop unlimitedly due to their high melting entropy and the enhanced anisotropic growth, in despite of the fact that TiB_2 platelets nucleate and grow as the primary phases. When it comes to the TiC secondary phases, because of a combination of their isotropic growth and high diffusion rate of C relative to B in metallic liquid, they grow rapidly once they begin to nucleate, and ceaselessly compete for growth space and free Ti atoms against TiB_2 primary phases. In this case, some fine TiB_2 platelets are surrounded by TiC irregular grains and stop to grow, thereby presenting the unique solidified microstructure where a number of fine TiB_2 platelets are embedded in irregular TiC grains, as showed in Fig. 3 and Fig. 5. Moreover, the dissolution of the molten stainless steel into the liquid ceramic occupies the growth space of TiB_2 primary phases mostly, and makes TiB_2 crystals more difficult to grow. Therefore, ultrafine-grained microstructures characterized by the thickness of TiB_2 primary phases smaller than 1 μm develop in either the intermediate or the nearby ceramic, as shown in Fig. 5. In addition, C atom diffuses into the molten stainless steel more widely than B atom does, just because of the high diffusion rate of C relative to B in ceramic liquid, thereby presenting the wider graded distribution of (Fe, Cr, Ni) enriched TiC_{1-x} phases than that of the others in the intermediate, as shown in Fig. 7 and Fig. 8. And the EDS results in Table 2 also certify this result

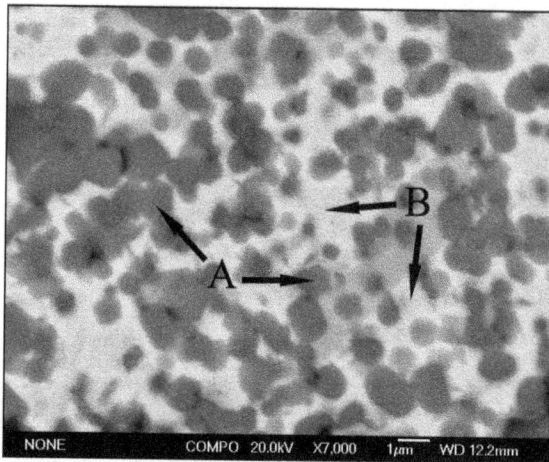

Fig. 7 (Fe, Cr, Ni) enriched TiC$_{1-x}$ carbides distributed at solidified stainless steel.

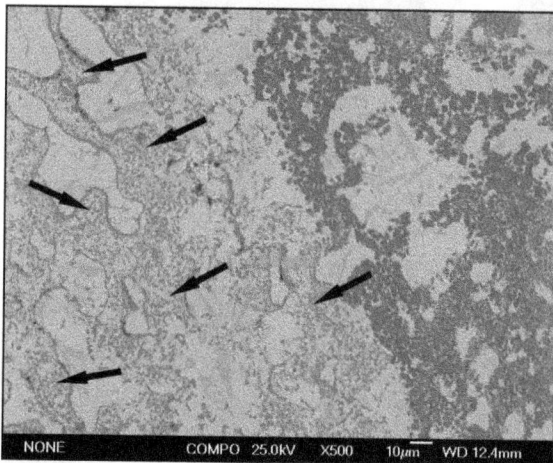

Fig. 8 Distribution of (Fe, Cr, Ni) enriched TiC$_{1-x}$ carbides from the intermediate to solidified stainless steel surface.

Table 2 Composition of the solidified stainless steel surface in Fig. 7 (unit: at%)

Position	Ti	C	Cr	Fe	Ni	Determined phase
A	44.86	33.51	6.82	7.80	7.01	(Fe, Cr, Ni) enriched TiC$_{1-x}$
B	6.01	—	18.35	68.78	6.86	Solidified stainless steel

mentioned above.

At the final stage of solidification, due to the dissolution of the molten stainless steel into the ceramic liquid, the residual liquid consisting mainly of the molten stainless steel spreads out at the surfaces of TiB$_2$ and TiC solids, and infiltrates the shrinkage cavities between the ceramic solids under a combination of capillary action and ultrahigh gravity field. So the densification of the intermediate layer and

the ceramic is promoted. Subsequently, the molten stainless steel happens to aggregate and develop the coarse metallic alloy in the intermediate, or solidify to form the boundaries of TiB$_2$ and TiC solids (as shown in Fig. 9), thereby presenting in the intermediate layer with the hybrid microstructures, as it reveals that different-size and different-morphology Fe–Cr–Ni alloy phases alternate with TiB$_2$ platelets and irregular TiC grains.

In conclusion, the formation mechanism of the laminated composite materials of TiB$_2$–TiC/1Cr18Ni9Ti fabricated by combustion synthesis in ultrahigh gravity field involves three stages: the formation of thermal explosion caused by a combination of mechanical activation and ultrahigh gravity field to yield high-temperature TiB$_2$–TiC ceramic liquid during the first stage; the formation of the molten stainless steel to dissolve TiB$_2$–TiC liquid, thereby yielding the diffusion-controlled concentration gradient from the ceramic liquid to the alloy liquid during the second stage; and the formation of rapid sequent solidification of the ceramic and the alloy during the third stage (as shown in Fig. 10). Therefore, the laminated composite materials are achieved in continuously graded composition and microstructure from the ceramic to the stainless steel. And, between

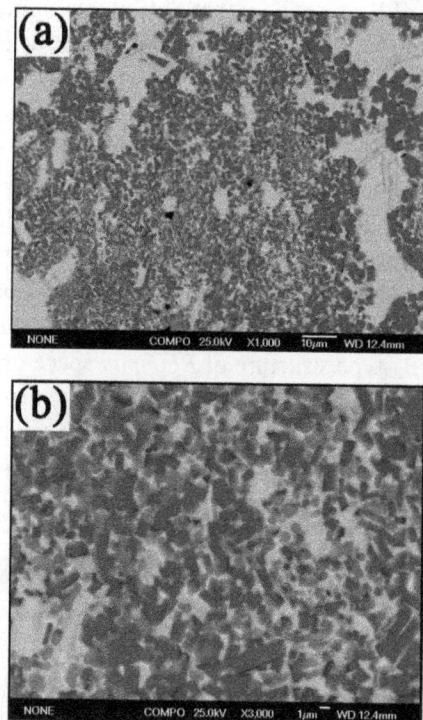

Fig. 9 Hybrid microstructure in the intermediate layer: (a) the segregation of the alloy phases; (b) the alloy phases distributed in the inter-region of TiB$_2$ and TiC.

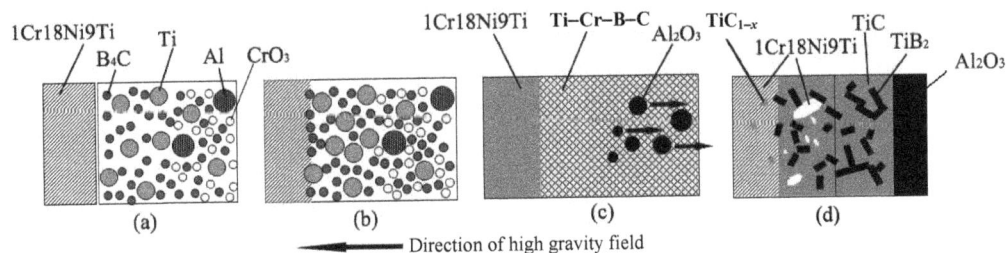

Fig. 10 Formation of the TiB$_2$–TiC/1Cr18Ni9Ti laminated composite by combustion synthesis in ultrahigh gravity field.

the ceramic and the alloy, the hybrid microstructure of multilevel and scale-span is presented with different-size, different-morphology Fe–Cr alloy phases alternating with TiB$_2$ platelets and irregular TiC grains in size from micrometer to nanometer.

Figure 11 shows the FESEM image of shear fracture surface of the ceramic/metal joint, and the shear strength of ceramic/metal joint was measured as 320 MPa. The experiments reveals that the shear fracture always takes place in the intermediate area near the stainless steel substrate where Al$_2$O$_3$ inclusions and cavities become the weak points and the crack sources in stress concentration. Hence, it can obtain higher joint strength by removing Al$_2$O$_3$ inclusions and cavities in combustion synthesis, so the higher high gravity field and the enough combusiton heat are necessary in these experiments.

4 Conclusions

By applying combustion synthesis in ultrahigh gravity field to prepare solidified TiB$_2$–TiC ceramic, the laminated composite materials of TiB$_2$–TiC ceramic to 1Cr18Ni9Ti stainless steel are achieved in

continuously graded microstructure via liquid dissolution and atomic interdiffusion. XRD, FESEM, and EDS results show that the ceramic is composed of a number of fine TiB$_2$ platelets, irregular TiC grains, and a few of Cr metallic binder. And the physical and mechanical properties reveal that the density, hardness, and fracture toughness of the ceramic are 4.3 g/cm^3, 18±1.5 GPa, and 13.5±1.8 MPa·m$^{1/2}$, respectively. The blend of TiB$_2$–TiC ceramic to stainless steel consists of three-layer structures, i.e., the TiB$_2$–TiC ceramic layer, the intermediate layer, and the stainless steel substrate. The key formation principle of the laminated composite materials of TiB$_2$-based ceramic to stainless steel by combustion synthesis in ultrahigh gravity field is considered to the following three elements: the formation of high-temperature TiB$_2$–TiC ceramic liquid during thermal explosion; the separation of Al$_2$O$_3$ drops and bubbles from ceramic liquid under high gravity field; the interdiffusion and filling between the ceramic liquid and the fused stainless steel surface. As a result, the laminated composite materials of TiB$_2$–TiC/1Cr18Ni9Ti are achieved in multilevel, scale-span hybrid microstructure, and the shear strength of ceramic/metal joint is measured as 320 MPa.

Acknowledgements

This work is sponsored by the National Natural Science Foundation of China (Grant No. 51072229).

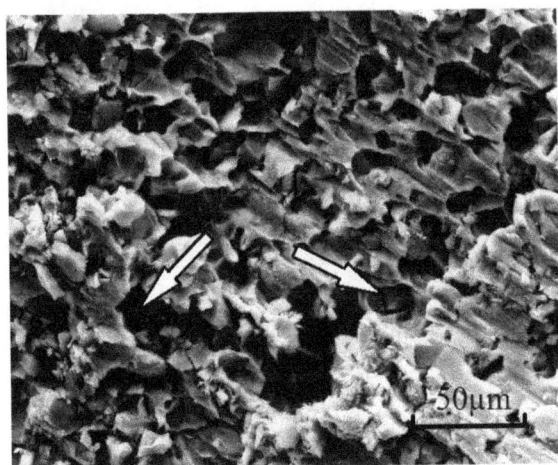

Fig. 11 FESEM image of shear fracture surface of the ceramic/metal joint.

References

[1] Zhang GJ, Jin ZZ, Yue XM. Effects of Ni addition on

mechanical properties of TiB$_2$/SiC composites prepared by reactive hot pressing (RHP). *J Mater Sci* 1997, **32**: 2093–2097.

[2] Lee TW, Lee CH. Microstructure and mechanical properties of TiB$_2$/TiAl composites produced by reactive sintering using a powder extrusion technique. *J Mater Sci Lett* 1999, **18**: 801–803.

[3] Vallauri D, Adrián ICA, Chrysanthou A. TiC–TiB$_2$ composites: A review of phase relationships, processing and properties. *J Eur Ceram Soc* 2008, **28**: 1697–1713.

[4] Degrave IE, Udalov YP. Composite powders of the TiC–TiB$_2$ system. *Glass Ceram* 2000, **57**: 396–398.

[5] Huang X, Zhang L, Zhao Z, *et al.* Microstructure transformation and mechanical properties of TiC–TiB$_2$ prepared by combustion synthesis in high gravity field. *Mat Sci Eng A* 2012, **553**: 105–111.

[6] Zhang Y, Feng D, He Z, *et al.* Progress in joining ceramics to metals. *J Iron Steel Res Int* 2006, **13**: 1–5.

[7] Huang X, Zhao Z, Zhang L. Layered composite of TiC–TiB$_2$ to Ti–6Al–4V in graded composition by combustion synthesis in high-gravity field. *J Phys: Conf Ser* 2013, **419**: 012027.

[8] Yin C, Chen YQ, Zhong SM, *et al.* Fractional-order sliding mode based extremum seeking control of a class of nonlinear system. *Automatica* 2014, **50**: 3173–3181.

[9] Yin C, Stark B, Chen YQ, *et al.* Adaptive minimum energy cognitive lighting control: Integer order vs fractional order strategies in sliding mode based extremum seeking. *Mechatronics* 2013, **23**: 863–872.

[10] Bever MB, Duwez PE. Gradients in composite materials. *Mater Sci Eng* 1972, **10**: 1–8.

[11] Huang X, Zhao Z, Zhang L, *et al.* The effects of ultra-high-gravity field on phase transformation and microstructure evolution of the TiC–TiB$_2$ ceramic fabricated by combustion synthesis. *Int J Refract Met H* 2014, **43**: 1–6.

[12] Morsi K. The diversity of combustion synthesis processing: A review. *J Mater Sci* 2012, **47**: 68–92.

[13] Chaudhari YA, Mahajan CM, Jagtap PP, *et al.* Structural, magnetic and dielectric properties of nano-crystalline Ni-doped BiFeO$_3$ ceramics formulated by self-propagating high-temperature synthesis. *J Adv Ceram* 2013, **2**: 135–140.

[14] Contreras L, Turrillas X, Vaughan GBM, *et al.* Time-resolved XRD study of TiC–TiB$_2$ composites obtained by SHS. *Acta Mater* 2004, **52**: 4783–4790.

[15] Mahmoodian R, Hassan MA, Hamdi M, *et al. In situ* TiC–Fe–Al$_2$O$_3$–TiAl/Ti$_3$Al composite coating processing using centrifugal assisted combustion synthesis. *Composites Part B* 2014, **59**: 279–284.

[16] Huang X, Zhao Z, Zhang L. Fusion bonding of solidified TiC–TiB$_2$ ceramic to Ti–6Al–4V alloy achieved by combustion synthesis in high-gravity field. *Mat Sci Eng A* 2013, **564**: 400–407.

[17] Aminikia B. Investigation of the pre-milling effect on synthesis of nanocrystalline TiB$_2$–TiC composite prepared by SHS method. *Powder Technol* 2012, **232**: 78–86.

[18] Kurz W, Fisher DJ. *Fundamentals of Solidification*, 4th edn. Switzerland: Trans Tech Publications Ltd, 1998.

10

Catalytic liquid phase oxidation of 1,4-dioxane over a Pt/CeO$_2$–ZrO$_2$–Bi$_2$O$_3$/SBA-16 catalyst

Pil-Gyu CHOI, Takanobu OHNO, Nashito FUKUHARA,
Toshiyuki MASUI, Nobuhito IMANAKA[*]

Department of Applied Chemistry, Faculty of Engineering, Osaka University, 2-1 Yamadaoka, Suita, Osaka 565-0871, Japan

Abstract: A Pt/CeO$_2$–ZrO$_2$–Bi$_2$O$_3$/SBA-16 (Santa Barbara Amorphous No. 16) catalyst was prepared by hydrothermal and wet impregnation methods for catalytic purification of 1,4-dioxane in water. SBA-16 has a number of mesopores and the average size of the pores is 9.4 nm. In the present catalyst, platinum and CeO$_2$–ZrO$_2$–Bi$_2$O$_3$ were successfully dispersed in the pores of the SBA-16 support, and the temperature dependence of the liquid phase oxidation of 1,4-dioxane was examined. The oxidation reaction proceeded effectively in the air atmosphere in the temperature range of 40–80 ℃.

Keywords: composite materials; oxidation; porous materials; 1,4-dioxane; catalyst

1 Introduction

1,4-dioxane is a cyclic ether that possesses satisfactory solubility in water as well as organic solvents. It exists in its liquid state at ordinary temperature, and its boiling point is 101 ℃ [1] which is almost equivalent to that of water (100 ℃). 1,4-dioxane has been used as a solvent for extraction, purification and chemical reaction [2–4], but it has been identified as a cancer-causing pollutant by animal testing [5]. Therefore, a stringent environmental quality standard for water pollution has been established for 1,4-dioxane: for example, it is 0.05 mg/L or below in Japan [6]. Unfortunately, however, 1,4-dioxane is a persistent substance not to be susceptible to hydrolysis and biodegradation [4,7]. Accordingly, it is significantly difficult to eliminate 1,4-dioxane with

microbial processes at wastewater treatment plants [8]. Furthermore, 1,4-dioxane is poorly adsorbed onto active carbon, and it is possible to be eliminated only in the presence of ozone by oxidation [9]. Therefore, an effective treatment process remains to be established, and it has been required to find a novel water treatment technology for degradation of 1,4-dioxane.

In this study, a porous catalyst was prepared for the degradation of 1,4-dioxane in water, where platinum and CeO$_2$–ZrO$_2$–Bi$_2$O$_3$ were dispersed in the pores of the SBA-16 (Santa Barbara Amorphous No. 16) support [10] which has pore size around 9.4 nm suitable for the insertion of platinum and CeO$_2$–ZrO$_2$–Bi$_2$O$_3$ particles. In the present catalyst, platinum works as a main oxidation catalyst and CeO$_2$–ZrO$_2$–Bi$_2$O$_3$ solid solution plays a promoter role to facilitate oxidation by supplying active oxygen from the inside bulk due to the oxygen storage and release properties [11–13]. When these components are

* Corresponding author.
E-mail: imanaka@chem.eng.osaka-u.ac.jp

supported inside the mesopores of SBA-16, it becomes possible to adsorb and decompose 1,4-dioxane at once. The reaction is carried out in air with vigorous stirring, and, thereby, the lattice oxygen of the CeO_2–ZrO_2–Bi_2O_3 promoter is continuously recovered by absorption of oxygen from the air through liquid water phase. In this study, therefore, catalytic purification performance of 1,4-dioxane in water was evaluated for the Pt/CeO_2–ZrO_2–Bi_2O_3/SBA-16 catalyst.

2   Experimental procedure

2. 1   Catalyst preparation

SBA-16 was prepared according to the hydrothermal method described in previous studies [10,14]. Pluronic F-127 (1.6 g) and 1,3,5-trimethylbenzene (1.1 cm^3) were dissolved into 0.2 mol/dm^3 hydrochloric acid (90 cm^3), and the mixture was stirred at 35 ℃ for 3 h. After stirring, tetraethyl orthosilicate (7.1 cm^3) was added to the solution, and the mixture was stirred again at 35 ℃ for 20 h. Then the mixture was poured into a Teflon bottle in a sealed brass vessel and heated at 140 ℃ for 24 h. The solid product was collected by filtration, washed and dried at room temperature for 12 h. Finally, the dried powder was calcined at 600 ℃ for 4 h in air to obtain SBA-16.

Subsequently, $Ce_{0.68}Zr_{0.17}Bi_{0.15}O_{1.925}$ solid solution was supported on SBA-16 by an impregnation method. The SBA-16 support synthesized above (0.40 g) was dispersed into a stoichiometric mixture of 1.0 mol/dm^3 $Ce(NO_3)_3$ (0.30 cm^3), 0.1 mol/dm^3 $ZrO(NO_3)_2$ (0.73 cm^3) and 0.5 mol/dm^3 $Bi(NO_3)_3$ (0.13 cm^3) aqueous solutions. Then, 6.8×10^{-2} mol/dm^3 nitric acid aqueous solution (50 cm^3) was added, and the mixture was stirred at room temperature for 6 h. The content of $Ce_{0.68}Zr_{0.17}Bi_{0.15}O_{1.925}$ was adjusted to be 16 wt%, which was the optimum amount for toluene oxidation [15]. After stirring, the solvent was vaporized at 180 ℃. The powder obtained was ground in a mortar and calcined at 600 ℃ for 1 h in air.

Finally, platinum was dispersed on the $Ce_{0.68}Zr_{0.17}Bi_{0.15}O_{1.925}$/SBA-16 powder using a platinum colloid stabilized with polyvinylpyrrolidone. The Pt content was adjusted to be 7 wt% for exactly the same reason as $Ce_{0.68}Zr_{0.17}Bi_{0.15}O_{1.925}$ mentioned above [15]. After the deposition, the solvent was evaporated at 180 ℃. The dried powder was ground in a mortar and calcined at 500 ℃ for 4 h in air to obtain a Pt (7 wt%)/$Ce_{0.68}Zr_{0.17}Bi_{0.15}O_{1.925}$(16 wt%)/SBA-16 (denoted as Pt/CZB/SBA-16 hereafter) catalyst.

2. 2   Characterization

The composition of the Pt/CZB/SBA-16 catalyst was confirmed to be in good agreement with the stoichiometric ratio of the starting materials using X-ray fluorescence analysis (XRF, Rigaku Supermini200). X-ray powder diffraction and small angle X-ray scattering (XRD and SAXS respectively, Rigaku SmartLab) patterns were measured in the 2θ range of 10°–70° and 0.06°–2.0°, respectively. The Pt particle size after the heat treatment at 500 ℃ for 4 h in air was determined using the following Scherrer equation:

$$D = \frac{k \cdot \lambda}{\beta \cdot \cos\theta}$$

where D is the mean size of the crystalline domains; k is a dimensionless shape factor which is equal to 0.9; λ is the X-ray wavelength which is 0.154 nm for Cu Kα; β is the full width at half maximum (FWHM); and θ is the Bragg angle. The Brunauer–Emmett–Teller (BET) surface area and the pore size distribution were measured at −196 ℃ (Micromeritics Tristar 3000) using N_2 adsorption. Scanning electron microscopy (SEM, Shimadzu SS-550) and transmission electron microscopy (TEM, Hitachi H-9000NAR) were conducted to observe the morphology and size of the particles.

The oxidation reaction of 1,4-dioxane was conducted in the air atmosphere in batch mode using a mechanically stirred 300 cm^3 three-necked flask. Prior to the reaction, an aqueous solution of 100 ppm 1,4-dioxane (10 cm^3) was poured into the flask, and Pt/CZB/SBA-16, CZB/SBA-16, Pt/SBA-16 or SBA-16 (0.3 g) was loaded. The reactor was then heated to reflux in an oil bath at a stable operational temperature in the range of 40–80 ℃. After the 4 h reaction, the catalyst was separated by centrifugation, and the supernatant liquid was analyzed using gas chromatograph mass spectrometry (GCMS, Shimadzu GCMS-QP2010 Plus) to evaluate the rate of decrease in 1,4-dioxane. In addition, time course of catalytic activity for Pt/CZB/SBA-16 was also measured at 80 ℃ for 2 h, 4 h and 6 h using an aqueous solution of 100 ppm 1,4-dioxane.

3 Results and discussion

The SBA-16 support has a large surface area of 927 m^2/g, but it significantly decreases to 475 m^2/g after introduction of Pt and CZB into the mesopores of SBA-16. Figure 1 shows the N$_2$ adsorption–desorption isotherms of SBA-16 and Pt/CZB/SBA-16. The pore size distribution profiles calculated from the N$_2$ adsorption isotherms using the International Union of Pure and Applied Chemistry (IUPAC) and the Barrett–Joyner–Halenda (BJH) analysis [16] are also depicted in the insets. According to Brunauer–Derning–Deming–Teller (BDDT) classifications, the pattern of these isotherms is classified to the type IV, which is typical for materials with mesopore structures. The average pore size of the Pt/CZB/SBA-16 catalyst is 9.0 nm, which is smaller than that of SBA-16 (9.4 nm). The BJH analysis was carried out again to identify whether the difference was led by experimental error, but a reproducible result was obtained. These results also suggest that the pores are filled with Pt and/or CZB particles.

Figure 2 depicts the XRD pattern of the Pt/CZB/SBA-16 catalyst. The detectable diffraction peaks in the pattern are well indexed to those of Pt, CZB and SBA-16. The six broad peaks at $2\theta = 28.4°$,

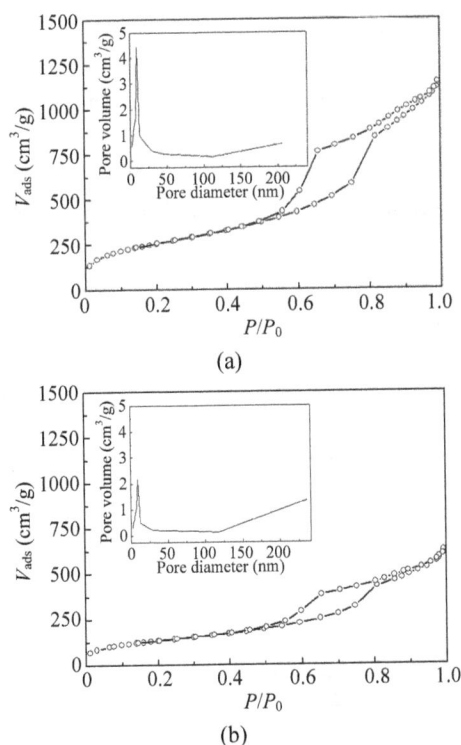

Fig. 2 XRD pattern of the Pt/CZB/SBA-16 catalyst.

32.8°, 46.7°, 56.3°, 58.5° and 69.7° are assigned to (111), (002), (022), (113), (222) and (004) reflections of cubic fluorite-type structure respectively, while three peaks observed at $2\theta = 39.6°$, 46.0° and 67.5° correspond to those of platinum. The diffraction peak at $2\theta = 22.7°$ is very broad due to the low crystallinity of the SBA-16 crystallites. The Pt particle size is calculated to be 8.9 nm from the Scherrer equation using the FWHM and angular position of the (111) peak.

Figure 3 shows the SAXS patterns of SBA-16 and Pt/CZB/SBA-16. The diffraction peak at $2\theta = 0.6°$ in Fig. 3(a) elucidates that mesopores of SBA-16 are successfully synthesized. After the loading of Pt and CZB, the intensity of the diffraction peak is significantly decreased in comparison with that for as-prepared SBA-16, and the diffraction angle shifts to higher direction indicating the decrease in the pore size of SBA-16. These results indicate that the Pt and/or CZB particles are loaded in the pores of SBA-16.

Figure 4 shows an SEM image of the SBA-16 support. The particle size of the SBA-16 support is approximately 2 μm. TEM images of the SBA-16 support and the Pt/CZB/SBA-16 catalyst are displayed in Fig. 5. The synthesized SBA-16 support possesses a

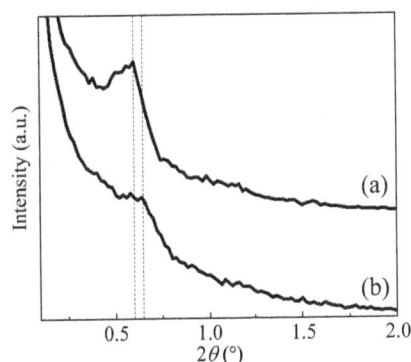

Fig. 1 N$_2$ adsorption–desorption isotherms of (a) SBA-16 and (b) Pt/CZB/SBA-16. The BJH pore size distribution profiles are shown in the insets.

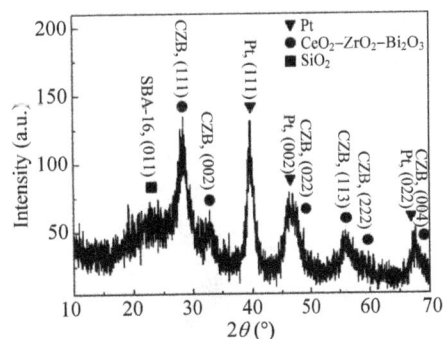

Fig. 3 SAXS patterns of (a) SBA-16 and (b) Pt/CZB/SBA-16.

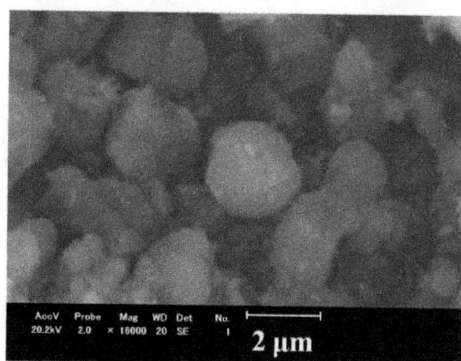

Fig. 4   SEM image of the SBA-16 support.

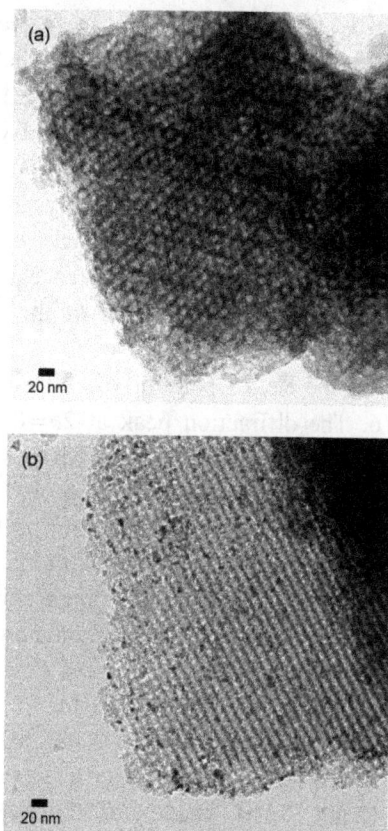

Fig. 5   TEM images of (a) SBA-16 support and (b) Pt/CZB/SBA-16 catalyst.

regular arrangement of mesopores, and the pore diameter is approximately 10 nm as shown in Fig. 5(a). After the introduction of the Pt and CZB particles, these particles are observed in Fig. 5(b) as dark spots in and on the SBA-16 support, although it is difficult to distinguish between Pt and CeO_2–ZrO_2–Bi_2O_3. These results are in agreement with the results of the SAXS measurement that Pt and CZB particles are well dispersed into the mesopores of SBA-16, as well as those of the BJH pore size distribution described above.

Finally, the results on the catalytic liquid phase

oxidation of 1,4-dioxane using SBA-16, CZB/SBA-16, Pt/SBA-16 and Pt/CZB/SBA-16 are tabulated in Tables 1 and 2. The supernatant liquid after the reaction was analyzed using GCMS. As a result, any by-products are not detected: only 1,4-dioxane is detected and its concentration is smaller than that before the reaction. Accordingly, it is considered that carbon dioxide and water are generated by oxidation of 1,4-dioxane. The residual percentage of 1,4-dioxane in Table 1 was calculated according to the following equation:

$$\frac{\text{Molar concentration of 1,4-dioxane after the reaction}}{\text{Molar concentration of 1,4-dioxane before the reaction}} \times 100\%$$

The rate of decrease in 1,4-dioxane after the oxidation reaction summarized in Table 2 was estimated by deduction of the residual percentage of 1,4-dioxane in the presence of CZB/SBA-16, Pt/SBA-16 and Pt/CZB/SBA-16 from that obtained in the blank test performed with SBA-16 in Table 1. In the blank test using SBA-16, effects of both vaporization and adsorption on the SBA-16 support are included. Accordingly, the rate of decrease in 1,4-dioxane in Table 2 shows the result just caused by oxidation.

As seen in Table 2, the oxidation catalysis of CZB/SBA-16, Pt/SBA-16 and Pt/CZB/SBA-16 is obviously recognized and the rate of decrease in 1,4-dioxane increases with a rise in the reaction temperature. Among the samples tested, Pt/CZB/SBA-16 shows the highest oxidation activity. These results demonstrate that the insertion of both Pt and CZB particles into the mesopores of the SBA-16 support enables high catalytic activity. In addition, time course of catalytic activity for Pt/CZB/SBA-16 was also measured at 80 ℃ for 2 h, 4 h and 6 h (Table 3). It clarifies   that   the   catalytic   reaction   proceeds

Table 1   Residual percentage of 1,4-dioxane after the 4 h reaction in the air atmosphere

Reaction temperature (℃)	Residual percentage of 1,4-dioxane (%)			
	SBA-16	CZB/SBA-16	Pt/SBA-16	Pt/CZB/SBA-16
40	74	67	65	60
60	79	68	64	49
80	75	53	50	41

Table 2   Rate of decrease in 1,4-dioxane after the 4 h reaction in the air atmosphere

Reaction temperature (℃)	Rate of decrease in 1,4-dioxane (%)		
	CZB/SBA-16	Pt/SBA-16	Pt/CZB/SBA-16
40	7	9	14
60	11	15	30
80	22	25	34

Table 3 Rate of decrease in 1,4-dioxane after the reaction at 80 °C for 2 h, 4 h and 6 h

Reaction time (h)	Residual percentage of 1,4-dioxane (%)		Rate of decrease in 1,4-dioxane (%)
	SBA-16	Pt/CZB/SBA-16	
2	73	60	13
4	75	41	34
6	76	38	38

continuously. Although the maximum degradation rate of decrease in 1,4-dioxane is 38% at 80 °C using Pt/CZB/SBA-16 at this time, the purification ability will be increased by the improvement of the oxygen supplying ability of the promoter by using CeO_2–ZrO_2–SnO_2 solid solutions [17] in the near future work.

4 Conclusions

A novel oxidation catalyst composed of Pt, CeO_2–ZrO_2–Bi_2O_3 and SBA-16 was prepared for liquid phase oxidation of 1,4-dioxane in wastewater. The oxidation reaction was progressed effectively by the loading of the Pt and CeO_2–ZrO_2–Bi_2O_3 particles into the mesopores of SBA-16. The catalytic activity was increased with increasing the reaction temperature and the catalytic reaction proceeded continuously with time course. As a result, the rate of decrease in 1,4-dioxane reached 38% after the reaction at 80 °C for 6 h in the air atmosphere.

References

[1] Budavari S, O'Neil MJ, Smith A, et al. The Merck Index, 11th edn. Rahway, NJ (USA): Merck & Co., Inc., 1989.

[2] Chitra S, Paramasivan K, Cheralathan M, et al. Degradation of 1,4-dioxane using advanced oxidation processes. Environ Sci Pollut R 2012, 19: 871–878.

[3] Adams CD, Scanlan PA, Secrist ND. Oxidation and biodegradability enhancement of 1,4-dioxane using hydrogen peroxide and ozone. Environ Sci Technol 1994, 28: 1812–1818.

[4] Fan W, Kubota Y, Tatsumi T. Oxidation of 1,4-dioxane over Ti-MWW in the presence of H_2O_2. ChemSusChem 2008, 1: 175–178.

[5] Yamazaki K, Ohno H, Asakura M, et al. Two-year toxicological and carcinogenesis studies of 1,4-dioxane in F344 rats and BDF1 mice-drinking studies. In Proceedings: Second Asia-Pacific Symposium on Environmental and Occupational Health 22–24 July, 1993: Kobe. Sumino K, Sato S, Shinkokai NG, Eds. Kobe University, 1994: 193–198.

[6] Ministry of the Environment, Government of Japan. Environmental quality standards for water pollution. Available at https://www.env.go.jp/en/water/wq/wp.pdf.

[7] Klečka GM, Gonsior SJ. Removal of 1,4-dioxane from wastewater. J Hazard Mater 1986, 12: 161–168.

[8] Makino R, Gamo M, Sato N, et al. Removal rate of 1,4-dioxane in a sewage treatment plant. J Jpn Soc Water Environ 2005, 28: 211–215.

[9] Aizawa T. Contamination of drinking water sources with novel pollutants. J Environ Conserv Engineering 2001, 30: 592–597.

[10] Zhao D, Huo Q, Feng J, et al. Nonionic triblock and star diblock copolymer and oligomeric surfactant syntheses of highly ordered, hydrothermally stable, mesoporous silica structures. J Am Chem Soc 1998, 120: 6024–6036.

[11] Masui T, Minami K, Koyabu K, et al. Synthesis and characterization of new promoters based on CeO_2–ZrO_2–Bi_2O_3 for automotive exhaust catalysts. Catal Today 2006, 117: 187–192.

[12] Imanaka N, Masui T, Koyabu K, et al. Significant low-temperature redox activity of $Ce_{0.64}Zr_{0.16}Bi_{0.20}O_{1.90}$ supported on γ-Al_2O_3. Adv Mater 2007, 19: 1608–1611.

[13] Imanaka N, Masui T, Yasuda K. Environmental catalysts for complete oxidation of volatile organic compounds and methane. Chem Lett 2011, 40: 780–785.

[14] Yotou H, Okamoto T, Ito M, et al. Novel method for insertion of Pt/CeZrO2 nanoparticles into mesoporous SBA-16 using hydrothermal treatment. Appl Catal A: Gen 2013, 458: 137–144.

[15] Masui T, Imadzu H, Matsuyama N, et al. Total oxidation of toluene on Pt/CeO_2–ZrO_2–Bi_2O_3/γ-Al_2O_3 catalysts prepared in the presence of polyvinyl pyrrolidone. J Hazard Mater 2010, 176: 1106–1109.

[16] Barrett EP, Joyner LG, Halenda PP. The determination of pore volume and area distributions in porous substances. I. Computations from nitrogen isotherms. J Am Chem Soc 1951, 73: 373–380.

[17] Yasuda K, Yoshimura A, Katsuma A, et al. Low-temperature complete combustion of volatile organic compounds over novel Pt/CeO_2–ZrO_2–SnO_2/γ-Al_2O_3 catalysts. Bull Chem Soc Jpn 2012, 85: 522–526.

11

Polymer optimization for the development of low-cost moisture sensor based on nanoporous alumina thin film

Manju PANDEY[a], Prabhash MISHRA[a], Debdulal SAHA[b], S. S. ISLAM[a,*]

[a]Nano-Sensor Research Laboratory, F/O Engineering and Technology, Jamia Millia Islamia (Central University), New Delhi, India
[b]Sensors and Actuators Division, Central Glass & Ceramic Research Institute, 196 Raja S. C. Mullick Road, Kolkata 700032, India

Abstract: Sol–gel processed alumina (Al_2O_3) thin film has been investigated for sensing moisture. The sensor was based on ordered nanoporous Al_2O_3 thin film, which consisted of gold electrodes on both sides of the film forming parallel plate capacitor. Alumina substrate was used for supporting thin film moisture sensor. Hydrophilicity was achieved by controlling the surface energy of the substrate and polymer (polyvinyl alcohol (PVA)) optimization was done for developing rigid thin film over it. A high change in capacitance was observed as the moisture level increased from 5 ppmV to 500 ppmV. Scanning electron microscopy (SEM) results revealed that pores were distributed uniformly throughout the sample, which enhanced the adsorption of water molecule over the film. X-ray diffraction (XRD) study clearly confirmed the gamma (γ) phase of alumina thin film. It was found that the sensitivity of our sensor was suitable for commercial application.

Keywords: sol–gel; dip-coating; polyvinyl alcohol (PVA); nanoporous; hydrophilicity; moisture sensor

1 Introduction

As the increasing demand for a successful packaging in food industry, a technique to detect trace level of moisture during micro packaging has become more important. As we know, failures are mainly due to contaminants such as ambient moisture present in atmosphere, so a highly sensitive and stable moisture sensor has gotten much attention [1]. There is an increasing interest in polymers for sensing applications [2]. A typical problem of polymers is their poor durability against water. In fact, polymeric materials used in moisture sensors are generally hydrophilic and sometimes even soluble in water. This problem is solved by sol–gel technique. Sol–gel is a chemical solution process used to make ceramic and glass materials at low temperature in the form of thin films, fibers or powders, in which sol is a colloidal or molecular suspension of solid particles of ions in a solvent. Sol–gel technology has received much attention in recent years in the field of sensing application. The reason behind is it is a low-cost,

---

* Corresponding author.
E-mail: safiul5996@gmail.com

easier-fabrication and precisely-composition-control process [3]. On the other hand, sol–gel is suitable for preparing porous material because when the solvent from the sol begins to evaporate and the particles or ions left behind begin to join together in a continuous network, it forms a lot of cavities [4]. The properties of sol–gel processed ceramic materials can be improved by the addition of different types of porosity-control additives. The frequently used porosity-control additives are some organic polymers especially polyvinyl alcohol (PVA) to the colloidal solution, because it leads to the achievement of a very good rigidity of the resulted thin film product [5]. Yoldas [6] followed the method for preparing alumina sol. Preliminary studies on electrical behavior of Al_2O_3 thin films in the presence of trace level moisture have been carried out [7,8]. Result shows that Al_2O_3 thin films could form a good porous layer with nano-sized pores for trace moisture sensors. In present research work, ceramic moisture sensors [7,9–11] established their good sensitivity in the range of 5–500 ppmV after optimizing the amount of polymer (PVA).

2 Experiment

2.1 Preparation of alumina sol solution

Alumina sol was prepared by Yoldas process [4]. Hydrolysis was performed by introducing an organometallic compound, Al-sec-butoxide ($AlC_{12}H_{27}O_3$), into de-ionized water (Al-sec-butoxide:water = 1:25). Then the solution was peptized by dropwise adding hydrochloric acid (acid/alkoxide molar ratio = 0.07) till it became transparent solution under continuous stirring at 90℃. The peptized solution was then refluxed for about 14–16 h. After refluxing, three different concentrations of PVA which acted as binder were added into 50 ml of refluxed sol. As we know, ceramic colloidal particles have the tendency of agglomeration in aqueous media. In general, aqueous suspensions can be stabilized by controlling the repulsive forces proceeding from a charged electric double layer, surrounding the particle (electrostatic stabilization), or from non-charged or charged polymers, adsorbed on the surface (steric and electrosteric stabilization, respectively) [12]. The electrokinetic or zeta potential measurements are used to estimate the electrostatic effect, which is responsible for the repulsive forces between the particles and

estimates the suspension stability. Studies on particle electrokinetic have been proved to be useful where colloid stability is involved [13]. So for the development of crack-free thin film over the gold coated substrate, the zeta potential (ζ) of the sol was controlled to be +22 mV by adding HCl for uniform dispersion of the sol particles as shown in Fig. 1. From the zeta potential data, lower pH values promise better dispersing efficiencies with respect to higher surface charges yielding more intensive repulsive forces. A pH range of 2.0–3.0 appears to be the most beneficial working range in dispersing process of boehmite (AlO(OH)), because at this range repulsive forces between colloidal particles are very high as compared to other pH range and simultaneously it increases the stability of the boehmite sol.

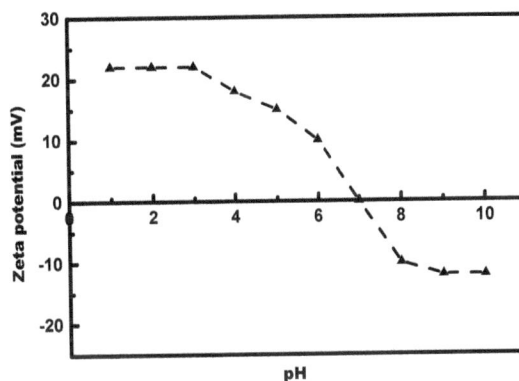

Fig. 1   Zeta potential curve for alumina sol.

2.2 Development of capacitive humidity sensor

The PVA mixed sol was coated on a gold-coated alumina substrate (20 mm × 20 mm × 1 mm) by dipping it in alumina sol at a speed of 10 cm/min and then pulling out at a speed of 1 cm/min. The process was repeated eight times. This was followed by drying and then sintering the films at 450℃ for 5 h with the heating and cooling rate of 100 ℃/h. A stable γ phase of alumina with high surface area [14] which is first priority for sensing mechanism was achieved after sintering and annealing the films. The second electrode using gold paste was developed by screen printing method on the alumina thin film coated substrate, and then finally fired at 950℃ for curing the top gold electrode as illustrated in Fig. 2. The film between the gold electrodes was used as transducing layer for sensing the ambient trace moisture. The effective dielectric of the porous layer underwent a large change when water vapour diffused to the inner regions of the pore structure and was adsorbed on the porous layer. The capacitive sensor developed in this work consisted of a non-porous substrate coated with gold on which the thin films were deposited. It was covered by another porous electrode of gold. Contacts were taken from the two parallel gold electrodes. The water vapour was free to diffuse through the porous electrode to the porous alumina film and showed increase in capacitance. Figure 3 shows the photograph of the fabricated capacitive moisture sensors.

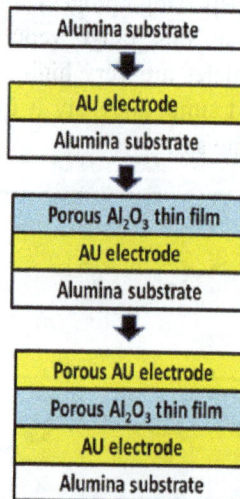

Fig. 2   Fabrication sequence of the moisture sensors based on alumina thin film.

Fig. 3   Photograph of the moisture sensor.

3   Results and discussion

3.1   Deposition of film by controlling the surface energy

Sol–gel thin film coating on substrate depends on

intrinsic properties of the coating and adhesion characteristics with the substrate material, which in turn, are controlled by the sol–gel processing parameters. Therefore, the interfacial adhesion properties of the film on substrate should be controlled by controlling the surface tension and contact angle between liquid and substrate using the sessile drop method. For developing rigid thin film over the substrate, it is very important that the difference between work of adhesion (W_a) and work of cohesion (W_{coh}) should be positive with large value, only then does the film adhere properly with the substrate which is controlled by optimizing contact angle. Optimized PVA quantity plays an important role in rigidity. For optimization, three samples were prepared by using different concentrations of PVA into the alumina sol and then contact angle was measured over the substrate using drop of prepared sol as shown in Table 1.

The wettability of liquids on ceramics is often evaluated in terms of work of adhesion, W_a, where W_a is usually determined from sessile drop experiment as shown in Fig. 4, in which surface tension σ_{lv} and contact angle θ are measured by using Young–Dupre equation:

$$W_a = \sigma_{lv}(1 + \cos\theta)$$

The liquid surface tension is controlled so that the liquid angle of contact with the substrate should maintain

$$W_a > W_{coh}$$

Table 1   Optimization of surface energy

Sample No.	PVA (g)	Contact angle θ	W_a (mN/m)	W_{coh} (mN/m)	$W_a - W_{coh}$ (mN/m)
1	0.5	74.40°	105.04	64.51	40.53
2	1.0	65.23°	93.73	41.87	51.86
3	1.5	37.64°	134.76	124.76	10.00

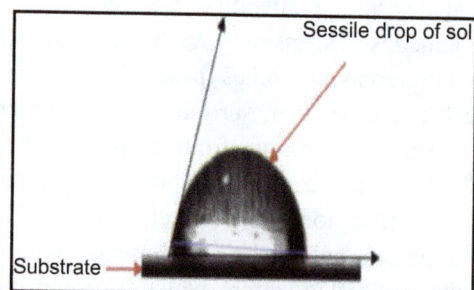

Fig. 4   Sessile-drop contact angle measurement.

3. 2   Surface area and pore size analysis

Porosity, surface area and pore size distribution are the basic quantities to specify a porous material. Brunauer, Emmet and Teller (BET) method was followed to measure surface area and Barrett–Joyner–Halenda (BJH) method for pore size distribution using surface area and pore size analyser (NOVA 2000e Model) as shown in Table 2. It yields well defined adsorption, desorption isotherms on most surfaces. The nitrogen adsorption isotherm gives the structural information about the network of micropores and mesopores. The formation of adsorption boundary curve is due to increase of pressure from zero to maximum value and desorption isotherm is for decrease of pressure from maximum to zero value. BET characterization was done after annealing the alumina films without gold electrode at 450 ℃ along with three different concentrations of PVA and then scratching them from the substrate.

Table 2   BET analysis for surface area, pore size and porosity present in γ phase of alumina thin film

Thin film (Al$_2$O$_3$)	Heat treatment (℃, h)	BET surface area (m^2/g)	Pore diameter (nm)	Porosity (%)
With 0.5 g PVA	450, 5	110.2	8.02	60
With 1.0 g PVA	450, 5	170.0	7.02	80
With 1.5 g PVA	450, 5	200.0	6.89	83

3. 3   Structural properties

X-ray diffraction (XRD) analysis was carried out by removing the annealed film from the substrate with a PAN analytical X-ray spectrometer (PW1830, Philips) with Cu Kα radiation ($\lambda = 1.54$ Å, $2\theta = 20°$–$80°$) at room temperature. All the peaks are corresponded to γ phase of alumina as shown in Fig. 5.

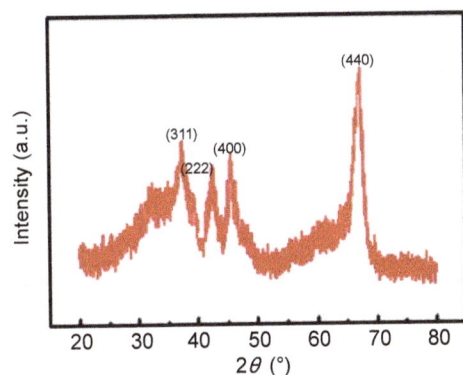

Fig. 5   XRD micrograph of γ-phase alumina film.

3. 4   Surface morphology of thin film

Field emission scanning electron microscope (FEI, Nova Nano SEM 450) was used for surface structure analysis as shown in Fig. 6. The micrographs show the effect of binder (PVA) concentration on the morphology of thin films. Figure 6(a) shows thin film

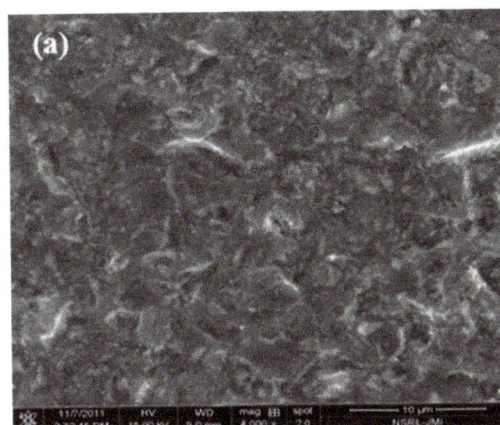

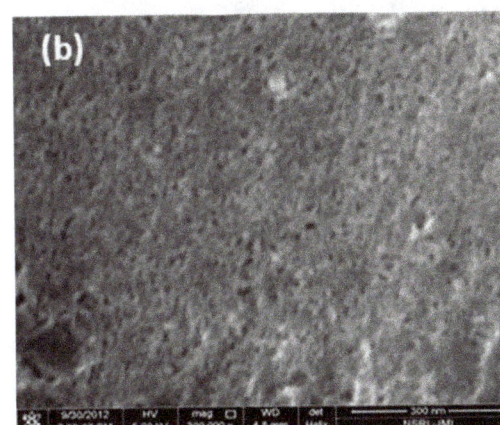

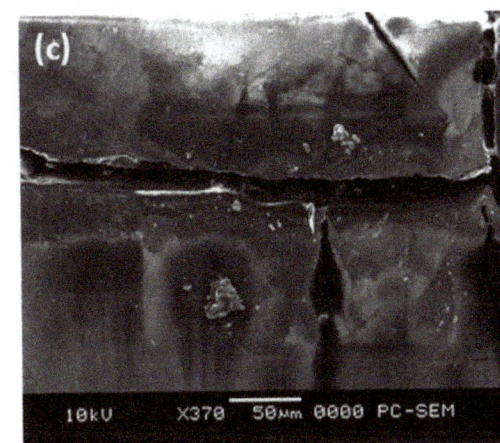

Fig. 6   Scanning electron micrographs of γ-phase alumina thin film with (a) 0.5 g PVA, (b) 1.0 g PVA, (c) 1.5 g PVA.

with less porosity, whereas Fig. 6(b) shows a film having high degree of porosity with a large enhancement of the active surface, responsible for the high sensitivity towards moisture and being best suitable for developing moisture sensor. Figure 6(c) shows a film which is full of cracks and is completely discarded for further process.

Roughness of the film (which contained optimum amount of binder that was 1.0 g in 50 ml sol) surface was checked by atomic force microscopy (AFM, Solver Next, NT-MDT) measurement. The contact mode measurement was followed with the tip radius of 20 nm. Roughness or porosity also depends on homogeneity and rigidity of thin film over the substrate. AFM image (Fig. 7) shows high roughness or porosity which is first priority in sensing application.

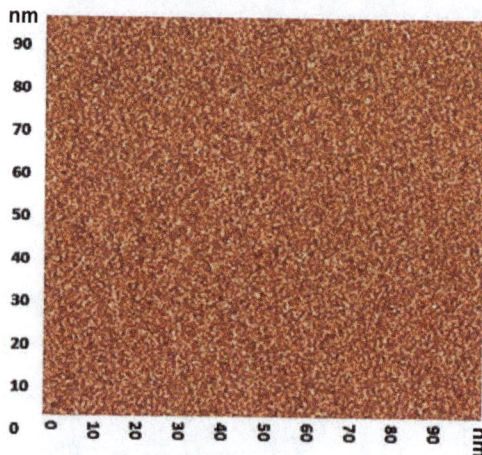

Fig. 7   AFM image of alumina thin film.

3.5   Sensing performance of sensor

The electrical characterization of the sensors was determined by placing the sensors in sample holder which was made up of stainless steel. Various concentrations of moisture were achieved by mixing dry nitrogen gas with a desired amount of moist air. The mixing gas flow rate was set to 50 ml/min. Dielectric analysis was performed by semiconductor characterization system (4200 SCS, Keithley) and trace moisture variation was monitored by a SHAW dew point meter (model No. SADPTR-R, UK) which is a commercial moisture sensor as illustrated in Fig. 8.

3.5.1   Sensitivity

The measurement has been performed from 5 ppmV to 500 ppmV at room temperature. At specific ppmV level of moisture, the capacitance has been recorded for 30 min. The sensitivity is defined by the slope or gradient of the capacitance to moisture curve.

The capacitance versus moisture level at 1 kHz (Fig. 9) shows the linear dependence between moisture and capacitance. The sensitivity is 1.11 pF/ppmV for the range of 5–500 ppmV.

3.5.2   Response and recovery time

Response and recovery characteristics have been observed by applying moist gas of 4–10 ppmV repeatedly. Figure 10 shows the response and recovery curve and it is observed that the response time is 20 s and recovery time is 30 s. Comparison between fabricated moisture sensor and commercial SHAW moisture sensor is shown in Table 3.

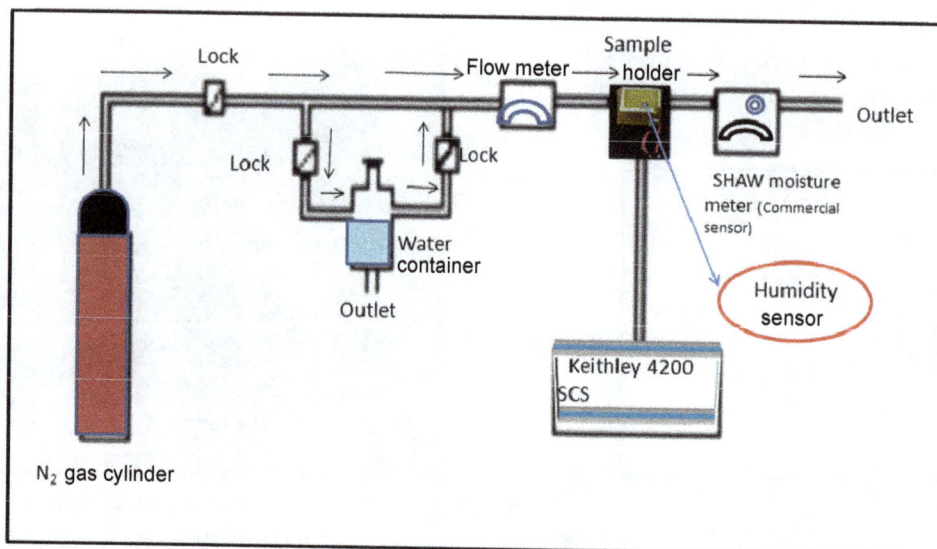

Fig. 8   Moisture sensing setup.

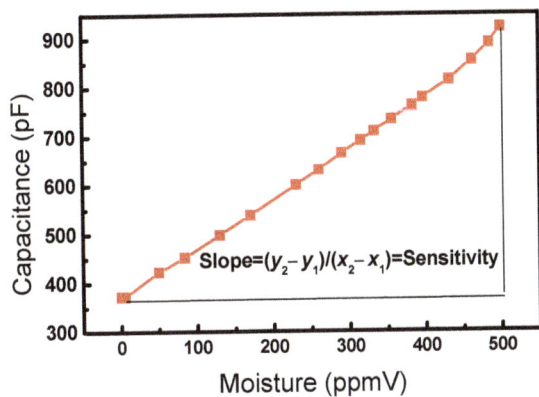

Fig. 9   Capacitance versus humidity curve.

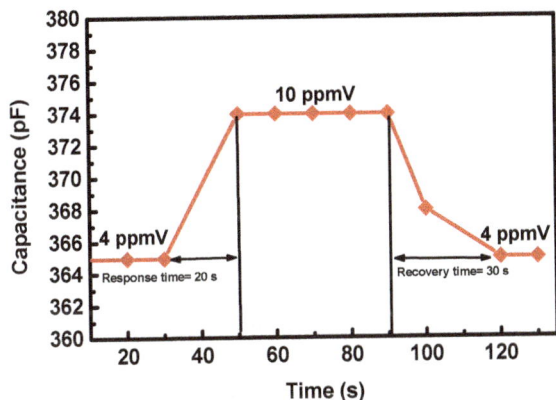

Fig. 10   Response and recovery time of humidity sensor.

Table 3   Comparison of output response of developed sensor with a commercial SHAW moisture analyzer

Type of sensor	Sensitivity (pF/ppmV)	Range (ppmV)	Response time (s)
SHAW moisture sensor	1.05	0–1000	258
γ-phase moisture sensor	1.11	5–500	20

4   Conclusions

The thin film of alumina prepared by sol–gel technology was investigated as a capacitive humidity sensor. Rigidity of thin film was achieved by binder optimization and controlling the contact angle over the substrate. It also found the linear variation of trace moisture from 5 ppmV to 500 ppmV. Sensitivity and resolution were also good due to the presence of high surface area as a result large change in capacitance with the absorption of moisture. The outcome of our studies accomplished that γ phase in alumina would be preferred for fabrication of trace moisture sensor.

References

[1] Chen Y-T, Hsu W-T, Najafi K, et al. Vacuum packaging technology using localized aluminum/silicon-to-glass bonding. J Microelectromech S 2002, **11**: 556–565.

[2] Harsányi G. Polymer Films in Sensors Applications. Budapest: Technomic Publishing Company, 1995.

[3] Mistry KK, Saha D, Sengupta K. Sol–gel processed Al_2O_3 thick film template as sensitive capacitive trace moisture sensor. Sensor Actuat B: Chem 2005, **106**: 258–262.

[4] Bagwell RB, Messing GL. Critical factors in the production of sol–gel derived porous alumina. Key Eng Mat 1996, **115**: 45–64.

[5] Ecsedi Z. Synthesis of tailored porosity materials using the sol–gel method. Chem Bull "POLITEHNICA" Univ (Timişoara) 2007, **52**: 14–17.

[6] Yoldas BE. Alumina gels that form porous transparent Al_2O_3. J Mater Sci 1975, **10**: 1856–1860.

[7] Chatterjee S, Basu S, Chattopadhyay D, et al. Humidity sensor using porous tape cast alumina substrate. Rev Sci Instrum 2001, **72**: 2792–2795.

[8] Islam T, Mistry KK, Sengupta K, et al. Measurement of gas moisture in ppm range using porous silicon and porous alumina sensor. Sens Mater 2004, **16**: 345–356.

[9] Basu S, Saha M, Chatterjee S, et al. Porous ceramic sensor for measurement of gas moisture sensor in the ppm range. Mater Lett 2001, **49**: 29–33.

[10] Basu S, Chatterjee S, Saha M, et al. Study of electrical characteristics of porous alumina sensors for detection of low moisture in gases. Sensor Actuat B: Chem 2001, **79**: 182–186.

[11] Saha D, Das S, Sengupta K. Development of commercial nanoporous trace moisture sensor following sol–gel thin film technique. Sensor Actuat B: Chem 2008, **128**: 383–387.

[12] Adamczyk Z, Weroński P. Application of the DLVO theory for particle deposition problems. Adv Colloid Interfac 1999, **83**: 137–226.

[13] Pettersson A, Marino G, Pursiheimo A, et al. Electrosteric stabilization of Al_2O_3, ZrO_2, and $3Y–ZrO_2$ suspensions: Effect of dissociation and type of polyelectrolyte. J Colloid Interface Sci 2000, **228**: 73–81.

[14] Pandey M, Tyagi K, Mishra P, et al. Nanoporous morphology of alumina films prepared by sol–gel dip coating method on alumina substrate. J Sol–Gel Sci Technol 2012, **64**: 282–288.

12

The effect of milling additives on powder properties and sintered body microstructure of NiO

L. Jay DEINER[a,*], Michael A. ROTTMAYER[b], Bryan C. EIGENBRODT[c]

[a]Department of Chemistry, New York City College of Technology, City University of New York, 300 Jay St., Brooklyn, NY 11201, USA
[b]The Air Force Research Labs, Wright-Patterson Air Force Base, OH 45433, USA
[c]Department of Chemistry, Villanova University, 800 E. Lancaster Ave., Villanova, PA 19085, USA

Abstract: The evolution of powder particle size, crystal structure, and surface chemistry was evaluated for micron scale NiO powders subjected to impact milling with commonly employed milling additives: methanol, Vertrel XF, and amorphous carbon. The effect of the different comminution protocols on sintered body microstructure was evaluated for high temperature sintering in inert atmosphere (N_2). X-ray photoelectron spectroscopy showed that NiO powder surface chemistry is surprisingly sensitive to milling additive choice. In particular, the proportion of powder surface defect sites varied with additive, and methanol left an alcohol or alkoxy residue even after drying. Upon sintering to intermediate temperatures (1100 ℃), scanning electron microscopy (SEM) showed that slurry milled NiO powders exhibit hindered sintering behaviors. This effect was amplified for NiO milled with methanol, in which sub-500 nm grain sizes dominated even after sintering to 1100 ℃. Upon heating to high temperatures (1500 ℃), simultaneous differential scanning calorimetry/thermogravimetric analysis (DSC/TGA) showed that the powders containing carbon residues undergo carbothermal reduction, resulting in a melting transition between 1425 and 1454 ℃. Taken together, the results demonstrated that when processing metal oxide powders for advanced ceramics, the choice of milling additive is crucial as it exerts significant control over sintered body microstructure.

Keywords: nickel oxide; impact milling; sintering; densification; grain growth

1 Introduction

Nickel oxide is a catalytically important material, especially in the context of emerging energy technologies like solid oxide fuel cells [1,2]. In these devices, the microstructure of the sintered NiO/YSZ cermet anode strongly affects the solid oxide fuel cell performance [3–5]. Specific attempts at controlling

NiO/YSZ cermet microstructure have focused on starting catalyst powder particle sizes [6], the ratio between NiO and YSZ content [7,8], and processing techniques including mechanical alloying through milling [9–12]. Nonetheless, there remain challenges to achieving an ideal anode structure, especially in the area of controlling grain growth and densification of the NiO component during high temperature sintering (> 1000 ℃). In principle, such growth could be controlled by tuning the physical (particle shape, size, size distribution, and extent of agglomeration [13,14]),

* Corresponding author.
E-mail: Ldeiner@citytech.cuny.edu

chemical (grain boundary chemistry [15], oxide impurities [16], and surface segregating dopants [17]), and electronic (surface oxidation state [18]) characteristics of the starting NiO powders. Continued development of synthetic and/or processing techniques for tuning the characteristics of NiO powders is thus necessary.

For the purpose of engineering nanoscale and micron scale powders, "bottom up" synthetic methods exhibit excellent structure/property control [19], but "top down" methods like high energy milling may be a more cost effective way to create large volumes of powders whose properties have been tuned to optimize sintered body microstructure. Variations in milling time and processing conditions provide the ability to control powder particle size, size distribution, particle shape, defect density, oxidation state, and surface chemistry [20–23]. In the present work, we investigate the use of three milling additives (amorphous carbon, methanol, and Vertrel XF) for tuning the particle size, particle size distribution, surface chemistry, and surface oxidation state of NiO powders. We correlate these powder properties with the microstructure of sintered NiO compacts and with the energetics of sintering. Previous work has shown that the presence of residues from processing additives hinders sintering in some cases [24] and facilitates sintering in others [25]. By correlating how additive-driven changes in the chemical and physical structures of the powders affect sintered body microstructure, this work aims to provide principles for choosing milling additives for large scale production of NiO powders engineered to produce sintered parts with controlled microstructures.

2 Materials and methods

The starting NiO material (Nickelous Oxide, Green, Powder, Baker Analyzed Reagent for Electronic Ceramics, Avantor Performance Materials, Center Valley, PA) was subjected to high energy ball milling using a SPEX 8000M Mixer/Mill (SPEX Sample Prep, Metuchen, NJ). All milling experiments were conducted in a zirconia vessel with two 12.7 mm spherical zirconia beads employed as milling media. The total milling time per sample was 480 min. As described in Table 1, the NiO samples were dry milled with an amorphous carbon additive or slurry milled with either methanol or Vertrel XF (1,1,1,2,2,3,4,5,5,5-decafluoropentane). For reference, NiO milled

Table 1 Summary of the samples and milling conditions employed for grinding NiO

Sample	Additive
NiO	None
NiO/C (1.5%)	Printex L amorphous carbon, 1.5 wt%
NiO/methanol	Methanol (Sigma Aldrich, Anhydrous, 99.8%)
NiO/Vertrel	Vertrel XF (1,1,1,2,2,3,4,5,5,5-decafluoropentane, DuPont)

with no additive was also included. In all milling experiments, the total mass of powders was 10 g. The powders milled with amorphous carbon contained 1.5 wt% carbon. For slurry milling, ~11 mL of liquid was added to the 10 g of NiO. The liquid was refilled every 90 min to compensate for evaporation.

The Brunauer–Emmett–Teller (BET) surface areas of the powders before and after milling were obtained with an ASAP 2020 Physisorption Analyzer (Micromeritics). The particle morphologies and approximate size distributions were analyzed before and after milling using scanning electron microscopy (SEM; Amray 1910, FE-SEM). Prior to SEM imaging, the powder samples were sputtered with 2 nm of Au/Pd (EMS 150RS, Electron Microscopy Sciences). Before and after milling, crystal structure was obtained via X-ray diffraction (Bruker, D8). Insight into the NiO surface oxidation state and presence of chemisorbed species before and after milling was obtained using X-ray photoelectron spectroscopy of the powder samples (Surface Science Instruments M-Probe, Service Physics, Inc.). A monochromatic Al Kα filament (1486.6 eV) was employed, and the emission current and operating voltage were fixed at 20 mA and 10 kV, respectively. All X-ray photoelectron spectroscopy data were shown after Shirley baseline correction. In accord with the literature, the binding energies were corrected for charging using the O 1s peak of bulk NiO at 529.5 eV as an internal calibration standard [26]. As a check of this binding energy correction method, we compared the binding energy values derived using the O 1s internal calibration method to those obtained by correcting for charging using the adventitious carbon peak at 284.5 eV [27,28]. To within the limits of ±0.2 eV, both methods of correction for charging led to consistent peak values in X-ray photoelectron spectroscopy.

The milled powders were pressed into pellets using a Carver press (Carver, Inc.) employing a pressure of 10000 psi for 5 min. The diameter of the die was 12.7 mm, and the mass of NiO used per pellet was 0.75 g. No organic binder was added during powder compaction. The resulting green pellet height was 1.61 mm. The pellets were then subjected to sintering in N_2 using a reducing furnace. The heating ramp was 1 ℃/min to the maximum dwell temperature of either 1100 or 1500 ℃. The samples were held for 4 h at the maximum dwell temperature, then cooled at a rate of 5 ℃/min. After sintering, the pellets were cleaved and analyzed in cross section with SEM (JSM 6060, Joel). Sample porosities were determined from the cross sectional images using the SEM software.

Simultaneous differential scanning calorimetry/ thermogravimetric analysis (DSC/TGA) measurements were performed with a TA Instruments SDT Q600 instrument. The measurements were performed by loading 20–45 mg of the NiO powders into an alumina sample pan. All measurements were performed under flowing nitrogen (Airgas, High Purity, 4.8 grade). The rate of flow was 100 mL/min. The temperature was stabilized at 40 ℃ for 10 min before ramping at a rate of 10 ℃/min to a maximum temperature of 1500 ℃. The samples were held at the maximum temperature for 30 min. Weight change and heat flow were measured during the temperature ramp and during the dwell time at the maximum temperature. In order to establish the reversibility of heat flow events for the NiO/C and NiO/methanol samples, additional experiments were performed in which the temperature was stabilized at 40 ℃ for 10 mim before ramping at a rate of 10 ℃/min to a maximum temperature of 1500 ℃. The samples were then cooled at a rate of 10 ℃/min down to a final temperature of 300 ℃. For all experiments, the heat flow was normalized by the real time sample weight and was presented as a weight corrected heat flow.

3 Results and discussion

3. 1 Powder characterizations

The BET surface areas and corresponding average particle diameters are shown in Table 2 for all milling procedures. The average particle diameters were estimated from the BET surface areas using Eq. (1) [29]:

$$D_{BET} = 6 / (\rho \times S_{BET}) \qquad (1)$$

where D_{BET} is the particle diameter estimated from BET; ρ is the powder density; and S_{BET} is the BET surface area. As seen in Table 2, the BET surface area

Table 2 BET surface areas and estimated particle sizes of NiO powders before and after 480 min of milling

Sample	Milling time (min)	BET surface area (m^2/g)	BET average particle size (nm)
NiO	0	3.5	255
NiO	480	12.4	72
NiO/C (1.5%)	480	16.8	54
NiO/methanol	480	4.6	198
NiO/Vertrel	480	4.9	183

of the as-received NiO powders was ~3.5 m^2/g. After 480 min of milling, all of the dry milled samples have BET surface areas > 10 m^2/g. In contrast, the powders that were slurry milled in Vertrel XF and methanol exhibit more modest BET surface area increases to ~4.9 and 4.6 m^2/g, respectively. The observation of a more rapid decrease in particle size for dry versus wet grinding is consistent with previous work, documenting structural changes of α-Fe_2O_3 as a function of milling environment [30]. In contrast, a more rapid decrease in particle size is observed for slurry milling of ZrO_2 as opposed to dry milling [31]. This apparent inconsistency is resolved when we consider that the zirconia study measured particle size using sedimentation. As such, agglomerates and primary particles are indistinguishable. It is expected that the dry milled samples of ZrO_2 contain smaller primary particles, but they are more heavily aggregated as compared to the corresponding slurry milled samples.

From Table 2, it is also clear that the surface area of the NiO dry milled with carbon is greater than NiO dry milled with no additive. If the BET surface area of NiO milled with carbon is estimated to be a weighted average of the BET surface areas of the starting Printex L carbon material (150 m^2/g) and NiO milled alone, the estimated value is 14.5 m^2/g, still less than the measured value shown in Table 2. As such, it is probable that the carbon acts as a grinding aid for NiO [32,33] while itself undergoing comminution during milling.

The SEM images of the powders provide insight into the particle size distributions of the milled samples (Fig. 1). The unmilled powders (Fig. 1(a)) are characterized by aggregated and irregular NiO particles, with few particles below 200 nm in size. After milling for 480 min with no additive (Fig. 1(b)), the extent of aggregation increases while the particle size distribution now includes many more particles in the 100 nm range. Milling for 480 min with carbon (Fig. 1(c)) also increases the frequency of small

(<100 nm) particles. Slurry milling in Vertrel XF (Fig. 1(d)) produces a more homogeneous particle size distribution with few particles less than 100 nm. Similarly, slurry milling in methanol produces a particle size distribution with few particles below 100 nm. These qualitative results support the BET surface area measurements which indicate only modest surface area increases with slurry milling.

Upon dry milling, the color of the NiO samples shifts from green to dark brown. Since stoichiometric NiO is bright green and non-stoichiometric Ni_xO_y is dark brown [34], it is suspected that dry milling either changes the bulk crystal structure of the NiO or changes the surface chemistry of the NiO. The former possibility can be excluded because X-ray diffraction does not provide any evidence for a change in bulk crystal structure upon milling for 480 min with or without additives (Fig. 2). All of the peaks present in the X-ray diffraction spectra of NiO milled for 480 min are in the same location as the peaks for unmilled NiO. These peaks are consistent with reported X-ray diffraction patterns for NiO, indicating that milling does not induce changes to the crystal structure of the bulk materials [35]. Peak broadening is observed for dry milled NiO, but not for the slurry milled NiO (Fig. 2). This broadening is attributed primarily to

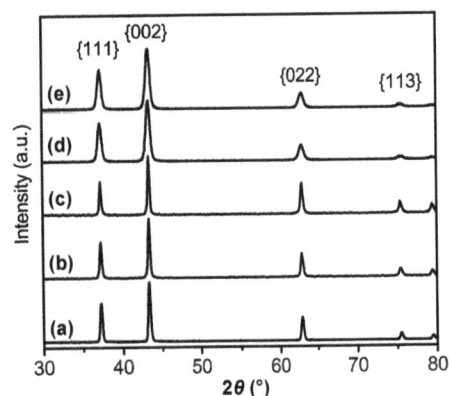

Fig. 2 X-ray diffraction data for: (a) unmilled NiO; (b) NiO slurry milled in methanol; (c) NiO slurry milled in Vertrel; (d) dry milled NiO; and (e) NiO dry milled with carbon.

Fig. 1 SEM images of NiO powders after: (a) no milling; (b) 480 min of milling with no additive; (c) 480 min of milling with 1.5% Printex L carbon; (d) 480 min of milling with Vertrel XF; and (e) 480 min of milling with methanol.

crystallite size reduction with possible contributions from milling induced strain [35]. This interpretation is consistent with the fact that significant particle size reduction is observed in BET for dry milled NiO, but not for slurry milled NiO.

While X-ray diffraction measurements indicate that milling does not change the bulk crystal structure of NiO, X-ray photoelectron spectroscopy provides clear evidence for milling-induced chemical modification of the NiO powder surface. The Ni 2p3/2 region of the unmilled NiO (Fig. 3(a)) displays a multiplet which is fitted by two peaks centered at 853.8 and 855.4 eV. It also displays a shake-up feature fit by peaks centered at 860.7 and 863.6 eV. The peak locations and peak widths are consistent with the literature reports for NiO [27,36]. In the milled NiO samples, the Ni 2p3/2 region displays subtle differences in the extent to which the multiplet splitting peaks are resolved. This is consistent with changes to the NiO surface either through modification of the Ni^{2+}/Ni^{3+} ratio or through the presence of adsorbates [26,37].

The origin of the chemical modifications to the NiO surface can be further understood by examination of the O 1s region of the X-ray photoelectron spectroscopy data (Fig. 3(b) and Table 3). Unmilled NiO displays two peaks, one at 529.4 eV (full width at half maximum (FWHM) = 1.2 eV) and the other at 531.0 eV (FWHM = 2.3 eV) (Fig. 3(b)(i)). These peaks are frequently observed for NiO. The 529.4 eV is assigned to oxygen in the NiO lattice [27,36]. In previous works, the 531.0 eV has been ascribed either to oxygen from hydroxyl or to other oxygen-containing moieties, most likely adsorbed at defect sites [38,39]. It has recently been shown that the 531.0 eV peak can

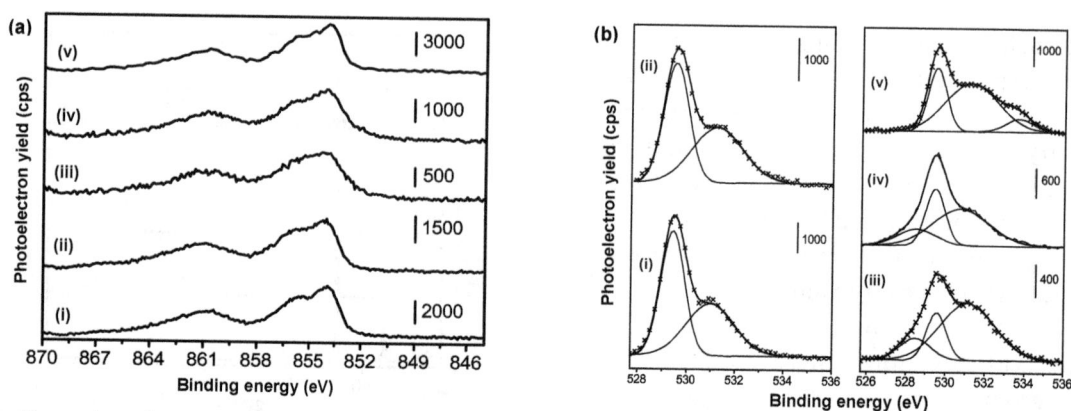

Fig. 3 X-ray photoelectron spectroscopy data for the (a) Ni 2p3/2 region and (b) O 1s region of: (i) unmilled NiO; (ii) NiO milled for 480 min with no additive; (iii) NiO milled for 480 min with carbon Printex L; (iv) NiO milled for 480 min with Vertrel XF; and (v) NiO milled for 480 min with methanol.

Table 3 Peak fits for the O 1s region of X-ray photoelectron spectroscopy data for NiO before and after undergoing high energy milling treatments

	Unmilled NiO	NiO milled with no additive	NiO milled with Printex carbon	NiO milled with Vertrel	NiO milled with methanol
Binding energy (eV)	529.4	529.5	529.5	529.4	529.4
FWHM (eV)	1.2	1.2	1.2	1.2	1.1
Assignment	Lattice oxygen	Lattice oxygen	Lattice oxygen	Lattice oxygen	Lattice oxygen
Binding energy (eV)	531.0	531.2	531.1	530.7	531.2
FWHM (eV)	2.3	2.5	2.8	2.8	3.1
Assignment	Defect oxygen	Defect oxygen	Defect oxygen	Defect oxygen	Defect oxygen, alkoxy, OH
Binding energy (eV)			528.5	528.5	
FWHM (eV)			1.9	2.2	
Assignment			Non-equilibrium oxygen	Non-equilibrium oxygen	
Binding energy (eV)					533.5
FWHM (eV)					1.6
Assignment					Alcohol

also be observed in the absence of adsorbed oxygen containing species, suggesting that it can be associated with the defect sites themselves [27,40]. In our unmilled NiO sample, the ratio between the area of the lattice oxygen peak and the 531.0 eV oxygen peak is 1.2:1. After the NiO is milled with no additive, the same two O 1s peaks are present, but the ratio of the lattice oxygen to the 531.0 eV oxygen peak is 1:1.1, indicating an increase in defect or adsorbate sites upon milling (Fig. 3(b)(ii)). When NiO is dry milled in the presence of carbon, the shape of the O 1s peak changes such that it can no longer be fitted solely by the states used to describe unmilled NiO. The peak at 531.1 eV has a larger FWHM (2.8 eV) as compared to the unmilled NiO and the NiO milled without any additive. This is likely due to a greater variety of defect sites arising on the particles milled with carbon. In addition, there is a new low binding energy feature centered at 528.5 eV (FWHM = 1.9 eV). This peak has been observed previously for the early stages of solution processed NiO film growth [41] and AgO film growth [42]. In both cases, it is ascribed to a transient oxygen

feature associated with oxygen bonded to partially reduced Ni. The O 1s region of NiO milled with Vertrel displays the same three states as NiO milled with carbon. However, the ratios of the integrated areas of these peaks differ. For NiO milled in carbon, the ratio of the integrated areas of the 528.5 eV peak to that of the 529.5 eV peak and 531 eV peak is 1:1.4:4.6. For NiO milled in Vertrel, the corresponding ratio is 1:1.8:3.4. This difference is consistent with the BET data that indicates that carbon acts as a milling aid, cleaving NiO at a more rapid rate and creating a larger number of surface defects. Finally, the O 1s region of NiO milled in methanol shows three peaks centered at 529.5, 531.2, and 533.7 eV. The peak at 529.5 eV is ascribed to the lattice oxygen of NiO, and the peak at 533.7 eV is ascribed to the intact alcohol [43,44]. While the peak at 531.2 eV is primarily associated with surface defects as described above, it may also contain contributions from the dissociation of methanol to alkoxide and OH. Support for the presence of alkoxide and OH is provided by the increased FWHM (3.1 eV) and high integrated area of the 531.2 eV peak, almost twice as large as the 529.5 peak. Since the BET data indicate that the extent of comminution of NiO in methanol is not that high, it seems likely that for NiO milled in methanol, the 531.2 eV peak is not solely composed of contributions from defect sites. The presence of C–H stretching peaks in diffuse reflectance infrared spectroscopy of the methanol milled NiO (data not shown) provides support for the hypothesis that alcohol and/or alkoxide moieties are present on the surface of NiO milled with methanol.

3. 2 Scanning electron microscopy and thermal analyses of sintering

SEM images of sintered pellets from milled and unmilled NiO show significant differences in microstructure as a function of milling additives (Fig. 4). For intermediate temperature sintering (1100 ℃) (Fig. 4(a)), the unmilled NiO sample shows notable but inhomogeneous particle size growth with some sub-micron particles remaining, but significant fusion of particles into grains greater than two microns. The porosity of the unmilled NiO sample sintered to 1100 ℃ is 1.6%. The sample milled with carbon undergoes an accelerated grain growth, maintaining almost no sub-micron particles and achieving a porosity of 1.7%, nearly the same as that of the unmilled NiO sample. In contrast, both of the slurry milled samples (NiO/Vertrel and NiO/methanol) display a hindered particle size growth maintaining smaller, more homogeneous particle sizes, and higher porosities of 2.3%. The combination of hindered particle size growth and higher porosity suggests that, after sintering to intermediate temperatures, the slurry milled samples bear a closer resemblance to unsintered pressed powder compact as compared to the dry milled or unmilled samples. Notably, for NiO milled in methanol, sintering to 1100 ℃ results in minimal particle size growth with almost all particles maintaining sizes below 500 nm. Despite this hindered particle size growth, for both NiO/methanol and NiO/Vertrel, there are regions where the small particles have fused together resulting in amorphous regions with indistinct grain boundaries. Finally, we note that preliminary investigations of NiO milled with no additive produce a structure with inhomogeneous pores and large irregular grains. Because of the non-optimal characteristics of this microstructure, we elected not to carry it forward in the future studies.

Differences in porosity and grain growth persist

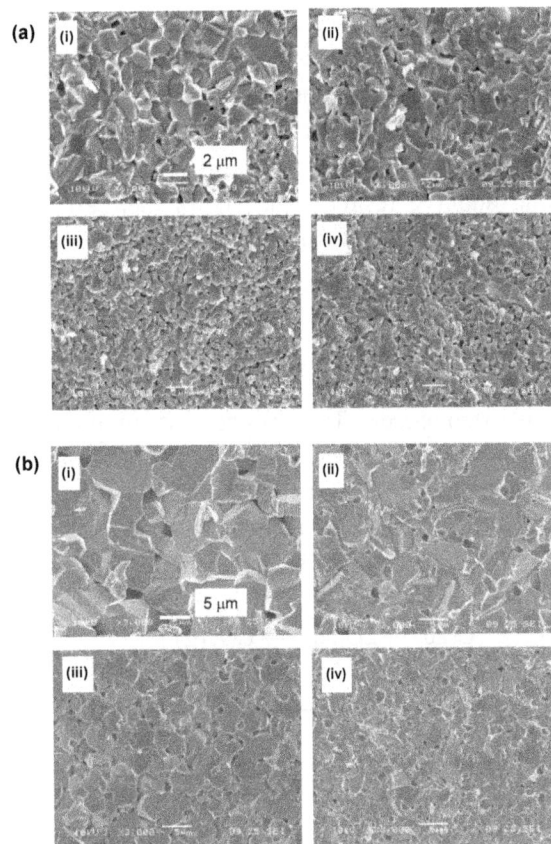

Fig. 4 SEM images of NiO powders sintered in N_2 to (a) 1100 ℃ and (b) 1500 ℃. Powders are: (i) NiO unmilled; (ii) NiO milled for 480 min with carbon; (iii) NiO milled for 480 min with methanol; and (iv) NiO milled for 480 min with Vertrel XF.

even when the NiO is sintered to 1500 ℃ (Fig. 4(b)), well into the range of final stage sintering [45]. After the 1500 ℃ sintering treatment, the unmilled NiO has grains in the range of 5 microns and above, with distinct grain boundaries and an open structure of micron scale pores. Between 1100 and 1500 ℃, total porosity of this sample decreases only slightly to a final value of 1.3%. NiO sample dry milled with carbon has almost no discernible grains and isolated pores ranging from nanoscale to micron scale. During the final stage of sintering, the total porosity decreases to 0.8%, a much more pronounced decrease than that seen for the unmilled sample. After sintering to 1500 ℃, NiO milled with methanol displays irregular particles ranging from 1 to 5 microns and pores ranging from less than 1–2 microns. As in the case of NiO milled with carbon, the porosity decreases noticeably to a final value of 1.4%. NiO milled in Vertrel has no discernible grains and a large number of isolated pores ranging in size from sub-micron to 2 microns. Similar to unmilled NiO, the porosity of NiO milled with Vertrel decreases only modestly between 1100 and 1500 ℃ (two tenths of a percent), resulting in a final porosity of 2.1%.

Simultaneous DSC/TGA confirms the presence of milling additive residues and shows their effects on heat flow during sintering (Fig. 5). In the thermogravimetric analysis trace (Fig. 5(a)), NiO milled with 1.5 wt% carbon shows the most pronounced mass loss, 5.1%, upon heating. The majority of mass is lost between 513 and 800 ℃ in a distinct two-step process. This two-step pattern of mass loss has been observed previously for NiO–YSZ, and has been attributed to the carbothermal reduction of NiO [46]. In this process, carbon abstracts surface oxygen and desorbs from the surface as represented by overall Eqs. (2) and (3):

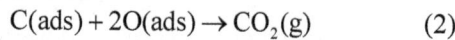

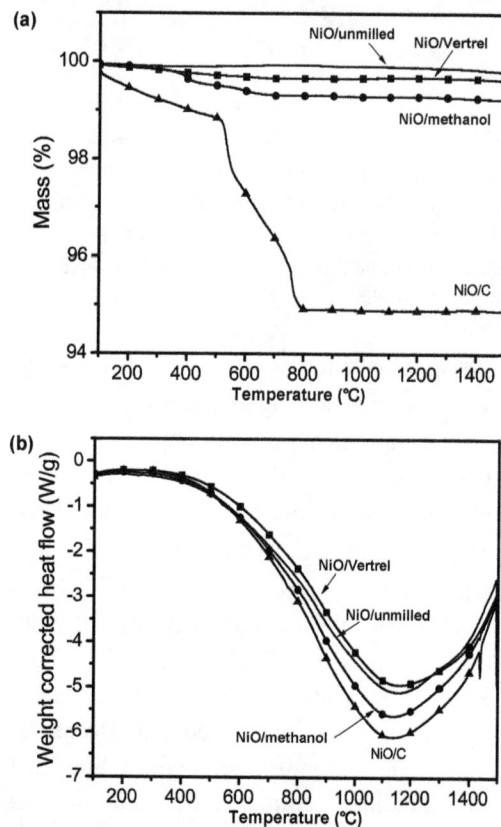

$$C(ads) + 2O(ads) \rightarrow CO_2(g) \qquad (2)$$

Fig. 5 (a) Thermogravimetric analysis and (b) differential scanning calorimetry for unmilled NiO (black line), NiO milled with carbon (black line + triangle symbols), NiO milled with methanol (black line + circle symbols), and NiO milled with Vertrel XF (black line + square symbols). In the differential scanning calorimetry traces, exothermic events produce an upward peak.

$$C(ads) + O(ads) \rightarrow CO(g) \qquad (3)$$

If 1.5 wt% carbon were to desorb fully by combination to CO_2 (Eq. (2)), a total mass loss of 5.5% would be expected. The observed mass loss of 5.1% suggests that 80% of the carbon desorbs as CO_2 and 20% desorbs as CO. These desorption events are endothermic as observed by the downward peaks at 536 and 761 ℃ in the weight corrected heat flow recorded during differential scanning calorimetry (Fig. 5(b)). These endothermic events are most obvious in the first derivative of the weight corrected heat flow (Fig. 6(a)).

Since the present thermal experiments take place under flowing $N_2(g)$, the only source of oxygen for the

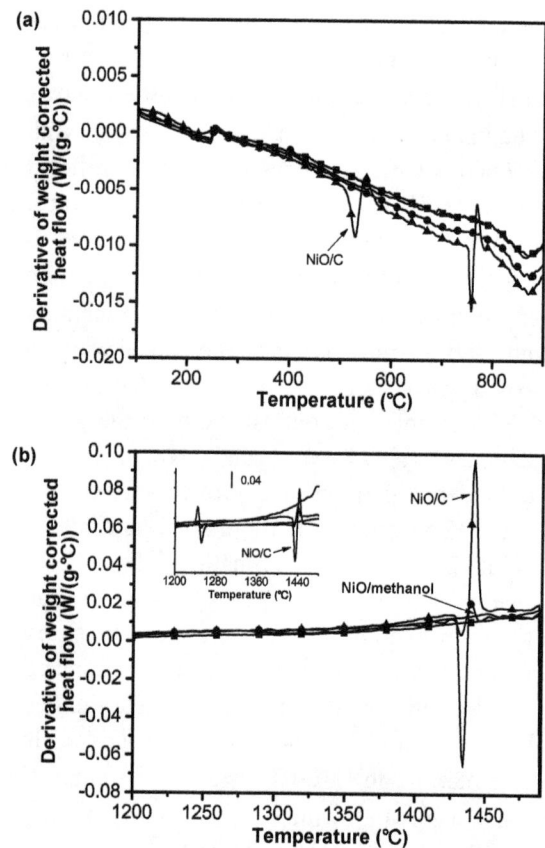

Fig. 6 Temperature derivative of the weight corrected heat flow from differential scanning calorimetry measurements of unmilled NiO (black line), NiO milled with carbon (black line + triangle symbols), NiO milled with methanol (black line + circle symbols), and NiO milled with Vertrel XF (black line + square symbols) for the temperature ranges of (a) 100–900 ℃ and (b) 1200–1500 ℃. Inset in (b): derivative of the weight corrected heat flow for NiO/C and NiO/methanol (as labeled) in both the forward and reverse scans. The NiO/methanol trace has been multiplied by four in order to be visible on the same scan as the NiO/C trace.

combination to CO or CO_2 is that bound to the NiO. Thus, when carbon is used as a milling additive, the NiO surface is partially reduced after 800 ℃. As a result, a distinct endothermic melting event is observed between 1425 and 1454 ℃ in the differential scanning calorimetry for the NiO milled with C (Figs. 5(b) and 6(b)). The interpretation of this event as melting is supported by its reversibility, manifested in an exothermic solidification event, observed from 1265 to 1230 ℃ during the cooling curve (Fig. 6(b) inset). That the melting event is attributable to metallic or reduced Ni formed by CO/CO_2 desorption is further supported by the observation that the melting peak is not observed for the same differential scanning calorimetry experiment performed in air where oxygen is present to replenish any abstracted from the surface (data not shown). Finally, the melting point of metallic nickel is 1455 ℃, quite close to the temperature observed in the present study [47].

NiO milled with methanol shows a mass loss of 0.7% (Fig. 5(a)) indicating that the alcohol milling residue is present at the beginning of the thermogravimetric analysis experiment. This observation is in accord with the X-ray photoelectron spectroscopy data which suggests the presence of a C–O containing species. As in the case of atomic carbon, the presence of the alcohol milling residue results in partial reduction of the NiO surface, and an endothermic melting transition between 1425 and 1454 ℃ (Figs. 5(b) and 6(b)). As in the case of NiO milled with carbon, this melting transition is shown to be reversible in the cooling curve (Fig. 6(b) inset). The partial surface reduction and resulting melting transition are in accord with previous studies of methanol adsorbed on NiO. In these works, upon heating, molecular desorption of the alcohol competes with surface decomposition to carbon, hydrogen, and oxygen [48]. The carbon-containing decomposition products may abstract surface oxygen upon desorption.

In contrast to NiO milled with amorphous carbon or methanol, the mass loss of NiO milled with Vertrel is significantly less (0.3 wt%, Fig. 5(a)) and there are no marked melting transitions or other significant endothermic or exothermic events. This suggests that Vertrel is a fairly inert milling additive that does not decompose appreciably on the NiO surface. This observation is in accord with the differences in the bond dissociation energies of C–H (338 kJ/mol) versus C–F (513 kJ/mol) [47].

High energy milling simultaneously influences NiO powder particle size distribution and surface chemistry. As such, the evolution of the powder microstructures upon sintering is a function of the concerted effects of differences in the powder particle size distributions and surface chemistries. Typically, fine powders with uniform size distributions display less rapid grain growth during intermediate stage sintering [49]. It is therefore reasonable that the relatively uniform particle size distributions of NiO powders milled in Vertrel and methanol result in smaller grain sizes after sintering to 1100 ℃ in comparison to unmilled NiO or NiO milled with carbon. These effects can be reinforced by the surface chemical differences imparted by the milling additives. Notably, it has recently been shown that the presence of surface bound C–OH or C–O–C groups hinder surface diffusion and hence sintering of TiO_2 nanoparticles [24]. In the present study, the NiO powders milled in methanol and shown by X-ray photoelectron spectroscopy and thermogravimetric analysis to have residual surface bound alkoxy or alcohol, display a hindered sintering behavior. The methanol milled powders sintered to 1100 ℃ exhibit less dramatic grain growth in comparison to the powders milled with Vertrel and carbon. However, in the present study, the mechanism of hindered sintering cannot be connected to hindered surface diffusion due to the presence of adsorbed species because thermogravimetric analysis shows that all additives desorb from the surface before 900 ℃. As such, the differences in the sintered microstructure are more likely to be due to the way that additives modify the surface oxidation state upon desorption. This is evident for the samples sintered to 1500 ℃, where carbothermal reduction by carbon and methanol decomposition products results in metallic nickel species that undergo melting by 1455 ℃. This melting transition is observed in differential scanning calorimetry and manifested in the extreme grain growth experienced between 1100 and 1500 ℃ for the powders milled in methanol and carbon. The melting transition may also be the driving force behind the significant decreases in porosity between 1100 and 1500 ℃, observed for the NiO milled with carbon and methanol. Unmilled NiO and NiO milled with an additive that does not induce carbothermal reduction (Vertrel XF) show far less pronounced decreases in porosity between 1100 and 1500 ℃.

4 Conclusions

The choice of milling additive was shown to have a significant effect on the particle size distribution and surface chemistry of NiO powders. Residues from all milling additives were detectable by X-ray photoelectron spectroscopy and thermogravimetric analysis. Even though all milling additives desorbed from the powders by 900 ℃, NiO milled with different additives and sintered to 1100 and 1500 ℃ displayed microstructural differences. Some of these differences may be due to the effects of the milling additives on the particle size distributions of the powders. However, the presence of distinct melting transitions for NiO milled with carbon and methanol suggests another mechanism by which additives may steer sintered body microstructure, even past the additive desorption temperature. Specifically, when sintering is performed in an oxygen free atmosphere, desorption of carbon-containing species may drive surface reduction. This change in surface oxidation state introduces a metallic Ni species which facilitates rapid grain growth and decrease in porosity. The presence of such a mechanism suggests that the interplay between milling additive and sintering environment cannot be disregarded, even for high temperature sintering of NiO.

Acknowledgements

The authors thank the N.Y.S. Graduate Research and Teaching Initiative (GRTI) for financial support. L. J. D. thanks the Air Force Summer Faculty Fellowship program for fellowship support. The authors thank the Advanced Imaging Facility of the College of Staten Island for SEM images, and Prof. William L'Amoreaux and Dr. Mike Bucaro for their help with SEM images.

References

[1] Adams TA, Nease J, Tucker D, et al. Energy conversion with solid oxide fuel cell systems: A review of concepts and outlooks for the short- and long-term. Ind Eng Chem Res 2013, 52: 3089–3111.

[2] Cowin PI, Petit CTG, Lan R, et al. Recent progress in the development of anode materials for solid oxide fuel cells. Adv Energ Mater 2011, 1: 314–332.

[3] Clemmer RMC, Corbin SF. The influence of pore and Ni morphology on the electrical conductivity of porous Ni/YSZ composite anodes for use in solid oxide fuel cell applications. Solid State Ionics 2009, 180: 721–730.

[4] Guo W, Liu J. The effect of nickel oxide microstructure on the performance of Ni–YSZ anode-supported SOFCs. Solid State Ionics 2008, 179: 1516–1520.

[5] Itoh H, Yamamoto T, Mori M, et al. Configurational and electrical behavior of Ni–YSZ cermet with novel microstructure for solid oxide fuel cell anodes. J Electrochem Soc 1997, 144: 641–646.

[6] Wang Y, Walter ME, Sabolsky K, et al. Effects of powder sizes and reduction parameters on the strength of Ni–YSZ anodes. Solid State Ionics 2006, 177: 1517–1527.

[7] Jiang SP. Sintering behavior of Ni/Y$_2$O$_3$–ZrO$_2$ cermet electrodes of solid oxide fuel cells. J Mater Sci 2003, 38: 3775–3782.

[8] Wilson JR, Barnett SA. Solid oxide fuel cell Ni–YSZ anodes: Effect of composition on microstructure and performance. Electrochem Solid-State Lett 2008, 11: B181–B185.

[9] Cho HJ, Choi GM. Effect of milling methods on performance of Ni–Y$_2$O$_3$-stabilized ZrO$_2$ anode for solid oxide fuel cell. J Power Sources 2008, 176: 96–101.

[10] Hong HS, Chae U-S, Choo S-T. The effect of ball milling parameters and Ni concentration on a YSZ-coated Ni composite for a high temperature electrolysis cathode. J Alloys Compd 2008, 449: 331–334.

[11] Restivo TAG, de Mello-Castanho SRH. Nickel–zirconia cermet processing by mechanical alloying for solid oxide fuel cell anodes. J Power Sources 2008, 185: 1262–1266.

[12] Tietz F, Dias FJ, Simwonis D, et al. Evaluation of commercial nickel oxide powders for components in solid oxide fuel cells. J Eur Ceram Soc 2000, 20: 1023–1034.

[13] Bowen P, Carry C. From powders to sintered pieces: Forming, transformations and sintering of nanostructured ceramic oxides. Powder Technol 2002, 128: 248–255.

[14] Chaim R, Levin M, Shlayer A, et al. Sintering and densification of nanocrystalline ceramic oxide powders: A review. Adv Appl Ceram 2008, 107: 159–169.

[15] Koch CC. The synthesis and structure of nanocrystalline materials produced by mechanical attrition: A review. Nanostruct Mater 1993, 2: 109–129.

[16] Jung S-H, Oh H-C, Kim J-H, et al. Pretreatment of zirconium diboride powder to improve densification. J Alloys Compd 2013, 548: 173–179.

[17] Castro RHR, Pereira GJ, Gouvêa D. Surface modification of SnO$_2$ nanoparticles containing Mg or Fe: Effects on sintering. Appl Surf Sci 2007, 253: 4581–4585.

[18] Phung X, Groza J, Stach EA, et al. Surface characterization of metal nanoparticles. Mat Sci Eng A 2003, 359: 261–268.

[19] Burda C, Chen X, Narayanan R, et al. Chemistry and properties of nanocrystals of different shapes. Chem Rev 2005, 105: 1025–1102.

[20] Eser O, Kurama S. The effect of the wet-milling process on sintering temperature and the amount of additive of SiAlON ceramics. *Ceram Int* 2010, **36**: 1283–1288.

[21] Ivanov E, Suryanarayana C. Materials and process design through mechanochemical routes. *J Mater Synth Proces* 2000, **8**: 235–244.

[22] Koch CC, Cho YS. Nanocrystals by high energy ball milling. *Nanostruct Mater* 1992, **1**: 207–212.

[23] Zhang DL. Processing of advanced materials using high-energy mechanical milling. *Prog Mater Sci* 2004, **49**: 537–560.

[24] Lu K, Liang Y, Li W. Hindered sintering behaviors of titania nanoparticle-based materials. *Mater Lett* 2012, **89**: 77–80.

[25] Faudot F, Gaffet E, Harmelin M. Identification by DSC and DTA of the oxygen and carbon contamination due to the use of ethanol during mechanical alloying of Cu–Fe powders. *J Mater Sci* 1993, **28**: 2669–2676.

[26] Peck MA, Langell MA. Comparison of nanoscaled and bulk NiO structural and environmental characteristics by XRD, XAFS, and XPS. *Chem Mater* 2012, **24**: 4483–4490.

[27] Biesinger MC, Payne BP, Lau LWM, *et al.* X-ray photoelectron spectroscopic chemical state quantification of mixed nickel metal, oxide and hydroxide systems. *Surf Interface Anal* 2009, **41**: 324–332.

[28] Barr TL, Seal S. Nature of the use of adventitious carbon as a binding energy standard. *J Vac Sci Technol A* 1995, **13**: 1239–1246.

[29] Jang HD, Kim S-K, Kim S-J. Effect of particle size and phase composition of titanium dioxide nanoparticles on the photocatalytic properties. *J Nanopart Res* 2001, **3**: 141–147.

[30] Sánchez LC, Arboleda JD, Saragovi C, *et al.* Magnetic and structural properties of pure hematite submitted to mechanical milling in air and ethanol. *Physica B* 2007, **389**: 145–149.

[31] Spearing DR, Huang JY. Zircon synthesis via sintering of milled SiO_2 and ZrO_2. *J Am Ceram Soc* 1998, **81**: 1964–1966.

[32] Zhou S, Chen H, Ding C, *et al.* Effectiveness of crystallitic carbon from coal as milling aid and for hydrogen storage during milling with magnesium. *Fuel* 2013, **109**: 68–75.

[33] Huang ZG, Guo ZP, Calka A, *et al.* Effects of carbon black, graphite and carbon nanotube additives on hydrogen storage properties of magnesium. *J Alloys Compd* 2007, **427**: 94–100.

[34] Richardson JT, Yiagas DI, Turk B, *et al.* Origin of superparamagnetism in nickel oxide. *J Appl Phys* 1991, **70**: 6977.

[35] Gonçalves NS, Carvalho JA, Lima ZM, *et al.* Size–strain study of NiO nanoparticles by X-ray powder diffraction line broadening. *Mater Lett* 2012, **72**: 36–38.

[36] Biesinger MC, Payne BP, Grosvenor AP, *et al.* Resolving surface chemical states in XPS analysis of first row transition metals, oxides and hydroxides: Cr, Mn, Fe, Co and Ni. *Appl Surf Sci* 2011, **257**: 2717–2730.

[37] Payne BP, Grosvenor AP, Biesinger MC, *et al.* Structure and growth of oxides on polycrystalline nickel surfaces. *Surf Interface Anal* 2007, **39**: 582–592.

[38] McIntyre NS, Cook MG. X-ray photoelectron studies on some oxides and hydroxides of cobalt, nickel, and copper. *Anal Chem* 1975, **47**: 2208–2213.

[39] Uhlenbrock S, Scharfschwerdt C, Neumann M, *et al.* The influence of defects on the Ni 2p and O 1s XPS of NiO. *J Phys: Condens Matter* 1992, **4**: 7973.

[40] Payne BP, Biesinger MC, McIntyre NS. Use of oxygen/nickel ratios in the XPS characterisation of oxide phases on nickel metal and nickel alloy surfaces. *J Electron Spectrosc Relat Phenom* 2012, **185**: 159–166.

[41] Jung J, Kim DL, Oh SH, *et al.* Stability enhancement of organic solar cells with solution-processed nickel oxide thin films as hole transport layers. *Sol Energy Mater Sol Cells* 2012, **102**: 103–108.

[42] Rocha TCR, Oestereich A, Demidov DV, *et al.* The silver–oxygen system in catalysis: New insights by near ambient pressure X-ray photoelectron spectroscopy. *Phys Chem Chem Phys* 2012, **14**: 4554–4564.

[43] Au CT, Hirsch W, Hirschwald W. Adsorption and interaction of methanol with zinc oxide: Single crystal faces and zinc oxide–copper catalyst surfaces studied by photoelectron spectroscopy (XPS and UPS). *Surf Sci* 1989, **221**: 113–130.

[44] Deiner LJ, Serafin JG, Friend CM, *et al.* Insight into the partial oxidation of propene: The reactions of 2-propen-1-ol on clean and O-covered Mo(110). *J Am Chem Soc* 2003, **125**: 13252–13257.

[45] Iida Y. Sintering of high-purity nickel oxide. *J Am Ceram Soc* 1958, **41**: 397–406.

[46] Arico E, Tabuti F, Fonseca FC, *et al.* Carbothermal reduction of the YSZ–NiO solid oxide fuel cell anode precursor by carbon-based materials. *J Therm Anal Calorim* 2009, **97**: 157–161.

[47] Haynes WN. *CRC Handbook of Chemistry and Physics.* Boca Raton, FL: CRC Press/Taylor and Francis, 2014.

[48] Natile MM, Glisenti A. Surface reactivity of NiO: Interaction with methanol. *Chem Mater* 2002, **14**: 4895–4903.

[49] Shiau F-S, Fang T-T, Leu T-H. Effects of milling and particle size distribution on the sintering behavior and the evolution of the microstructure in sintering powder compacts. *Mater Chem Phys* 1998, **57**: 33–40.

13

Influence of variables on the synthesis of CoFe₂O₄ pigment by the complex polymerization method

P. N. MEDEIROS[a,*], Y. F. GOMES[a], M. R. D. BOMIO[a], I. M. G. SANTOS[b], M. R. S. SILVA[b], C. A. PASKOCIMAS[a], M. S. LI[c], F. V. MOTTA[a]

[a]*Department of Materials Engineering, Federal University of Rio Grande do Norte, Campus Lagoa Nova, CEP 59078-900-Natal/RN, Brazil*
[b]*Department of Chemistry, Federal University of Paraíba, Cidade Universitária, CEP 58051-900-João Pessoa/PB, Brazi*
[c]*Institute Physics of São Carlos, USP, CEP 13566-590, São Carlos, São Paulo, Brazil*

Abstract: Synthetic inorganic pigments are most widely used in ceramic applications due to their excellent chemical and thermal stability and their lower toxicity to both human and environment as well. In the present work, black ceramic pigment $CoFe_2O_4$ has been synthesized by the complex polymerization method (CPM) with good chemical homogeneity. In order to study the influence of variables on the process of obtaining pigment through CPM, $2^{(5-2)}$ fractional factorial design with resolution III was used. The variables studied in the mathematical modeling were: citric acid/metal concentration, pre-calcination time, calcination temperature, calcination time, and calcination rate. Powder pigments were characterized by X-ray diffraction (XRD), scanning electron microscopy (SEM), and UV–visible (UV–Vis) spectroscopy. Based on the results, the formation of cobalt ferrite phase ($CoFe_2O_4$) with spinel structure was verified. The color of pigments obtained showed dark shades, from black to gray. The model adjusted to the conditions proposed in this study due to the determination coefficient of 99.9% and variance (R^2) showed that all factors are significant at the confidence level of 95%.

Keywords: pigment; complex polymerization method (CPM); fractional factorial design

1 Introduction

Ceramic pigments are powders composed of inorganic oxides consisting of a ceramic matrix of crystalline nature and stable with respect to color when dissolved in glasses or ceramic glazes at high temperatures. The color of each pigment is the result of the addition of chromophore ions (usually transition metals) in an inert matrix (oxide or oxide systems) [1,2].

Usually, a ceramic pigment is a metal transition complex oxide obtained by calcination process with three main characteristics: (a) thermal stability, maintaining its identity when temperature increases; (b) chemical stability, maintaining its identity when fired with glazes or ceramic matrices; and (c) high tinting strength when dispersed and fired with glazes or ceramic matrices. Other characteristics such as high dispensability in vehicles, high refractive index (in order to avoid transparency and to increase its tinting

---

* Corresponding author.
E-mail: yfeliciano@gmail.com

strength), acid and alkali resistance, and low abrasive strength are also suitable [3].

$CoFe_2O_4$ is typically synthesized using inorganic precursors (primarily nitrate), followed by calcination at high temperature to obtain the spinel structure. A synthesis route frequently investigated in the literature for obtaining $CoFe_2O_4$ at low temperatures is the complex polymerization method (CPM); this study used fractional factorial design to optimize the synthesis process of cobalt aluminate ($CoFe_2O_4$) at operating conditions to determine a means to save time, thus reducing the number of experiments and analyzing the input variables that would influence the process simultaneously [42].

The main coloring methods of ceramics are based on dyes or pigments. In general, a dye or soluble colorant is a colored substance that interacts with the matrix to which it is applied and is usually soluble in the application media called matrix or substrate.

Black pigments are generally produced from the mixture or pure oxides of metals, and the most important oxides in this class of dyes are: cobalt, iron, chromium, and nickel [4,5]. Black pigments are most widely used in the ceramic industry, representing approximately 25% of the total consumption, and are obtained from two main crystalline structures: hematite and spinel [6,7]. Spinel-type pigments are characterized by being stable under the effect of various factors and are widely used in the decoration of ceramic products [8,9].

Cobalt ferrite is a black pigment widely used in the ceramic industry due to its excellent properties such as chemical and thermal stability [10]. Among ferritic materials, $CoFe_2O_4$ has the cubic form of partially inverse spinel [11–13].

Various chemical methods, such as conventional ceramic method [2,14,15], combustion synthesis, co-precipitation [16], sol–gel method [2,17], polymeric precursor [1,18], microwave synthesis [19], and others, have been used to synthesize inorganic pigments. The complex polymerization method (CPM) based on the Pechini method offers the possibility of preparing complexes of good homogeneity at molecular scale and a good stoichiometric control. The temperatures required are lower than in conventional methods, as in reactions between materials in the solid state or decompositions [20–24]. Many variables can influence the synthesis of oxides by CPM [25–27]. Therefore, an experimental design should be used to investigate the factors that most influence the results.

The experimental design consists of a set of tests established with statistical criteria aimed at determining the influence of variables on the results of a given system or process. The benefits are the reduced number of trials without compromising the quality of information, simultaneous study of several variables, and representation of the process studied by mathematical expressions [28–31].

In this study, a $2^{(5-2)}$ fractional factorial design was used to evaluate the influence of variables on the synthesis of $CoFe_2O_4$ by CPM. It optimized the process of cobalt ferrite ($CoFe_2O_4$) synthesis at operating conditions to determine a means to save time, thus reducing the number of experiments and analyzing the input variables that would influence the process simultaneously.

2 Experimental

2.1 Materials and reagents

Cobalt nitrate ($CoN_2O_6 \cdot 6H_2O$, Aldrich, 98%), iron nitrate ($Fe(NO_3)_3 \cdot 9H_2O$, Synth, 99%), citric acid ($C_6H_8O_7 \cdot H_2O$, Synth, 99.5%), and ethylene glycol ($C_2H_6O_2$, Synth, 99%) were used to prepare the cobalt ferrite precursor resins ($CoFe_2O_4$).

2.2 Experimental design

The fractional factorial design increases the amount of information obtained and reduces the number of experiments with the aim of studying the influence of variables on the synthesis of cobalt ferrite through CPM; a series of experiments were carried out according to a statistical experimental design using the STATISTIC Software 7.0 [32].

To build the design matrix of experiments, five factors were chosen, each with two levels, resulting in 32 combinations. However, a $2^{(5-2)}$ fractional factorial design with resolution III was used. When using a $2^{(5-2)}$ experiment with eight combinations (only one fourth of the entire experiment), the number of experiments, the time spent on the experiments, and the consumption of reagents were reduced. The five factors were: citric acid/metal concentration, pre-calcination time, calcination temperature, calcination time, and calcination rate, assessed on two levels (−1 and +1) and three replicates at the center point (0), totaling 11 experiments. In response, the mean reflectance percentage in the visible range (400–

700 nm) for each sample synthesized was used. The syntheses were carried out according to the scheme shown in Table 1.

The $2^{(5-2)} = 2^3$ fractional factorial approach reduced the number of experiments to 11, including eight factorial and three central points.

The values of wavelengths (Y) related to higher peak absorbance of pigments obtained by UV–visible (UV–Vis) spectroscopy analysis are the response variables. Based on this analysis [42], the general second-order or quadratic equation may be generated, as represented in Eq. (1):

$$Y = \beta_0 + \sum_{j=1}^{k} \beta_j x_j + \sum\sum_{i<j} \beta_{ij} x_i x_j + \sum_{j=1}^{k} \beta_{jj} x_j^2 + \varepsilon \tag{1}$$

where Y is the expected response wavelength; x_i is the encoded or non-encoded value of factors (citric acid/metal concentration, pre-calcination time, calcination temperature, calcination time, and calcination rate); β_0 is a constant; i is the leading coefficient for each variable; and ij is the effect of the interaction of coefficients [42].

The values of polynomial coefficients and the response surface of the second-order model were obtained using the STATISTICA Software 7.0, and the model was validated for the processing conditions used in this study.

2. 3 Cobalt ferrite synthesis

The materials used to obtain cobalt ferrite were citric acid, iron nitrate, cobalt nitrate, and ethylene glycol. Initially, citric acid was dissolved in distilled water under stirring and heating at about 70 ℃. After dissolution, iron nitrate was added and then cobalt nitrate, with citric acid/metal ratio according to the experimental design. Then, ethylene glycol was added

at a weight ratio of 60:40 in relation to citric acid. The heating temperature was kept at about 75 ℃ for the polymerization of the esterification reaction. The reduction of the solution volume was expected to occur for the removal of excess solvent resulting in a viscous material, the polymeric resin. This gel was submitted to pre-calcination in a muffle furnace at 350 ℃ by the time determined in the statistical planning. Subsequently, the material was deagglomerated and approximately 2 g was removed for thermal analysis. The rest of powders were calcined at 700 ℃, 800 ℃, and 900 ℃ according to the experimental design.

Pigments were characterized by X-ray diffraction (XRD) on a Shimadzu XRD-7000 diffractometer using Cu Kα radiation. The morphology of pigment particles after calcination was observed in a scanning electron microscope (SEM) from Hitachi High Technologies. UV–Vis spectroscopy analyses were performed on a Shimadzu UV–visible brand, with accessory reflectance UV-2550, with wavelength in the region between 190 and 900 nm

3 Results and discussion

3. 1 Characterization of CoFe$_2$O$_4$ particles

Reflectance measurements were used to characterize the optical behavior of the pigments obtained in the visible region. Typical reflectance curves are illustrated in Fig. 1. The results obtained in response to the experimental design are shown in Table 2. The results for all CoFe$_2$O$_4$ powders synthesized show absorption band in the entire visible wavelength range, which is typical of low-reflectance systems, indicating the formation of dark-colored pigments [6]. The charge transfer between Co–O and Fe–O along with d–d electronic transitions of Co^{2+} and Fe^{3+} in multiple

Table 1 $2^{(5-2)}$ fractional factorial design with three replicates at the center point

Experiment	Citric acid/ metal concentration	Pre-calcination time (h)	Calcination temperature (℃)	Calcination time (h)	Calcination rate (℃/min)
1	2:1 (−1)	1 (−1)	700 (−1)	6 (1)	11 (1)
2	4:1 (1)	1 (−1)	700 (−1)	2 (−1)	5 (−1)
3	2:1 (−1)	3 (1)	700 (−1)	2 (−1)	11 (1)
4	4:1 (1)	3 (1)	700 (−1)	6 (1)	5 (−1)
5	2:1 (−1)	1 (−1)	900 (1)	6 (1)	5 (−1)
6	4:1 (1)	1 (−1)	900 (1)	2 (−1)	11 (1)
7	2:1 (−1)	3 (1)	900 (1)	2 (−1)	5 (−1)
8	4:1 (1)	3 (1)	900 (1)	6 (1)	11 (1)
9	3:1 (0)	2 (0)	800 (0)	4 (0)	8 (0)
10	3:1 (0)	2 (0)	800 (0)	4 (0)	8 (0)
11	3:1 (0)	2 (0)	800 (0)	4 (0)	8 (0)

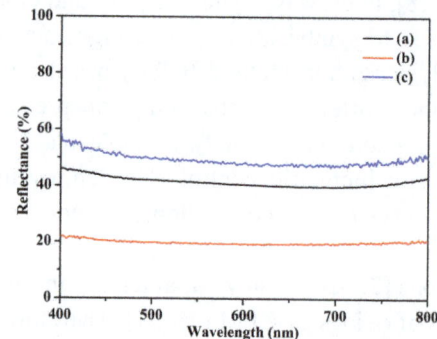

Fig. 1 Reflectance spectra of CoFe$_2$O$_4$ particles obtained by experiments (a) 4, (b) 7, and (c) 9.

Table 2   Results of the reflectance percentage of CoFe$_2$O$_4$ particles

Experiment	Reflectance (%)
1	25.60
2	14.60
3	28.12
4	41.65
5	37.09
6	48.77
7	19.78
8	46.06
9	49.46
10	49.35
11	49.68

coordination ensures complete absorption throughout the visible spectrum [12].

According to the literature, the reflectance curve of theoretical black color would present 0% reflectance at all wavelengths. However, in practice, it is observed that a pigment with the best possible black color shows reflectance values close to 0% in the visible wavelength range [33,34].

In Fig. 2, photomicrographs of cobalt ferrite pigments show black color by using the obtained statistical design values of factors (citric acid/metal concentration, pre-calcination time, calcination temperature, calcination time, and calcination rate) and the route from polymeric precursors.

All CoFe$_2$O$_4$ samples were characterized by XRD. Figure 3 shows the XRD patterns obtained from experiments 4, 7, and 9. According to the statistical design (Table 1), in general, in all powders, the major phase of the CoFe$_2$O$_4$ spinel structure was identified

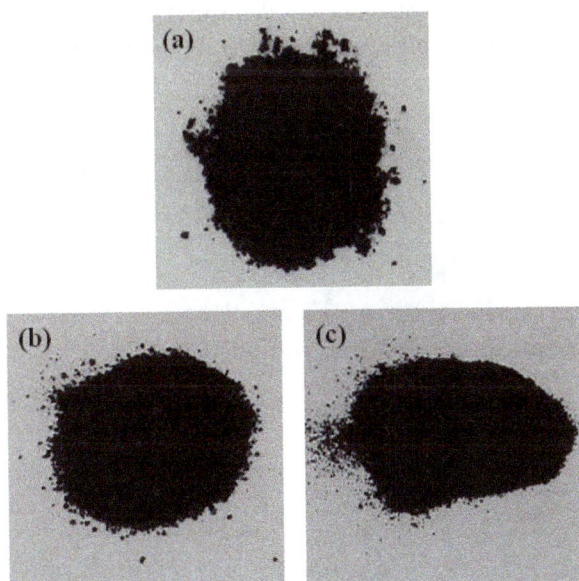

Fig. 3   XRD patterns of CoFe$_2$O$_4$ particles obtained by experiments (a) 4, (b) 7, and (c) 9.

according to the JCPDS 22-1086 form [35–39].

In micrographs obtained for CoFe$_2$O$_4$ samples, a strong agglomeration is evidenced in all compounds, which is attributed to both the synthesis method used and the magnetic characteristic of the material, which is due to the attraction between particles tending to form clusters [40]. Another factor that can lead to the formation of partially sintered agglomerates is additional heating due to the combustion of the organic material during calcination [41]. Figure 4 shows the morphology of particles obtained from experiments 4, 7, and 9.

3. 2   Experimental design

The experimental design was analyzed using the STATISTICA software and analysis of variance (ANOVA), which evaluate whether the effect and interaction between the factors investigated are of significance with regard to the experimental error. The significance of the main factors and their interactions were evaluated by the F-test with 95% confidence

Fig. 2   Photomicrographs of cobalt ferrite pigments showing black color by using the obtained statistical design experiments (a) 4, (b) 7, and (c) 9.

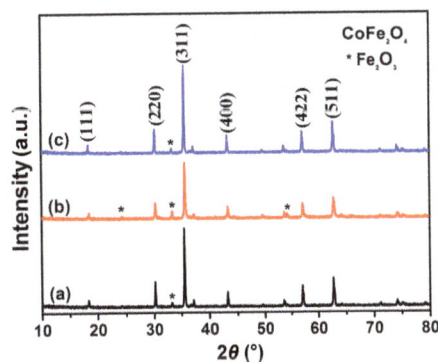

Fig. 4   SEM micrographs of CoFe$_2$O$_4$ particles obtained by experiments (a) 4, (b) 7, and (c) 9.

level (Table 3) [42].

The results of the ANOVA analysis are presented in Table 3, showing that all the effects and interactions between the factors are significant ($P < 0.05$).

The estimated effects and coefficients are listed in Table 4. The experimental error and the main factors and their interactions were evaluated by the F-test with 95% confidence level as reported by Montgomery [42]

Table 3   Analysis of variance (ANOVA) for the suggested model

Source	DF	SS	MS	F	P
(1) Citric acid/metal concentration	1	204.971	204.970	7619.44	0.000131
(2) Pre-calcination time (h)	1	11.386	11.386	423.26	0.002354
(3) Calcination temperature (℃)	1	217.799	217.799	8096.33	0.000123
(4) Calcination time (h)	1	191.375	191.375	7114.05	0.000141
(5) Calcination rate (℃/min)	1	156.840	156.840	5830.26	0.000171
(3)×(4)	1	12.301	12.301	457.26	0.002180
(4)×(5)	1	307.322	307.322	11424.17	0.000088
(1)×(3)×(5)	1	614.789	614.789	22853.75	0.000044
Pure error	2	0.054	0.0269		
Total	10	1716.836			

DF: degree of freedom; SS: sum of square; MS: mean square; and F: F-test.

Table 4   Estimated effects of the experimental design

	Effect	Pure error	t(2)	P	Confidence level 95%
Mean/ interaction	49.4960	0.094694	522.693	0.000004	(49.0886; 49.9034)
(1) Citric acid/metal concentration	10.1235	0.115976	87.289	0.000131	(9.6245; 10.6225)
(2) Pre-calcination time (h)	2.3860	0.115976	20.576	0.002354	(1.8870; 2.8850)
(3) Calcination temperature (℃)	10.4355	0.115976	89.980	0.000123	(9.9365; 10.9345)
(4) Calcination time (h)	9.7820	0.115976	84.345	0.000141	(9.2830; 10.2810)
(5) Calcination rate (℃/min)	8.8555	0.115976	76.356	0.000171	(8.3565; 9.3545)
(3)×(4)	−2.4800	0.115976	−21.384	0.002180	(−2.9790; −1.9810)
(4)×(5)	−12.396	0.115976	−106.884	0.000088	(−12.8950; −11.8970)
(1)×(3)×(5)	−33.5725	0.222078	−151.175	0.000044	(−34.5280; −32.6170)

$R^2 = 0.99997$; adjustment: 0.99984; MS pure error = 0.026901.

and Mason et al. [32]. Table 4 shows the main effects and interactions of independent variables and as response, the reflectance percentage for a linear model, considering the interactions between variables with 95% confidence level. Data were obtained considering the pure error.

As illustrated in the Pareto diagram (Fig. 5), effects are statistically significant on the response variables. It is observed that the determination coefficient (R^2) is approximately 99.99%, indicating that a linear model is able to represent the relationship between effects and response.

In the Pareto chart, the five input variables with values greater than 87.28939 ($P = 0.05$) at the right of the line are significant; the calcination temperature has greater statistical significance compared to the other factors studied. The negative value of −151.175 indicates higher values of the reflectance range to the lower level with respect to the calcination time. The calcination rate shows positive numeric value of 76.35612, indicating an increase in the reflectance value when the citric acid/metal concentration reaches its maximum level (Fig. 5).

In Fig. 6, the values predicted by the model are represented by the straight line, while observed values are represented by points. Predicted values versus observed values using the model equation were obtained from data of the response surface of UV–Vis absorbance. Effects with less than 95% significance according to the F-test used in this study are not reported. Observed values are close to predicted values, demonstrating that the model is adequate (Fig. 6).

Based on the significant effects, the following model (Eq. (2)) is proposed:

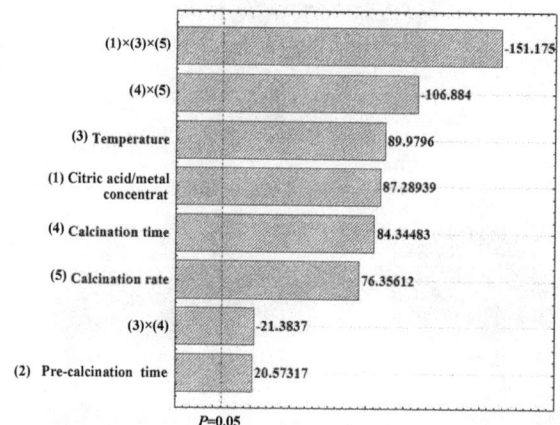

Fig. 5 Pareto diagram of the $2^{(5-2)}$ fractional factorial design showing the influence of the factors studied.

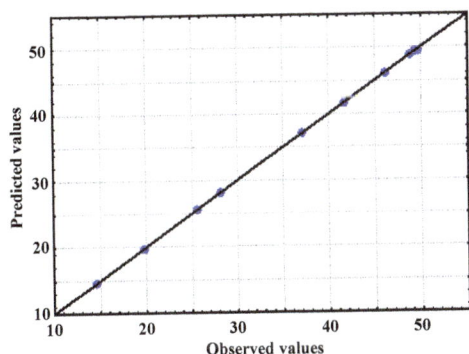

Fig. 6 Values predicted by the model versus observed values, according to the reflectance values.

$$Y = 49.496 + (4.891 \times \text{Calcination time}) +$$
$$(4.42775 \times \text{Calcination rate}) -$$
$$(1.241 \times \text{Calcination time}) -$$
$$(6.198 \times \text{Calcination time} \times \text{Calcination rate}) -$$
$$(16.78625 \times \text{Calcination rate}) + 11.4725 \qquad (2)$$

The wavelength model equation is expressed to the following interval (−1, 0, +1), or variables citric acid/metal concentration, pre-calcination time, calcination temperature, calcination time, and calcination rate.

Equation (2) was used to predict the wavelength, in which the model function is valid for the defined interval, i.e., for variables citric acid/metal concentration, pre-calcination time, calcination temperature, calcination time, and calcination rate. In the valid range from −1 to +1 for the experimental conditions used in this study, the determination coefficient of the proposed mathematical model can explain about 99.9% of the variance (R^2) [32,42].

The results of experiments show that all variables influence in synthesis of cobalt ferrite by CPM; however, the interaction between citric acid/metal concentration, calcination temperature, and calcination rate is more significant.

4 Conclusions

The results obtained in this study indicated that the CPM was favorable for the synthesis of $CoFe_2O_4$ pigment with spinel structure. XRD showed that the $CoFe_2O_4$ phase was obtained for all samples, some of which showed Fe_2O_3 and Co_3O_4 as secondary phases. The spectra in the UV–Vis region indicated that the synthesized pigment showed darker shades, with color predominantly from black to gray.

According to the $2^{(5-2)}$ fractional factorial design, it was observed that all main effects were significant, as well as the interactions between effects, at confidence level of 95%. The linear model showed optimal adjustment with $R^2 = 99.99\%$, being also statistically significant. The design of experiments used showed that all variables studied influence the production of cobalt ferrite pigment by CPM.

Acknowledgements

The authors thank the financial support of the Brazilian research financing institutions: RECAM (Rede de Catalisadores Ambientais), CNPq, and CAPES.

References

[1] Cunha JD, Melo DMA, Martinelli AE, et al. Ceramic pigment obtained by polymeric precursors. Dyes Pigments 2005, 65: 11–14.

[2] Gorodylova N, Kosinová V, Dohnalová Ž, et al. New purple-blue ceramic pigments based on $CoZr_4(PO_4)_6$. Dyes Pigments 2013, 98: 393–404.

[3] Monrós G. Pigment, Ceramic, Encyclopedia of Color Science and Technology. Springer, 2014.

[4] Escardino A, Mestre S, Barba A, et al. Kinetic study of black Fe_2O_3–Cr_2O_3 pigment synthesis: I, influence of synthesis time and temperature. J Am Ceram Soc 2003, 86: 945–950.

[5] Zaichuk AV, Belyi YI. Black ceramic pigments based on open-hearth slag. Glass Ceram+ 2012, 69: 99–103.

[6] Costa G, Della VP, Ribeiro MJ, et al. Synthesis of black ceramic pigments from secondary raw materials. Dyes Pigments 2008, 77: 137–144.

[7] Hajjaji W, Seabra MP, Labrincha JA. Evaluation of metal-ions containing sludges in the preparation of black inorganic pigments. J Hazard Mater 2011, 185: 619–625.

[8] Maslennikova GN. Pigments of the spinel type. Glass Ceram+ 2001, 58: 216–220.

[9] Mestre S, Palacios MD, Agut P. Solution combustion synthesis of (Co,Fe)Cr_2O_4 pigments. J Eur Ceram Soc 2012, 32: 1995–1999.

[10] Yüngevis H, Ozel E. Effect of the milling process on the properties of $CoFe_2O_4$ pigment. Ceram Int 2013, 39: 5503–5511.

[11] Liu X-M, Fu S-Y, Xiao H-M, et al. Synthesis of nanocrystalline spinel $CoFe_2O_4$ via a polymer-pyrolysis

route. *Physica B* 2005, **370**: 14–21.

[12] Cavalcante PMT, Dondi M, Guarini G, *et al.* Colour performance of ceramic nano-pigments. *Dyes Pigments* 2009, **80**: 226–232.

[13] Jia Z, Ren D, Zhu R. Synthesis, characterization and magnetic properties of CoFe$_2$O$_4$ nanorods. *Mater Lett* 2012, **66**: 128–131.

[14] Costa G, Ribeiro MJ, Hajjaji W, *et al.* Ni-doped hibonite (CaAl$_{12}$O$_{19}$): A new turquoise blue ceramic pigment. *J Eur Ceram Soc* 2009, **29**: 2671–2678.

[15] Gargori C, Cerro S, Galindo R, *et al.* New vanadium doped calcium titanate ceramic pigment. *Ceram Int* 2011, **37**: 3665–3670.

[16] Llusar M, Zielinska A, Tena MA, *et al.* Blue-violet ceramic pigments based on Co and Mg Co$_{2-x}$Mg$_x$P$_2$O$_7$ diphosphates. *J Eur Ceram Soc* 2010, **30**: 1887–1896.

[17] Ricceri R, Ardizzone S, Baldi G, *et al.* Ceramic pigments obtained by sol–gel techniques and by mechanochemical insertion of color centers in Al$_2$O$_3$ host matrix. *J Eur Ceram Soc* 2002, **22**: 629–637.

[18] Melo D, Vieira FTG, Costa TCC, *et al.* Lanthanum cobaltite black pigments with perovskite structure. *Dyes Pigments* 2013, **98**: 459–463.

[19] Blosi M, Albonetti S, Gatti F, *et al.* Au–Ag nanoparticles as red pigment in ceramic inks for digital decoration. *Dyes Pigments* 2012, **94**: 355–362.

[20] Kakihana M. Invited review "sol–gel" preparation of high temperature superconducting oxides. *J Sol–Gel Sci Technol* 1996, **6**: 7–55.

[21] Bernardi MIB, De Vicente FS, Li MS, *et al.* Colored films produced by electron beam deposition from nanometric TiO$_2$ and Al$_2$O$_3$ pigment powders obtained by modified polymeric precursor method. *Dyes Pigments* 2007, **75**: 693–700.

[22] Cho W-S, Kakihana M. Crystallization of ceramic pigment CoAl$_2$O$_4$ nanocrystals from Co–Al metal organic precursor. *J Alloys Compd* 1999, **287**: 87–90.

[23] Razpotnik T, Maček J. Synthesis of nickel oxide/zirconia powders via a modified Pechini method. *J Eur Ceram Soc* 2007, **27**: 1405–1410.

[24] Mariappan CR, Galven C, Crosnier-Lopez M-P, *et al.* Synthesis of nanostructured LiTi$_2$(PO$_4$)$_3$ powder by a Pechini-type polymerizable complex method. *J Solid State Chem* 2006, **179**: 450–456.

[25] Vieira FTG, Melo DS, de Lima SJG, *et al.* The influence of temperature on the color of TiO$_2$:Cr pigments. *Mater Res Bull* 2009, **44**: 1086–1092.

[26] Chai Y-L, Chang Y-S, Chen G-J, *et al.* The effects of heat-treatment on the structure evolution and crystallinity of ZnTiO$_3$ nano-crystals prepared by Pechini process. *Mater Res Bull* 2008, **43**: 1066–1073.

[27] Mohammadia MR, Fray DJ. Low temperature nanostructured zinc titanate by an aqueous particulate sol–gel route: Optimisation of heat treatment condition based on Zn:Ti molar ratio. *J Eur Ceram Soc* 2010, **30**: 947–961.

[28] Box GEP, Hunter WG, Hunter JS. *Statistic for Experiments: Design, Innovation, and Discovery*, 2nd edn. New York: Wiley, 2005.

[29] Rautio J, Perämäki P, Honkamo J, *et al.* Effect of synthesis method variables on particle size in the preparation of homogeneous doped nano ZnO material. *Microchem J* 2009, **91**: 272–276.

[30] Rosario AV, Pereira EC. Comparison of the electrochemical behavior of CeO$_2$–SnO$_2$ and CeO$_2$–TiO$_2$ electrodes produced by the Pechini method. *Thin Solid Films* 2002, **410**: 1–7.

[31] Maran JP, Manikandan S. Response surface modeling and optimization of process parameters for aqueous extraction of pigments from prickly pear (*Opuntia ficus-indica*) fruit. *Dyes Pigments* 2012, **95**: 465–472.

[32] Mason RL, Gunst RF, Hess JL. *Statistical Design and Analysis of Experiments: With Aplications to Engineering and Science*, 2nd edn. New Jersey: John Wiley & Sons, 2003.

[33] Sánchez MYH, Baena OJR. Síntesis y caracterización colorimétrica de un pigmento negro tipo espinela. CONAMET/SAM-2008.

[34] Shen L, Qiao Y, Guo Y, *et al.* Preparation of nanometer-sized black iron oxide pigment by recycling of blast furnace flue dust. *J Hazard Mater* 2010, **177**: 495–500.

[35] Sajjia M, Oubaha M, Prescott T, *et al.* Development of cobalt ferrite powder preparation employing the sol–gel technique and its structural characterization. *J Alloys Compd* 2010, **506**: 400–406.

[36] Verma KC, Singh VP, Ram M, *et al.* Structural, microstructural and magnetic properties of NiFe$_2$O$_4$, CoFe$_2$O$_4$ and MnFe$_2$O$_4$ nanoferrite thin films. *J Magn Magn Mater* 2011, **323**: 3271–3275.

[37] Airimioaei M, Ciomaga CE, Apostolescu N, *et al.* Synthesis and functional properties of the Ni$_{1-x}$Mn$_x$Fe$_2$O$_4$ ferrites. *J Alloys Compd* 2011, **509**: 8065–8072.

[38] Lavela P, Tirado JL. CoFe$_2$O$_4$ and NiFe$_2$O$_4$ synthesized by sol–gel procedures for their use as anode materials for Li ion batteries. *J Power Sources* 2007, **172**: 379–387.

[39] Simões AZ, Riccardi CS, Dos Santos ML, *et al.* Effect of annealing atmosphere on phase formation and electrical characteristics of bismuth ferrite thin films. *Mater Res Bull* 2009, **44**: 1747–1752.

[40] Gimenes R, Baldissera MR, da Silva MRA, *et al.* Structural and magnetic characterization of Mn$_x$Zn$_{1-x}$Fe$_2$O$_4$ ($x = 0.2$; 0.35; 0.65; 0.8; 1.0) ferrites obtained by the citrate precursor method. *Ceram Int* 2012, **38**: 741–746.

[41] Popa M, Calderon-Moreno JM. Lanthanum cobaltite nanoparticles using the polymeric precursor method. *J Eur Ceram Soc* 2009, **29**: 2281–2287.

[42] Montgomery DC. *Design and Analysis of Experiments*, 5th edn. New York: Wiley, 2001.

14

Selective liquid phase oxidation of cyclohexane over Pt/CeO$_2$–ZrO$_2$–SnO$_2$/SiO$_2$ catalysts with molecular oxygen

Nobuhito IMANAKA[*], Toshiyuki MASUI, Kazuya JYOKO

Department of Applied Chemistry, Faculty of Engineering, Osaka University, 2-1 Yamadaoka, Suita, Osaka 565-0871, Japan

Abstract: Partial oxidation of cyclohexane into cyclohexanone and cyclohexanol (KA-oil) is an industrially significant reaction for producing precursors for the synthesis of ε-caprolactam and adipic acid, which are the building blocks of nylon. However, to date, the cyclohexane conversion ratio has usually been limited to less than 6% to prevent further oxidation of the cyclohexanol and cyclohexanone targets. In this study, we report that Pt/CeO$_2$–ZrO$_2$–SnO$_2$/SiO$_2$, in which CeO$_2$–ZrO$_2$–SnO$_2$ provide reactive oxygen molecules from inside the bulk, can act as efficient catalysts. Optimization of the catalyst composition and reaction conditions provided a cyclohexane conversion ratio of 24.1% and a total selectivity for cyclohexanol and cyclohexanone of 83.4% at 130 ℃ in 0.5 MPa (4.9 atm) air for 7 h over a 5wt%Pt/16wt%Ce$_{0.68}$Zr$_{0.17}$Sn$_{0.15}$O$_{2.0}$/SiO$_2$ catalyst. This catalyst has significant advantages over conventional catalysts because the reaction proceeds at a lower pressure, and there is no need for toxic radical initiators or free-radical scavengers.

Keywords: composite materials; oxidation; cyclohexane; KA-oil; catalyst

1 Introduction

The selective oxidation of cyclohexane (CyH) to cyclohexanone (Cy=O, denoted as "K") and cyclohexanol (CyOH, denoted as "A"), which is a radical chain reaction, is a significant process in the chemical industry, because these are important intermediates for the manufacture of nylon-6 and nylon-6,6 [1]. The conventional industrial process for CyH oxidation to a mixture of K and A (known as KA-oil) involves the use of soluble cobalt or manganese salts as homogeneous catalysts that are radical generation agents, and the reaction is conducted at 140–180 ℃ and 0.8–2 MPa (7.9–19.8 atm) in air for 15–60 min [1]. KA-oil is produced by this method with 70%–85% selectivity; however, the CyH conversion ratio is usually limited to less than 6% to prevent further oxidation of the target KA-oil.

A number of alternative processes that employ heterogeneous catalysts have been proposed for selective CyH oxidation using molecular oxygen with a particular focus on improving the CyH conversion ratio without a significant decrease in the selectivity for KA-oil. Catalysts consisting of gold nanoparticles supported on zeolites such as ZSM-5 (Zeolite Socony Mobil No. 5) [2] and MCM-41 (Mobil Composition of Matter No. 41) [3] can convert CyH to KA-oil at 150 ℃ in 1 MPa (9.9 atm) O$_2$ atmosphere for 2–6 h with a 6%–19% conversion ratio and 92%–100% selectivity, even in the absence of free-radical

* Corresponding author.
E-mail: imanaka@chem.eng.osaka-u.ac.jp

scavengers [4,5]. An Au/SiO$_2$ catalyst, in which gold nanoparticles were embedded in amorphous silica, also exhibited a high catalytic activity of 22.7% for CyH conversion and 80.6% selectivity for KA-oil at 150 ℃ and 1.5 MPa (14.8 atm) O$_2$ for 3 h, although toxic and explosive *tert*-butyl hydroperoxide (TBHP) was used as a radical initiator [6]. Size-controlled Au clusters on hydroxyapatite exhibited a CyH conversion ratio of 6.7%–14.9% with 94%–99% selectivity for KA-oil at 150 ℃ under 1 MPa (9.9 atm) O$_2$ for 4 h [7], although the addition of a small amount of TBHP was also essential in this case to initiate the reaction.

Mesoporous chromium and iron terephthalates (Cr- and Fe-MIL-101, respectively), which are categorized as metal–organic frameworks, have demonstrated high catalytic activity and selectivity for the oxidation of CyH [8]. The Cr-MIL-101/TBHP system produces KA-oil with 92% selectivity at a 25% CyH conversion ratio, while Fe-MIL-101/TBHP/O$_2$ gives a mixture of cyclohexyl hydroperoxide, K, and A with 99% total selectivity (49% selectivity for KA-oil) and a 38% conversion ratio at 70 ℃ in air (1 atm) for 8 h. However, the addition of TBHP is also essential in these reactions. Fe-filled carbon nanotubes are also active catalysts that have yielded a CyH conversion ratio of about 37% at 125 ℃ and 1.5 MPa (14.8 atm) O$_2$ for 8 h, although the selectivity to KA-oil was about 30% due to over-oxidation [9].

This situation has prompted the examination of novel and advanced CyH oxidation catalysts that can realize high CyH conversion ratios with high selectivity for KA-oil in a liquid phase reaction. We have previously prepared novel catalysts to realize the complete oxidation of volatile organic compounds under moderate conditions [10,11]. These catalysts involved a combination of platinum and a solid promoter to supply reactive oxygen molecules, which was important in order to allow oxidation of hydrocarbons in the gas phase [11,12]. Complete oxidation of ethylene, toluene, and acetaldehyde was realized at temperatures as low as 55, 110, and 140 ℃, respectively, over a 10wt%Pt/16wt%Ce$_{0.68}$Zr$_{0.17}$Sn$_{0.15}$O$_{2.0}$/γ-Al$_2$O$_3$ catalyst [13].

In the present study, we have designed advanced catalysts that are optimized for CyH oxidation to KA-oil in the liquid phase without the need for radical initiators or free-radical scavengers. To realize such advanced catalysts, we have focused on a combination of Pt, Ce$_{0.68}$Zr$_{0.17}$Sn$_{0.15}$O$_{2.0}$, and SiO$_2$ (silica gel). In this catalyst, platinum acts as the main oxidation agent and the Ce$_{0.68}$Zr$_{0.17}$Sn$_{0.15}$O$_{2.0}$ solid solution is the promoter from which reactive oxygen molecules are provided. Silica gel is employed as a catalyst support instead of γ-Al$_2$O$_3$ because it has superior adsorption capabilities for C$_6$ and C$_7$ hydrocarbons [14]. In this study, $x$wt%Pt/16wt%Ce$_{0.68}$Zr$_{0.17}$Sn$_{0.15}$O$_{2.0}$/SiO$_2$ (hereafter denoted as Pt(x)/CZS/SiO$_2$, where $1 \leqslant x \leqslant 10$) catalysts were prepared and their oxidation activity for the conversion of CyH into K and A was investigated. Furthermore, the catalyst composition was optimized to simultaneously realize high activity and high selectivity.

2 Experiment

2.1 Preparation of the Pt/CeO$_2$–ZrO$_2$–SnO$_2$/SiO$_2$ catalysts

The 16wt%CZS/SiO$_2$ support was synthesized by co-precipitation and impregnation methods. SnC$_2$O$_4$ was dissolved in a mixture of 1.0 mol·L^{-1} Ce(NO$_3$)$_3$ and 0.1 mol·L^{-1} ZrO(NO$_3$)$_2$ aqueous solutions in a stoichiometric ratio, and the mixture was then impregnated into commercially available silica gel (Fuji Silysia Chemical, CARiACT Q-3). The CZS content was adjusted to 16 wt% of the total support to optimize the amount of promoter available so as to provide the highest oxidation activity [13]. The pH of the aqueous mixture was adjusted to 11 by dropwise adding aqueous ammonia (5%). After stirring for 12 h at room temperature, the resulting precipitate was collected by filtration, washed several times with deionized water, and then dried at 80 ℃ for 12 h. The sample was ground in an agate mortar and then calcined at 600 ℃ for 1 h in the ambient atmosphere. Supported platinum catalysts ($x$wt%Pt/16wt%Ce$_{0.68}$Zr$_{0.17}$Sn$_{0.15}$O$_{2.0}$/SiO$_2$) ($1 \leqslant x \leqslant 10$) were prepared by impregnating the 16wt%CZS/SiO$_2$ support with a 4 wt% platinum colloid aqueous solution stabilized with polyvinylpyrrolidone (Tanaka Kikinzoku Kogyo Co., Ltd.). After impregnation, the sample was dried at 80 ℃ for 6 h and then calcined at 500 ℃ for 4 h. As references, 5wt%Pt/$y$wt%Ce$_{0.68}$Zr$_{0.17}$Sn$_{0.15}$O$_{2.0}$/SiO$_2$ catalysts (hereafter denoted as Pt(5)/CZS(y)/SiO$_2$, where $1 \leqslant y \leqslant 30$) were also prepared to identify the feature of CZS.

2.2 Cyclohexane oxidation

The cyclohexane (CyH) oxidation reaction was conducted in batch mode using a mechanically stirred 50 mL autoclave made from SUS316 stainless steel. Prior to the reaction, CyH (5.0 g) and the catalyst (10 mg) were loaded into the autoclave, and dry air was supplied to the reactor. The reactor was sealed and heated to a stable operational temperature. After the reaction, the products were collected by decantation and analyzed using gas chromatograph mass spectrometry (GCMS; Shimadzu GCMS-QP2010 Plus). The turnover frequency (TOF) is defined as the number of cyclohexane molecules that a catalyst can convert to KA-oil per surface Pt site per unit of time. The amount of Pt atom exposed on the surface (Pt_s) was calculated from the amount of CO chemisorption, assuming that the metallic particles are spherical and one CO molecule adsorbs on one surface platinum atom ($CO/Pt_s = 1$). Accordingly, TOF can be calculated from the following equation:

$$TOF(h^{-1}) = \frac{KA\text{-oil synthesis rate}\,(mol \cdot g^{-1} \cdot h^{-1})}{Amount\ of\ CO\ adsorption\,(mol \cdot g^{-1})}$$

2.3 Characterization

The sample compositions were analyzed using X-ray fluorescence spectrometry (XRF; Rigaku ZSX-100e). The crystal structures of the catalysts were identified by X-ray powder diffraction (XRD; Rigaku SmartLab) analysis using Cu Kα radiation (40 kV, 30 mA). The Brunauer–Emmett–Teller (BET) specific surface areas were measured by nitrogen adsorption at −196 ℃ (Micromeritics Tristar 3000). Transmission electron microscopy (TEM) was performed at an accelerating voltage of 300 kV (Hitachi H-9000NAR). The metal dispersion analysis was carried out using a pulse method at –50 ℃ with 10%CO–He (0.03 mL). The Pt metal dispersion was defined as the following equation, where the amount of surface Pt atoms was calculated from the CO chemisorption values as mentioned above.

Dispersion (%) =

$$\frac{Amount\ of\ Pt\ atoms\ exposed\ on\ the\ surface\ (Pt_s)}{Total\ amount\ of\ Pt\ atoms\ (Pt)} \times 100$$

3 Results and discussion

3.1 Characterization of the catalysts

The chemical composition of the samples was confirmed using XRF. The BET specific surface area

and platinum dispersion data for the catalysts are summarized in Table 1. The BET specific surface area decreases with increasing platinum content, which may be due to the much larger density of Pt than that of the oxide promoter and support. The platinum dispersion in the catalysts decreases sharply when the amount of platinum becomes higher than 5 wt%, which suggests significant aggregation of the platinum particles. To confirm this, the degree of aggregation of platinum on the Pt(x)/CZS/SiO2 (x = 1, 5, and 7) catalysts was observed using TEM. Figure 1 shows that there are no notable differences between the Pt(1)/CZS/SiO2 and Pt(5)/CZS/SiO2 samples, which both have a particle

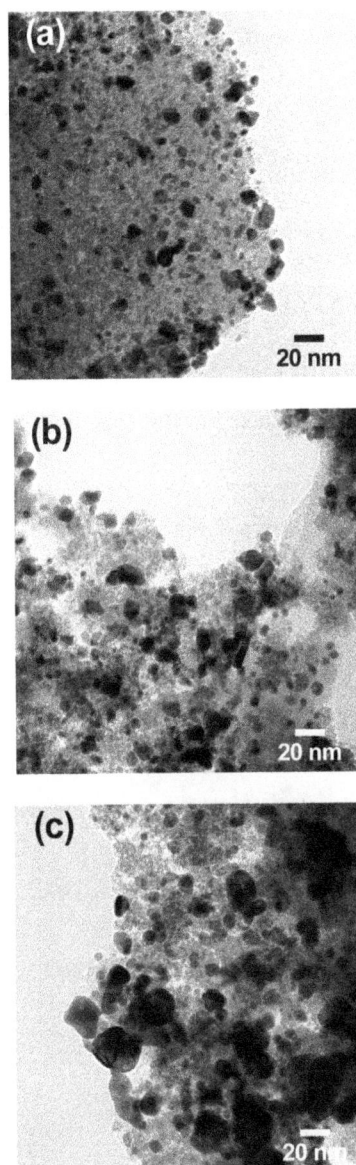

Fig. 1 TEM images of (a) Pt(1)/CZS/SiO2, (b) Pt(5)/CZS/SiO2, and (c) Pt(7)/CZS/SiO2 catalysts for KA-oil synthesis.

Table 1   Catalytic performance of catalysts at 130 ℃ for 7 h in 0.5 MPa air atmosphere

Catalyst*	Surface area ($m^2 \cdot g^{-1}$)	Amount of CO chemisorption ($\mu mol \cdot g^{-1}$)	Total amount of Pt atoms ($\mu mol \cdot g^{-1}$)	Dispersion (%)	Conversion (mol%)	KA-oil yield (mmol)	KA-oil synthesis rate[†] ($mmol \cdot g^{-1} \cdot h^{-1}$)	Selectivity (mol%)		TOF[§] (h^{-1})
								Cyclohexanol	Cyclohexanone	
Pt(1)/CZS/SiO$_2$	322	11.4	51.3	22.2	15.0	7.84	112	58.7	29.3	9845
Pt(3)/CZS/SiO$_2$	320	33.2	154	21.6	17.5	8.86	127	56.0	29.2	3810
Pt(5)/CZS/SiO$_2$	313	48.9	256	19.1	24.1	11.9	171	55.2	28.2	3485
Pt(7)/CZS/SiO$_2$	290	29.1	359	8.1	15.5	7.77	111	55.4	29.0	3820
Pt(10)/CZS/SiO$_2$	275	36.4	513	7.1	12.6	6.42	92	55.6	30.1	2518

*Pt content was determined by XRF. [†]Reaction conditions: 10 mg catalyst, 5.0 g cyclohexane, 130 ℃, 0.5 MPa, and 7 h. [§]TOF was calculated from the rate of KA-oil (cyclohexanol and cyclohexanol) synthesis divided by the amount of CO molecules chemisorbed on the Pt surfaces.

size of approximately 10 nm. In contrast, a number of large platinum particles are observed for the Pt(7)/CZS/SiO$_2$ catalyst, which is consistent with the results of the BET surface area measurements.

Figure 2 shows XRD patterns for the Pt(x)/CZS/SiO$_2$ ($1 \leqslant x \leqslant 10$) catalysts. Only peaks corresponding to platinum, cubic fluorite-type oxide, and amorphous SiO$_2$ appear, and no crystalline impurities are observed. The positions of the peaks assigned to the cubic fluorite structure are the same for all of the samples, which indicates that the platinum is simply supported on the CZS/SiO$_2$ surface, and does not form a solid solution with CZS or SiO$_2$.

3. 2   Catalytic performance

The catalytic performance for the selective oxidation of CyH over the Pt(x)/CZS/SiO$_2$ catalysts was investigated in an air atmosphere (Table 1). Highly efficient and quantitative CyH oxidation to KA-oil could be accomplished at 130 ℃ for 7 h under a 0.5 MPa (4.9 atm) air atmosphere for all samples. Even

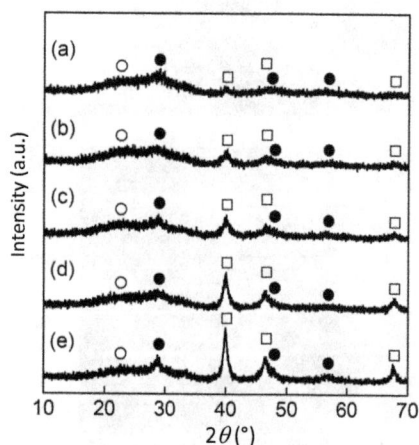

Fig. 2   XRD patterns for (a) Pt(1)/CZS/SiO$_2$, (b) Pt(3)/CZS/SiO$_2$, (c) Pt(5)/CZS/SiO$_2$, (d) Pt(7)/CZS/SiO$_2$, and (e) Pt(10)/CZS/SiO$_2$ catalysts (□: Pt, ●:CZS, ○: SiO$_2$).

without the use of radical initiators and free-radical scavengers, the results for both conversion and selectivity are excellent. In particular, the reaction pressure (0.5 MPa or 4.9 atm) employed in the catalytic system is significantly lower than that used with conventional methods (1–1.5 MPa or 9.9–14.8 atm) [4–9]. The highest activity is obtained for Pt(5)/CZS/SiO$_2$, for which a conversion ratio of 24.1% and a selectivity of 83.4% for KA-oil are achieved. The slight decrease in the conversion ratio for high Pt concentrations may be due to the large amount of aggregation of the Pt particles.

Figure 3(a) shows the dependence of the CyH conversion ratio and the KA-oil selectivity on the reaction pressure for the Pt(5)/CZS/SiO$_2$ catalyst, where the reaction temperature and time are fixed at 130 ℃ and 7 h, respectively. The CyH conversion ratio is improved by the application of pressure and the highest selectivity for KA-oil is obtained at 0.5 MPa (4.9 atm). However, the selectivity decreases as the pressure is increased further due to the over-oxidation of CyH into by-products such as pentanoic acid, glutaric acid, and carbon dioxide. Figure 3(b) shows the influence of the reaction temperature; the conversion ratio becomes saturated at 130 ℃ and is virtually constant at higher temperatures. However, the total selectivity for KA-oil decreases with increasing temperature. Figure 3(c) depicts the effect of the reaction time on the CyH conversion ratio and KA-oil selectivity for the Pt(5)/CZS/SiO$_2$ catalyst. The CyH oxidation reaction progresses from 3 to 12 h, and a conversion ratio of 24.1% and a selectivity of 83.4% for the two desired oxygenates are achieved at 7 h. Similar to previously reported results [6], the oxidation reaction does not appear to slow down with time, which indicates that there is no loss of catalytic activity. However, the selectivity for KA-oil decreases with increasing reaction time due to over-oxidation of CyH. The detailed results are summarized in Table 2.

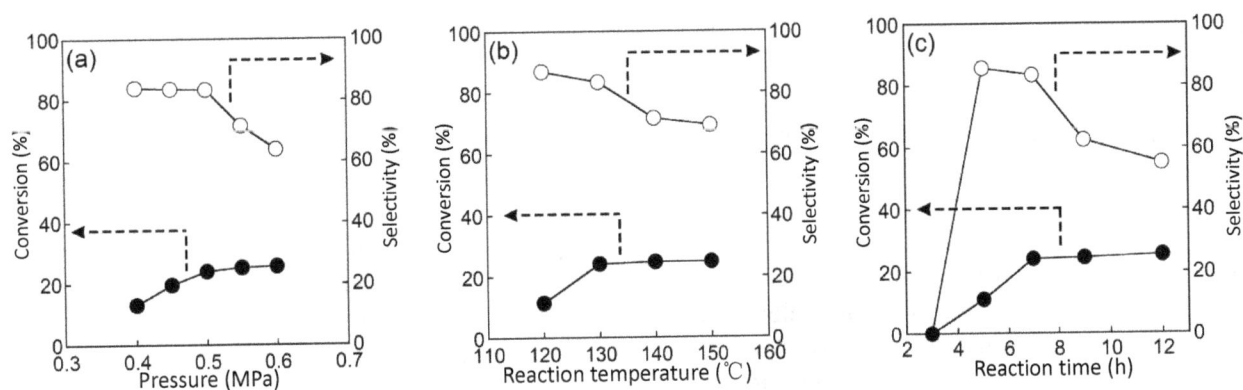

Fig. 3 Catalytic performance of the Pt(5)/CZS/SiO$_2$ catalyst. (a) Effect of reaction pressure on CyH conversion and selectivity for KA-oil (CyOH and Cy=O). Reaction conditions: 10 mg catalyst, 5.0 g cyclohexane, 130 °C, and 7 h. (b) Effect of reaction temperature. Reaction conditions: 10 mg catalyst, 5.0 g cyclohexane, 0.5 MPa (4.9 atm), and 7 h. (c) Effect of reaction time. Reaction conditions: 10 mg catalyst, 5.0 g cyclohexane, 130 °C, and 0.5 MPa (4.9 atm).

Table 2   Catalytic performance of the Pt(5)/CZS/SiO$_2$ catalyst under various reaction conditions

Pressure (MPa)	Temperature (°C)	Reaction time (h)	Conversion (mol%)	KA-oil yield (mmol)	KA-oil synthesis rate[‡] (mmol·g^{-1}·h^{-1})	Selectivity(mol%) Cyclo-hexanol	Cyclo-hexanone	TOF[§] (h^{-1})
0.3			0	0	0	0	0	0
0.4			13.2	6.74	96	56.6	29.4	1968
0.45			19.7	9.90	141	55.3	29.3	2889
0.5	130	7	24.1	11.9	171	55.2	28.2	3485
0.55			25.3	11.9	170	51.3	27.8	3470
0.6			25.8	9.81	140	40.0	24.0	2863
	120		11.5	5.94	85	56.5	30.4	1733
0.5	130	7	24.1	11.9	171	55.2	28.2	3485
	140		24.7	10.5	150	51.0	20.6	3066
	150		24.8	10.2	146	48.8	20.5	2980
		3	0	0	0	0	0	0
		5	11.1	5.64	113	54.1	31.5	2306
0.5	130	7	24.1	11.9	171	55.2	28.2	3485
		9	24.5	9.10	101	39.1	23.4	2065
		12	25.5	8.36	70	35.6	19.6	1424

[‡]Reaction conditions: 10 mg catalyst, 5.0 g cyclohexane. [§]TOF was calculated from the rate of KA-oil (cyclohexanol and cyclohexanol) synthesis divided by the number of CO molecules chemisorbed on the Pt surfaces.

To identify the feature of CZS, Pt(5)/CZS(y)/SiO$_2$ $1 \leqslant y \leqslant 30$ catalysts were also prepared and their catalytic activities were evaluated. Dependence of the catalytic performance of the Pt(5)/CZS(y)/SiO$_2$ catalysts on the amount of CZS is shown in Fig. 4, where the reaction conditions are 10 mg catalyst, 5.0 g cyclohexane, 130 °C, 0.5 MPa (4.9 atm), and 7 h. The CyH conversion increases with increasing the amount of CZS. The KA-oil selectivity is almost maintained until $y = 16$, but it gradually decreases when an excess amount of CZS is loaded on the catalyst due to over-oxidation. From these results, the feature of CZS is evidenced that appropriate amount of CZS addition is effective to increase the CyH conversion without decreasing the KA-oil selectivity.

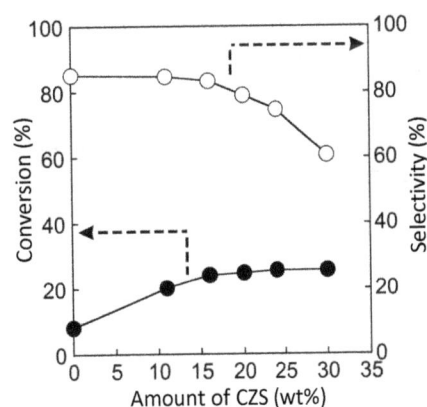

Fig. 4 Dependence of the catalytic performance of Pt(5)/CZS(y)/SiO$_2$ ($0 \leqslant y \leqslant 30$) catalysts on the amount of CZS. Reaction conditions: 10 mg catalyst, 5.0 g cyclohexane, 130 °C, 0.5 MPa (4.9 atm), and 7 h.

Figure 5 presents the relationship between the CyH conversion ratio and the selectivity for KA-oil under various reaction conditions for all of the catalysts in the present study. The selectivity for KA-oil decreases very gradually with increasing CyH conversion ratio until it reaches 24.1%. However, when the conversion ratio becomes greater than 24.1%, the selectivity decreases sharply due to over-oxidation. Therefore, it is crucial to limit the conversion ratio to just below this threshold in order to satisfy the requirements of both a high conversion ratio and high selectivity. Based on the results shown in Fig. 4, the optimum reaction conditions for obtaining the highest yield of KA-oil are 0.5 MPa (4.9 atm), 130 ℃, and 7 h with the Pt(5)/CZS/SiO$_2$ catalyst, for which 24.1% CyH conversion ratio and 83.4% selectivity are simultaneously realized. Furthermore, even though the catalysis test is repeated five times under the same conditions, the catalyst remains stable and no change is found in either the activity or selectivity.

3.3　Reaction mechanism

The CyH oxidation reaction proceeds via the free radical chain mechanism indicated in reactions (1) to (5) [15,16]:

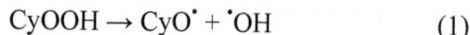

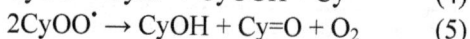

$$CyOOH \rightarrow CyO^{\bullet} + {}^{\bullet}OH \tag{1}$$
$$CyO^{\bullet} + CyH \rightarrow CyOH + Cy^{\bullet} \tag{2}$$
$$Cy^{\bullet} + O_2 \rightarrow CyOO^{\bullet} \tag{3}$$
$$CyOO^{\bullet} + CyH \rightarrow CyOOH + Cy^{\bullet} \tag{4}$$
$$2CyOO^{\bullet} \rightarrow CyOH + Cy=O + O_2 \tag{5}$$

The chain reaction is initiated by homolytic bond cleavage of cyclohexyl hydroperoxide (CyOOH) according to reaction (1). Cerium ions can exhibit redox activity between Ce^{3+} and Ce^{4+}, and thereby this initiation is catalyzed via the Haber–Weiss cycle [17]. The $CyO^{\bullet}$ radical produced by reaction (1) reacts with CyH via reaction (2) to form the cyclohexyl radical (Cy$^{\bullet}$). This then rapidly reacts with molecular O$_2$ to form the cyclohexyl peroxy radical (CyOO$^{\bullet}$) according to reaction (3), and the Cy$^{\bullet}$ radical is then continuously regenerated in reaction (4). The formation of CyOH and Cy=O has been attributed to chain termination by mutual destruction of two CyOO$^{\bullet}$ radicals according to reaction (5) [16,17].

The selectivity of CyOOH typically decreases with increasing conversion ratio [4]. Under the present reaction conditions, CyOOH is not detected in the products, because it decomposes into CyOH and Cy=O by the radical chain reaction. However, in the reaction mechanism shown in (1) to (5), CyOOH must be produced as the starting material. Since the present process does not involve a radical initiator, CyOOH is expected to be produced by catalytic oxidation of CyH [18,19], so that CyOOH formation is the rate-determining step [20]. In fact, there is an induction period for about 3 h, then the activity increases almost linearly until 7 h, and finally, it saturates after longer than 7 h due to over-oxidation, as seen in Fig. 3(c). This oxidation reaction is facilitated by the readily reducible $Ce_{0.68}Zr_{0.17}Sn_{0.15}O_{2.0}$, which can provide active oxygen species from the catalyst bulk. This oxygen then reacts with CyH at the interface of the Pt and $Ce_{0.68}Zr_{0.17}Sn_{0.15}O_{2.0}$ phases [11,13]. The Pt(5)/CZS/SiO$_2$ catalyst has a significantly high oxidation activity, so that there is no need for a radical initiator. However, for high CyH conversion ratios, both Cy=O and CyOH are further oxidized, and the selectivities for Cy=O and CyOH are consequently reduced. As a result, there are optimum reaction conditions for balancing both the selectivity and conversion ratio.

4　Conclusions

Novel Pt/CeO$_2$–ZrO$_2$–SnO$_2$/SiO$_2$ catalysts were successfully prepared for the partial oxidation of CyH into CyOH and Cy=O. The composition was optimized to simultaneously achieve both a high conversion ratio and a high selectivity; a cyclohexane conversion ratio of 24.1% and a selectivity of 83.4% for KA-oil were realized at 0.5 MPa (4.9 atm), 130 ℃, and 7 h using the 5wt%Pt/16wt%Ce$_{0.68}$Zr$_{0.17}$Sn$_{0.15}$O$_{2.0}$/SiO$_2$ catalyst. This catalytic process has significant advantages in that

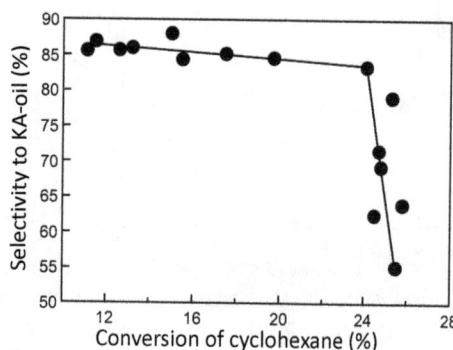

Fig. 5 Relationship between CyH conversion ratio and selectivity for KA-oil. Reaction conditions: 10 mg catalyst, 5.0 g cyclohexane, 0.4–0.6 MPa (3.9–5.9 atm), 120–150 ℃, and 5–12 h.

not only are radical initiators and free-radical scavengers not required, but also the reaction pressure is significantly lower than that typically applied in conventional reactions, due to the combination of catalytic oxidation by platinum, oxygen provision from the bulk of the CZS promoter, and the adsorption of C6 hydrocarbons onto SiO_2. These characteristics make $5wt\%Pt/16wt\%Ce_{0.68}Zr_{0.17}Sn_{0.15}O_{2.0}/SiO_2$ an efficient partial oxidation catalyst for conversion of CyH.

Acknowledgements

The authors deeply appreciate Professor Dr. S. Minakata for providing suggestions. We also thank Dr. T. Sakata and Professor Dr. H. Yasuda for technical assistance with the TEM measurements.

References

[1] Musser MT. Cyclohexanol and cyclohexanone. In *Ullmann's Encyclopedia of Industrial Chemistry, Vol. 11.* Weinheim: Wiley-VCH, 2012: 49–60.

[2] Kokotailo GT, Lawton SL, Olson DH, *et al.* Structure of synthetic zeolite ZSM-5. *Nature* 1978, **272**: 437–438.

[3] Kresge CT, Leonowicz ME, Roth WJ, *et al.* Ordered mesoporous molecular sieves synthesized by a liquid-crystal template mechanism. *Nature* 1992, **359**: 710–712.

[4] Zhao R, Ji D, Lv G, *et al.* A highly efficient oxidation of cyclohexane over Au/ZSM-5 molecular sieve catalyst with oxygen as oxidant. *Chem Commun* 2004, **40**: 904–905.

[5] Lü G, Zhao R, Qian G, *et al.* A highly efficient catalyst Au/MCM-41 for selective oxidation cyclohexane using oxygen. *Catal Lett* 2004, **97**: 115–118.

[6] Wang C, Chen L, Qi Z. One-pot synthesis of gold nanoparticles embedded in silica for cyclohexane oxidation. *Catal Sci Technol* 2013, **3**: 1123–1128.

[7] Liu Y, Tsunoyama H, Akita T, *et al.* Aerobic oxidation of cyclohexane catalyzed by size-controlled Au clusters on hydroxyapatite: Size effect in the sub-2 nm regime. *ACS Catal* 2011, **1**: 2–6.

[8] Maksimchuk NV, Kovalenko KA, Fedin VP, *et al.* Cyclohexane selective oxidation over metal–organic frameworks of MIL-101 family: Superior catalytic activity and selectivity. *Chem Commun* 2012, **48**: 6812–6814.

[9] Yang X, Yu H, Peng F, *et al.* Confined iron nanowires enhance the catalytic activity of carbon nanotubes in the aerobic oxidation of cyclohexane. *ChemSusChem* 2012, **5**: 1213–1217.

[10] Imanaka N, Masui T. Advanced materials for environmental catalysts. *Chem Rec* 2009, **9**: 40–50.

[11] Imanaka N, Masui T, Yasuda K. Environmental catalysts for complete oxidation of volatile organic compounds and methane. *Chem Lett* 2011, **40**: 780–785.

[12] Yasuda K, Masui T, Imanaka N. Complete oxidation of volatile organic compounds at moderate temperatures. In *Hazardous Materials*: *Types, Risks and Control*. Brar SK, Ed. New York: Nova Science Publishers, 2011: 424–431.

[13] Yasuda K, Yoshimura A, Katsuma A, *et al.* Low-temperature complete combustion of volatile organic compounds over novel $Pt/CeO_2–ZrO_2–SnO_2/\gamma\text{-}Al_2O_3$ catalysts. *B Chem Soc Jpn* 2012, **85**: 522–526.

[14] Suzuki H. Recovery of hydrocarbons from natural gas. *Oil Gas Business Environ Res* 1978, **11**: 95–103.

[15] Hereijgers BPC, Weckhuysen BM. Aerobic oxidation of cyclohexane by gold-based catalysts: New mechanistic insight by thorough product analysis. *J Catal* 2010, **270**: 16–25.

[16] Tolman CA, Druliner JD, Nappa MJ, *et al.* Alkane oxidation studies in du Pont's central research department. In *Activation and Functionalization of Alkanes*. Hill CL, Ed. Chichester: Wiley, 1989: 303–360.

[17] Sheldon RA, Kochi JK. *Metal-Catalyzed Oxidations of Organic Compounds*. New York: Academic Press, 1981.

[18] Ramanathan A, Hamdy MS, Parton R, *et al.* Co-TUD-1 catalysed aerobic oxidation of cyclohexane. *Appl Catal A*: *Gen* 2009, **355**: 78–82.

[19] Xu LX, He CH, Zhu MQ, *et al.* A highly active Au/Al_2O_3 catalyst for cyclohexane oxidation using molecular oxygen. *Catal Lett* 2007, **114**: 202–205.

[20] Pohorecki R, Bałdyga J. Moniuk W, *et al.* Kinetic model of cyclohexane oxidation. *Chem Eng Sci* 2001, **56**: 1285–1291.

15

Role of MgF$_2$ addition on high energy ball milled kalsilite: Implementation as dental porcelain with low temperature frit

Pattem Hemanth KUMAR[a,*], Abhinav SRIVASTAVA[a], Vijay KUMAR[a], Nandini JAISWAL[a], Pradeep KUMAR[b], Vinay Kumar SINGH[a]

[a]*Department of Ceramic Engineering, Indian Institute of Technology (BHU), Varanasi, India*
[b]*Department of Chemical Engineering, Indian Institute of Technology (BHU), Varanasi, India*

Abstract: Porcelain fused to metal (PFM) has received great attention over the last few years due to its importance in the dentistry. Kalsilite (K$_2$O·Al$_2$O$_3$·SiO$_2$) is a high thermal expansion porcelain, suitable for bonding to metals. However, kalsilite is a metastable phase which gets converted into crystalline leucite upon heating. In the current work feasibility of developing stable kalsilite phase, dispersion of MgF$_2$ in it as an additive and using mechanochemical synthesis are studied. Micro fine dental material has been formulated by mixing prepared kalsilite with low temperature frit (LTF) in different ratio. The crystalline phases evolved in fired powders are characterized by powder X-ray diffraction (XRD) technique. Kalsilite with different ratio of LTF has been cold pressed and heat treated to examine its coefficient of thermal expansion (CTE), flexural strength, apparent porosity (AP), bulk density (BD) and microstructure. Results indicate that MgF$_2$ addition and high milling duration help in kalsilite stabilization. Temperature also plays an important role in this stabilization, and at 1100 ℃ single phase kalsilite formation is observed. Present outcomes demonstrate that it is easily possible to synthesize a stable single phase kalsilite with desirable properties.

Keywords: kalsilite; porcelain fused to metal (PFM); dental ceramic; mechanochemical synthesis; thermal expansion; X-ray diffraction (XRD)

1 Introduction

Ceramic materials have been widely used in porcelain fused to metal (PFM) and all ceramic restoration systems over the last decade. They typically have high coefficient of thermal expansion (CTE) and high flexural strength. Kalsilite (KAlSiO$_4$) mineral has a network of tetrahedral Si and Al elements with charge balancing alkali metal ions [1]. CTE of kalsilite is $16×10^{-6}$(℃)$^{-1}$ [2]. Kalsilite is a significant constituent

* Corresponding author.
E-mail: phemanth111@gmail.com

in PFM and ceramic restoration systems [3]. It is used as the precursor of leucite [4]. Becerro *et al.* [4] have previously reported that kalsilite being a high thermal expansion ceramic, is suitable for bonding to metals. Kalsilite however crystallizes as a metastable phase when synthesizing leucite [5,6].

Kalsilite has been synthesized previously by various techniques such as hydrothermal method [7], sol–gel method [8,9] and solid-state method [10,11]. Accompanying with other synthesis methods, mechanochemical process is economical and suitable to prepare pure materials with a micro fine particle size [12–14]. This synthesis involves chemical reactions of solids under the action of mechanical forces. Driving force for mechanochemical synthesis is the strain energy stored in the fine powders. The mechanical energy produces structural imperfections in the powder particles during grinding and increases the reactivity of ground materials.

There is lack of research in the stabilization and implementation of this material as PFM. So in order to produce fine stoichiometric kalsilite, mechanochemical synthesis was employed. The aim to improve the stability of kalsilite phase using 2 wt% MgF_2 as an additive has been studied in the present investigation. The ground powders were heat treated at different temperatures and characterized via powder X-ray diffraction (XRD) technique. Optimised kalsilite powders were then mixed with different weight ratio of low temperature frit (LTF) and characterized for CTE, flexural strength, apparent porosity, bulk density and morphology.

2 Experimental procedure

2.1 Materials

Analytical reagent (AR) grade aluminium oxide (Al_2O_3), potassium carbonate (K_2CO_3) and silica (SiO_2) were used as raw materials for kalsilite formulation. Sodium carbonate (Na_2CO_3), potassium nitrate (KNO_3), silica (SiO_2), potassium carbonate (K_2CO_3), borax ($Na_2H_3BO_4 \cdot 10H_2O$), feldspar ($K_2O \cdot Al_2O_3 \cdot 6SiO_2$) and magnesium fluoride (MgF_2) were used for preparation of LTF. All materials were AR grade and procured from Loba Chemie Pvt. Ltd (Mumbai, India)

2.2 Preparation of kalsilite and LTF

Potassium carbonate, alumina and silica were weighed in stoichiometric ratio of kalsilite 1:1:1. Weighed mixtures of materials with and without the incorporation of 2 wt% MgF_2 were prepared in the same manner. These mixtures were then pulverized in a high energy planetary ball mill (supplied by V.B. Ceramics, Chennai, India) for 3 h and 6 h with the rate of 250 rpm. Milling was done in a 250 ml zirconia cylindrical jar. The grinding balls made of zirconia having diameter 10 mm were used as a hard grinding medium. Ball to powder weight ratio was kept 4:1. The milling operation of mixtures was carried out continuously at room temperature for 3 h and 6 h. The milled powders were heat treated in an electric furnace at 900–1100 ℃ at a heating rate of 6 ℃/min and

soaked for 1 h. The furnace was equipped with SiC heating element and a program PID528 manufactured by Selectron Process Controls Pvt Ltd., India. This program has the temperature control accuracy of ±1 ℃. To prepare LTF, starting components were mixed in an agate mortar. The mixture was melt in an alumina crucible at 1350 ℃ for 60 min. The molten frit was quenched in deionized water, air dried and then ball milled to pass a 350 mesh BSS.

2.3 Preparation of samples

The samples prepared for characterization contained different weight percentage of kalsilite. Optimization in choosing the prepared kalsilite was based on its maximum phase formation. Different formulations are presented in Table 1. Rectangular test bars were prepared using a uniaxial hydraulic press under 200 MPa. These bars were heated in a VITA VACUMAT 40T according to a standard dental veneering firing cycle pre-programmed by VITA. It consists of five steps from room temperature to 800 ℃. These five steps of firing cycle are: pre-heating at 500 ℃ for 2 min, heating from 500 ℃ to 800 ℃ for 6 min, 1 min soaking at 800 ℃ followed by cooling to 600 ℃ in 1 min.

2.4 Characterizations

The crystalline phases were identified by powder XRD technique. XRD patterns were observed using a portable XRD machine (Rigaku, Japan) using Cu Kα radiation employing Ni filter operating at 30 mA and 40 kV. Phase identification was carried out by comparing the respective powder XRD patterns with the standard database stated by JCPDS (PDF-2 Database 2003).

The thermal expansion of the material was

Table 1 Batch composition of different weight percentage of kalsilite and LTF samples

Sample coding	Firing temperature (℃)	Kalsilite (wt%)	LTF (wt%)
$K_{1000\text{-}20/80}$	1000	20	80
$K_{1000\text{-}25/75}$	1000	25	75
$K_{1000\text{-}30/70}$	1000	30	70
$K_{M1000\text{-}20/80}$	1000	20	80
$K_{M1000\text{-}25/75}$	1000	25	75
$K_{M1000\text{-}30/70}$	1000	30	70

determined by dilatometer (supplied by VB Ceramic Consultants, India) in the temperature range 20–550 ℃ at 6 ℃/min. The dilatometer was equipped with SiC heating element with a control accuracy of ±1 ℃. It had Nippon PID programmable digital temperature indicator cum controller. The samples for CTE measurement were cut and polished uniformly to the size of 45 mm × 15 mm × 10 mm.

Flexural strength measurements were done according to ASTM C78/C78M using universal testing machine Instron, 3344 (Germany). The specimens were bent in a three-point crossways fit with 20 mm span between the two supports (three-point bending). The load and the corresponding deflections were recorded. The flexural strength was calculated using the following equation:

$$F = \frac{3PL}{2bd^2} \qquad (1)$$

where F is the flexural strength (kg/cm^2); P is the maximum applied load; L is the span length; b is the width of specimen; and d is the depth of specimen.

Apparent porosity (AP) and bulk density (BD) of all the kalsilite and LTF samples were determined according to ASTM C20-00. All specimens were polished using emery papers of grades 1/0, 2/0, 3/0 and 4/0 (Sia, Switzerland) followed by polishing on a velvet cloth using diamond paste of grade 1/4-OS-475 (HIFIN). These polished specimens were chemically etched with 40% hydrofluoric acid for 10 s and then washed with distilled water. Finally they were dried and gold sputtered. Micrographs were recorded with the help of a scanning electron microscope (INSPECT 50 FEI).

3　Results and discussion

3.1　Phase analysis of the ground kalsilite at different temperatures

Figure 1 shows the XRD patterns of the sample milled for 3 h at different firing temperatures. Diffraction peaks are well matched to JCPDS Card No. 87-1707. It can be seen that XRD pattern of precursor fired at 900 ℃ contains hexagonal kalsilite as a major phase alike to JCPDS Card No. 87-1707. As temperature increases from 1000 ℃ to 1100 ℃, kalsilite is present as a major phase in addition to small amount of leucite phase. Unit-cell parameters are $a = b = 5.157$ Å and $c =$

8.706 Å and more similar to those given by Becerro et al. [4] ($a = 5.166$ Å and $c = 8.7123$ Å). Presence of broad peaks in XRD patterns shows the small crystallite size of the kalsilite.

Figure 2 shows the XRD patterns of the sample containing 2 wt% of MgF$_2$ milled for 3 h. It can be noted that XRD patterns of precursor fired at 900 ℃ and 1000 ℃ contains kalsilite as a major phase with hexagonal crystal structure. Furthermore when heat treatment is increased to 1100 ℃, kalsilite occurs as a major phase with some peaks of leucite crystalline phase. It can be noted that formation of leucite crystalline phase occurs at high temperature. Formation of leucite phase is less in the sample containing MgF$_2$ in comparison to that of the sample without MgF$_2$. It is probable that the presence of MgF$_2$ additive suppresses the crystallization of leucite.

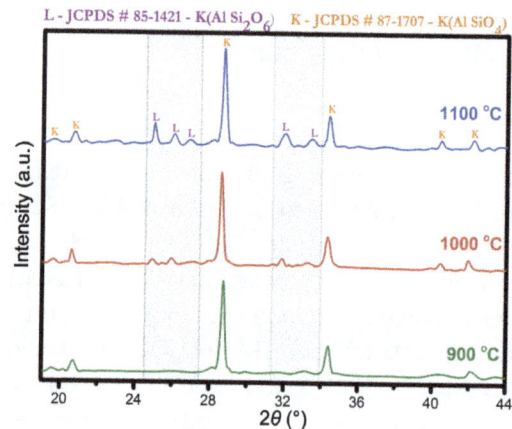

Fig. 1　XRD patterns of the sample milled for 3 h at different firing temperatures.

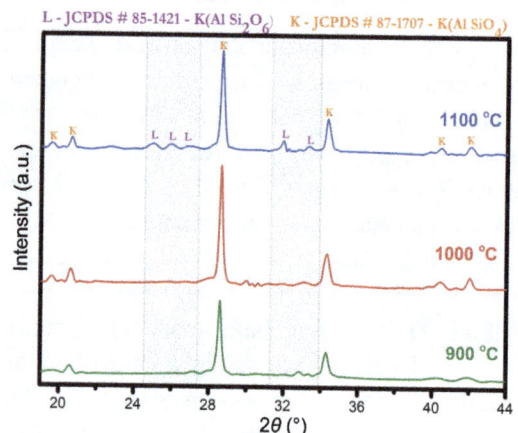

Fig. 2　XRD patterns of the sample containing 2 wt% of MgF$_2$ milled for 3 h at different firing temperatures.

Figure 3 shows the XRD patterns of the sample milled for 6 h. X-ray lines of the sample fired at 900 ℃ contain kalsilite as a major phase with small peaks of leucite phase. At temperatures 1000 ℃ and 1100 ℃, leucite is present as a main phase with small peaks of kalsilite crystalline phase. It might be due to high surface energy of ground sample. The smaller the particle size is, the larger the surface area and hence the higher the surface energy will be. Therefore, it can be concluded that the heat for formation of leucite is higher than that of kalsilite. It has also been previously reported by Zhang et al. [5] that activation energy for formation of leucite is higher than that of kalsilite. Kalsilite, therefore, crystallizes first as a metastable phase and then reacts with SiO_2 tetrahedra, which leads to formation of leucite phase. In the present work, the sample milled for 6 h has very fine particles in comparison to sample milled for 3 h. In former case therefore, most of the kalsilite coverts into leucite at higher temperatures.

Figure 4 shows the XRD patterns of the sample containing 2 wt% of MgF_2 milled for 6 h. It shows hexagonal kalsilite as a major phase with small peaks of leucite at 900 ℃, 1000 ℃ and 1100 ℃. Diffraction peaks are well matched to JCPDS Card Nos. 87-1707 and 85-1421. It can be concluded that MgF_2 promotes the formation of kalsilite. Zhang et al. [6] studied the effect of addition of CaF_2 on the crystallization of leucite. They found that CaF_2 decreases the crystallization temperature of leucite. Ionic radius of F^- ion (1.33 Å) is very close to radius of O^{2-} ion (1.4 Å); therefore, it replaces the O^{2-} ion and forms the Si–F or Al–F bonds [6]. Due to strong electrostatic attraction between Ca^{2+} and O^{2-} ions, network becomes loose. It

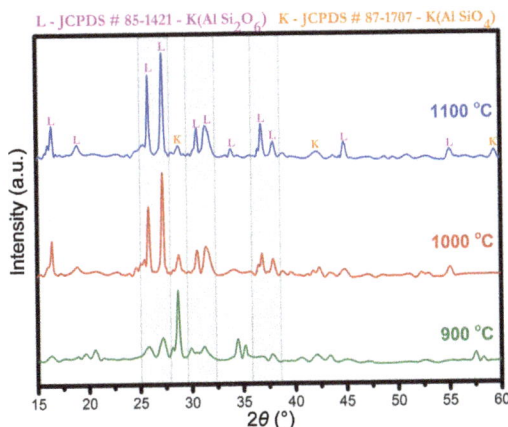

Fig. 4   XRD patterns of the sample containing 2 wt% of MgF_2 milled for 6 h at different firing temperatures.

helps the conversion of metastable kalsilite phase into stable leucite phase at low temperature.

In the present work, whereas, MgF_2 inhibits the formation of leucite phase and stabilizes the kalsilite metastable phase. It may be due to small amount of free silica present in the matrix. When F^- ion replaces O^{2-} ion in K^+, free Mg^{2+} reacts with free silica forming low temperature eutectic phase, enstatite ($MgO·SiO_2$). It decreases any silica present around grain boundary around the kalsilite and hence supresses the formation of leucite.

3.2   Coefficient of thermal expansion

Thermal compatibility of PFM from room temperature to the glass transition temperature can be assessed by measuring the average expansion coefficient of metal and porcelain in the range of 20–550 ℃. Its thermal expansion is essential to assure good bonding of the ceramic to metal. Figure 5 shows the CTE curves of kalsilite with different weight percentage of LTF, $K_{1000-20/80}$, $K_{1000-25/75}$ and $K_{1000-30/60}$. It is found that the addition of kalsilite to LTF increases the CTE of the whole matrix. It may be due to formation of leucite crystalline phase which has high CTE. Glass transition temperatures (T_g) of $K_{1000-20/80}$, $K_{1000-25/75}$ and $K_{1000-30/70}$, are determined to be 415 ℃, 425 ℃ and 440 ℃, respectively.

Figure 6 shows the CTE curves of kalsilite–2 wt% MgF_2 with different weight percentage of LTF, $K_{M1000-20/80}$, $K_{M1000-25/75}$ and $K_{M1000-30/70}$. It is found that the addition of MgF_2 to kalsilite decreases the CTE of final mixture. These CTE values are about 40% less than that of samples without MgF_2. This may be due to

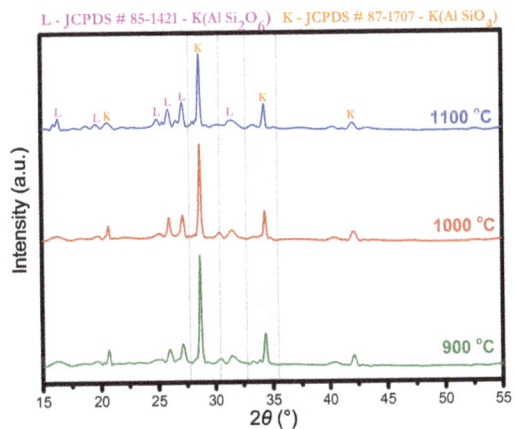

Fig. 3   XRD patterns of the sample milled for 6 h at different firing temperatures.

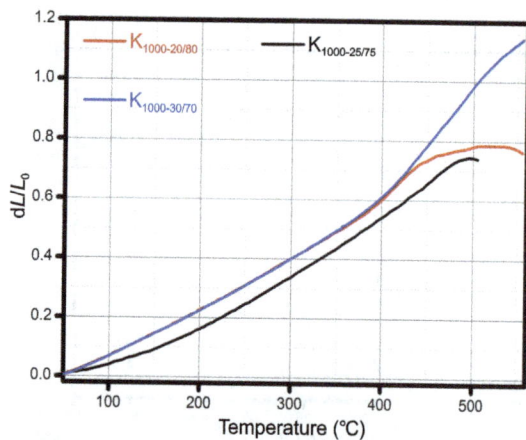

Fig. 5   CTE curves of the samples with different weight percentage of kalsilite and LTF.

the formation of low CTE major kalsilite phase. These results are in conformity with the results of XRD where the sample with MgF_2 has kalsilite as a major phase, i.e., MgF_2 suppresses the formation of leucite. The glass transition temperatures (T_g) of $K_{M1000-20/80}$, $K_{M1000-25/75}$ and $K_{M1000-30/70}$ are determined to be 415 ℃, 425 ℃ and 430 ℃, respectively. These prepared materials are suitable for PFM, as their CTE values of 14.0×10^{-6}–14.8×10^{-6} (℃)$^{-1}$ are close to standard CTE of nickel–chrome alloy (13.9×10^{-6} (℃)$^{-1}$).

3.3   Flexural strength

Figure 7 shows the flexural strength of K_{1000} and K_{M1000} with different weight percentage of kalsilite in LTF. Flexural strength increases with increasing the micro fine kalsilite in matrix. Homogenous dispersion of micro fine kalsilite grains within the glassy matrix leads to enhance the mechanical strength. K_{M1000} has higher flexural strength than K_{1000}. Subsequently the synthesised samples show better sinterability, low

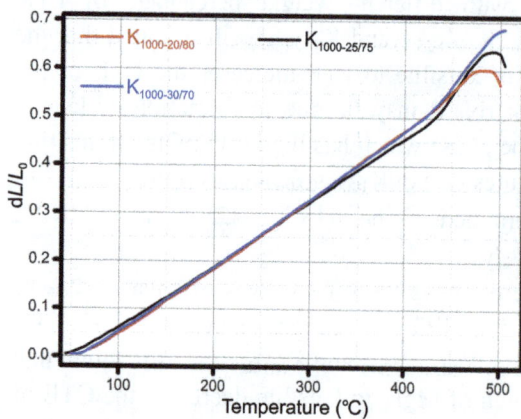

Fig. 6   CTE curves of the samples with different weight percentage of kalsilite–2 wt% MgF_2 and LTF.

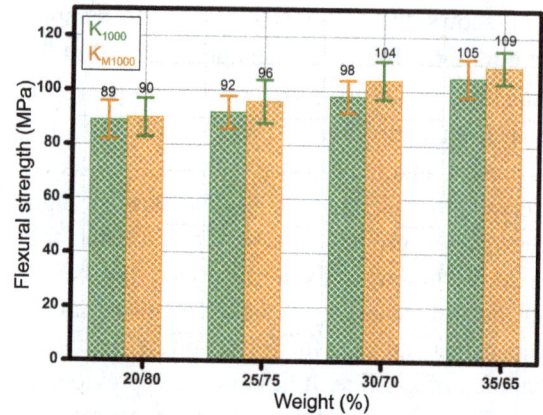

Fig. 7   Flexural strength of K_{1000} and K_{M1000} with different weight percentage of kalsilite and LTF.

porosity and high flexural strength. The flexural strength is more or less similar for all samples. However, the mixed samples with 20 wt%, 25 wt%, 30 wt% and 35 wt% kalsilite expressively result in higher flexural strength. This is due to presence of micro fine particles, i.e., larger surface area and hence less porosity.

3.4   Bulk density and apparent porosity

Figure 8 shows the variation of BD and AP with different weight percentage of kalsilite and LTF. It can be noted from Fig. 8 that BD increases with increasing the content of kalsilite followed by a continuous decrease of AP. Micro fine kalsilite particles dispersed throughout the glassy matrix help in improvement of packing density of the samples.

3.5   Microstructure evaluation through SEM

The microstructures of kalsilite with (K_{M1000}) and without (K_{1000}) MgF_2 with different weight percentage

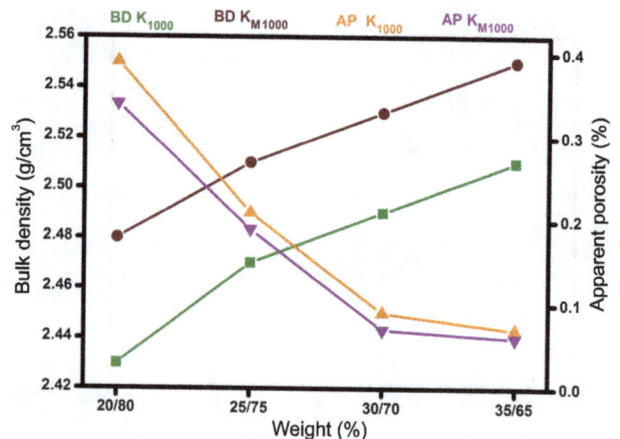

Fig. 8   BD and AP of the samples with different weight percentage of kalsilite and LTF.

of LTF are shown in Fig. 9. There is no visible micro-crack appearance due to phase transformation. Micrographs show a very dense structure which is also in conformity with the AP and BD plots. This also results in the high flexural strength.

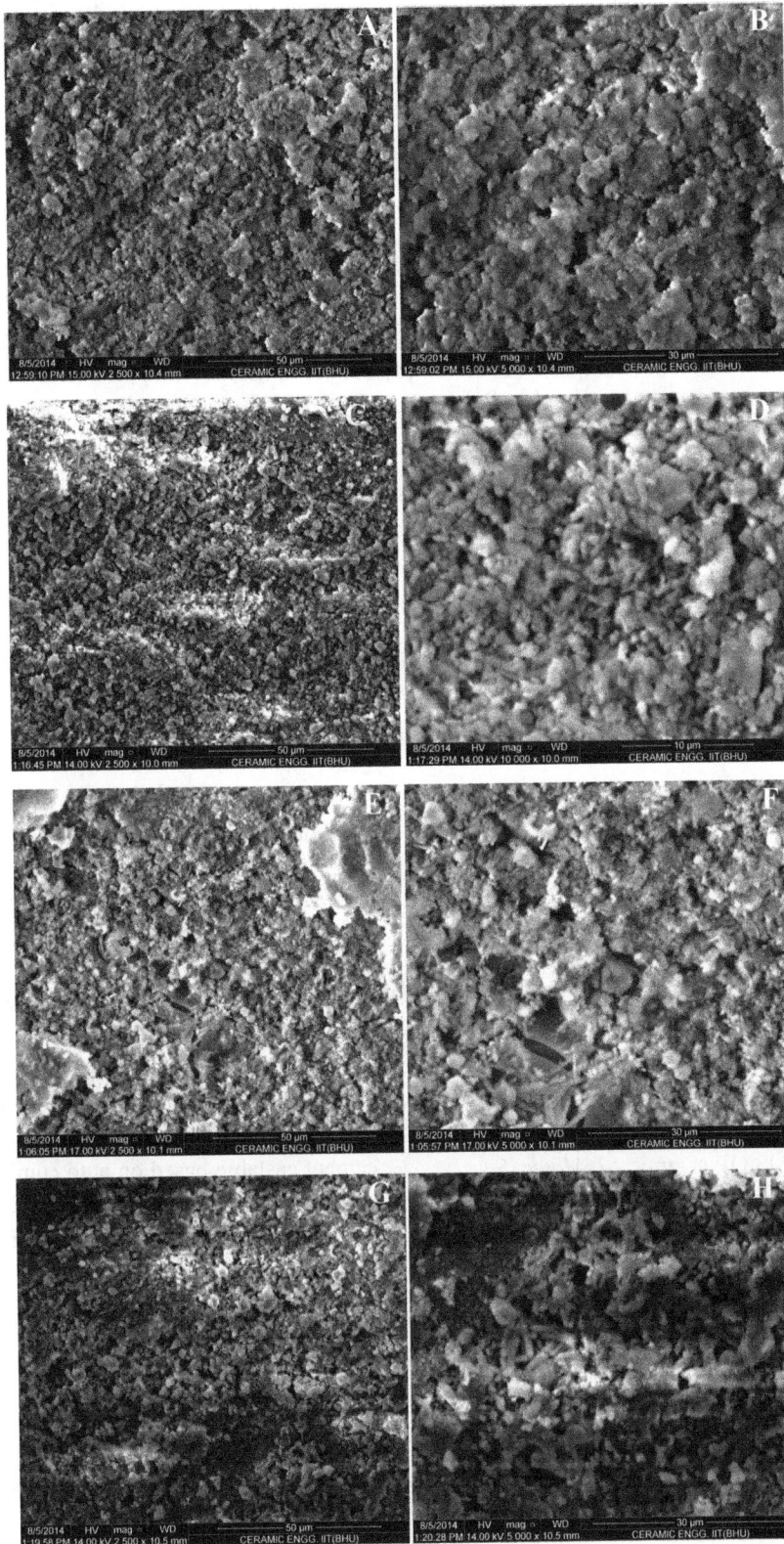

Fig. 9   Scanning electron micrographs of kalsilite with (K_{M1000}) and without (K_{1000}) MgF$_2$ with different weight percentage of LTF in different magnification: (a, b) $K_{M1000-25/75}$; (c, d) $K_{M1000-30/70}$; (e, f) $K_{1000-25/75}$; (g, h) $K_{1000-30/70}$.

4 Conclusions

Micro fine kalsilite has been successfully synthesised by high energy ball milling. Addition of MgF_2 supresses the formation of leucite phase and stabilizes the kalsilite phase. Kalsilite–2 wt% MgF_2 with different weight percentage of LTF, shows the CTE value of 14.0×10^{-6} (℃)$^{-1}$ to 14.8×10^{-6} (℃)$^{-1}$. This value is close to standard CTE of nickel–chrome alloy (13.9×10^{-6} (℃)$^{-1}$). Samples with 20 wt%, 25 wt% and 30 wt% kalsilite have higher flexural strength due to presence of micro fine kalsilite particles in the matrix. Micrographs show very dense structure with no visible cracks. This makes present material suitable for application in porcelain fused to metal.

Acknowledgements

The authors gratefully acknowledge the financial support of DST (TDT division, reference No. DST/SSTP/UP/197 (G) 2012), Ministry of Science & Technology, New Delhi, India.

References

[1] Andou Y, Kawahara A. The renfinement of the structure of synthetic kalsilite. *Mineral J* 1984, **12**: 153–161.

[2] Ota T, Takebayashi T, Takahashi M, *et al*. High thermal expansion KAlSiO$_4$ ceramic. *J Mater Sci* 1996, **31**: 1431–1433.

[3] Becerro AI, Escudero A, Mantovani M. The hydrothermal conversion of kaolinite to kalsilite: Influence of time, temperature, and pH. *Am Mineral* 2009, **94**: 1672–1678.

[4] Becerro AI, Mantovani M, Escudero A. Hydrothermal synthesis of kalsilite: A simple and economical method. *J Am Ceram Soc* 2009, **92**: 2204–2206.

[5] Zhang Y, Lv M, Chen D, *et al*. Leucite crystallization kinetics with kalsilite as a transition phase. *Mater Lett* 2007, **61**: 2978–2981.

[6] Zhang Y, Wu J, Rao P, *et al*. Low temperature synthesis of high purity leucite. *Mater Lett* 2006, **60**: 2819–2823.

[7] Kopp OC, Harris LA, Clark GW. The hydrothermal conversion of muscovite to kalsilite and an iron-rich mica. *Am Mineral* 1961, **46**: 719–727.

[8] Bogdanoviciene I, Jankeviciute A, Pinkas J, *et al*. Study of alumposilicate porcelain: Sol–gel preparation, characterization and erosion evaluated by gravimetric method. *Mater Res Bull* 2008, **43**: 2998–3007.

[9] Bogdanoviciene I, Jankeviciute A, Pinkas J, *et al*. Sol–gel synthesis and characterization of kalsilite-type alumosilicates. *Mater Sci* 2007, **13**: 214–218.

[10] Dimitrijevic R, Dondur V. Synthesis and characterization of KAlSiO$_4$ polymorphs on the SiO$_2$–KAlO$_2$ join. *J Solid State Chem* 1995, **115**: 214–224.

[11] Heller-Kallai L, Lapides I. Thermal reactions of kaolinite with potassium carbonate. *J Therm Anal Calorim* 2003, **71**: 689–698.

[12] Baláž P, Achimovičová M, Baláž M, *et al*. Hallmarks of mechanochemistry: From nanoparticles to technology. *Chem Soc Rev* 2013, **42**: 7571–7637.

[13] Kumar PH, Srivastava A, Kumar V, *et al*. Effect of high-energy ball milling and silica fume addition in BaCO$_3$–Al$_2$O$_3$. Part I: Formation of cementing phases. *J Am Ceram Soc* 2014, DOI: 10.1111/jace.13173.

[14] Srivastava A, Singh VK, Kumar V, *et al*. Low cement castable based on auto combustion processed high alumina cement and mechanochemically synthesized cordierite: Formulation and properties. *Ceram Int* 2014, **40**: 14061–14072.

16

Load-dependent indentation behavior of β-SiAlON and α-silicon carbide

Prasenjit BARICK[*], Dulal Chandra JANA, Bhaskar Prasad SAHA

Centre for Non-Oxide Ceramics, International Advanced Research Centre for Powder Metallurgy and New Materials, PO: Balapur, RCI Road, Hyderabad 500005, Andhra Pradesh, India

Abstract: A comparative study was carried out on the load-dependent indentation behavior with respect to hardness and induced cracks of β-SiAlON and α-silicon carbide ceramics. It is observed that silicon carbide (SiC) exhibits lower transition load, early cracking and severely crushed indentation sites, whereas β-SiAlON shows higher transition load and damage-free indentation zone even at the maximum applied load (294.19 N). Crack density is higher for α-SiC with comparison to β-SiAlON at each load. SiC exhibits both main and secondary radial types of cracking from low indentation load (0.98 N). Cracks are often associated with branching at higher load (> 9.80 N) for α-SiC. β-SiAlON exhibits cracks which are mainly radial types initiated at 4.90 N load. These opposing behaviors of β-SiAlON and α-SiC are attributed to their difference in hardness, toughness, and brittleness index. Higher brittleness of α-SiC results in early and severe cracking around its indentations. β-SiAlON shows less cracking due to its lower brittleness and higher toughness. The increased size of indentation-induced cracks of α-SiC is higher than that of β-SiAlON due to the rapid crack propagation in α-SiC with transgranular fracture behavior.

Keywords: SiAlON; silicon carbide (SiC); indentation; hardness; crack

1 Introduction

Hardness is the response of a material under specific set of conditions. Hardness of a material depends upon several parameters, including grain size, indenter load and loading rate, indenter geometry, surrounding environment, and accuracy in measurements [1–3]. Generally, hardness decreases with the increase of indenter load and indentation size, known as indentation size effect (ISE) [1,3–8]. This load-dependent hardness phenomenon occurs possibly due to the differential elastic recovery of indentations, cracking and stress relaxation, work hardening, dislocation pinning at the indentation area, and frictional effect between indenter facets and test specimen [1,3].

In general, harder and less tough ceramics, such as boron carbide and silicon carbide, are more brittle than less hard materials with higher fracture toughness, such as zirconia toughened alumina (ZTA) and β-SiAlON.

Previously, extensive research works were carried out on metallic as well as ceramic materials to

----
* Corresponding author.
E-mail: prasenjit_bhu@rediffmail.com

establish the know-how on load-dependent hardness behavior. The science of ISE was explained by several models, such as Meyer's law, Hays–Kendall approach, elastic recovery model, and proportional specimen resistance (PSR) model [7,9]. Wilantewicz et al. [3] have conducted ISE experiment on AlON, AD995 CAP3 Al$_2$O$_3$, pressureless-sintered SiC (Hexoloy SA), SiC–N, and SiC–B, using Knoop indenter within the load range of 0.49 N to 137.3 N; it has been shown that for each material, hardness drops by certain percentage (26%, 38%, 36%, 32%, and 31% respectively), and hardness continues to decrease with the increase of load beyond 19.6 N. It has also been shown that crack-dominated indentation area exhibits less hardness than that of less cracked indentation zones for silicon carbide and glasses [3,6]. Quinn and Quinn [10] have shown that beyond transition load, the slope of load–hardness plot becomes asymptotic, and severe crack formation starts beyond this particular load. In another study, the ISE of AlON and MgAl$_2$O$_4$ spinel was examined concluding that AlON shows rapid decrease in Vickers hardness than that of MgAl$_2$O$_4$ spinel [11]. Spinel exhibits wider load range (1–3.8 N) before transition load reaches, where deformation rather than fracture is dominant [11].

Silicon carbide (SiC) is a brittle material with hardness of 22–35 GPa and fracture toughness of 2.96–5.8 MPa·m$^{1/2}$ [12,13]; SiC finds applications in the areas of protective system, space mirror, and abrasives [14–16]. β-SiAlON is relatively less brittle than α-SiC, with hardness of 14–17 GPa and fracture toughness of 6.5–7.88 MPa·m$^{1/2}$ [17,18]; β-SiAlON is one of the promising materials for cutting tools, turbine vanes and blades, and turbocharger rotors [19–21]. Hardness has great importance in their applications for both the ceramics. Generally, hardness determines the damage resistance and stiffness of a material; fracture toughness governs the crack propagation behavior of a material under load. Hence, a systematic study was carried out to gain insights about the load-dependent indentation behavior, such as load vs. hardness and indentation-induced cracking of these two ceramics.

Since β-SiAlON and α-SiC maintain wide spacing at hardness and toughness scales, it is convincing to consider that the current experimental results can be considered as the reference for other structural ceramics lying within the aforesaid hardness and toughness ranges.

Twelve different loads in the range of 0.49–294.19 N with Vickers indenter were used to carry out the systematic study. The minimum load 0.49 N was chosen, because below this load, indentation size was difficult to measure using inbuilt optical microscope of hardness tester as indentations were smaller in size. The maximum applied indentation load was 294.19 N. This wide load range was chosen because of comparison purpose with respect to cracking behavior between the two ceramics. Beyond 29.4 N load, it was not possible to measure the diagonal size due to severe damage at indentation sites for α-SiC. Hence, load–hardness behavior comparison was made within the load range of 0.49–29.4 N. However, indentations were generated using load upto 294.19 N to show the difference between their indentation-induced cracking characteristics.

2 Experimental procedure

2.1 Materials

The raw premix powders corresponding to β-SiAlON composition were supplied by International Syalons (Newcastle) Limited (United Kingdom) used to produce β-SiAlON ceramics. RTP-SiC powders ($D_{50} = 0.8$ μm) were supplied by Saint-Gobain Ceramics (Norway) used to fabricate monolithic SiC. β-SiAlON was densified at 1750 ℃ through reaction sintering (liquid phase) of the premix powders, consisting of α-Si$_3$N$_4$, AlN, Al$_2$O$_3$ and Y$_2$O$_3$. Dense α-SiC ceramic was made by solid-state sintering of SiC powders at 2150 ℃. Physical and mechanical characteristics of the sintered products are shown in Table 1. Figures 1 and 2 show the scanning electron micrographs (SEM) of β-SiAlON and α-SiC, respectively.

Table 1   Physical and mechanical properties of β-SiAlON and α-SiC ceramics

	Property	β-SiAlON	α-SiC
Physical	Bulk density (g/cm^3)	3.27	3.16
	Relative density (%)	98–99	98–99
	Apparent porosoty (%)	0.16	—
	Water absorption (%)	0.05	—
Mechanical	$HV_{0.3}$ (kg/mm^2)	1646.00±150.14	2729.00±133.59
	Fracture toughness (MPa·m$^{1/2}$)	8.22±0.66	3.20±0.14
	3 point bend strength (MPa)	700±160	800±25

Fig. 1   SEM image for β-SiAlON.

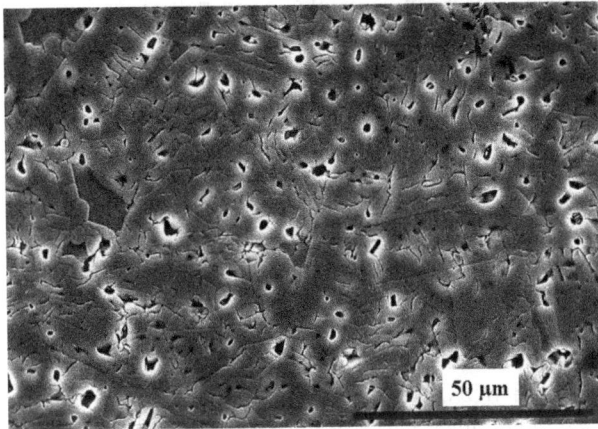

Fig. 2   SEM image for α-SiC.

respectively. β-SiAlON exhibits elongated grains; α-SiC shows equiaxed grains.

2. 2   Test specimen preparation

The samples were sectioned from 50 mm × 50 mm × 6 mm sintered blocks using high-speed precision cutting machine (ISOMET 4000, Buehler Ltd., USA). Then samples were hot mounted on a bakelite platform at 100 kg/cm² pressure and 120 ℃ temperature using hydraulically operated mounting press (BAINMOUNT, Chennai Metco, India). The sectioned specimens were polished with diamond slurry using variable-speed grinder–polisher (EcoMet 4000, Buehler Ltd., USA). First, the sample was made into maximum flatness using 30 μm diamond gritted disk. Subsequently, polishing was carried out using diamond slurry with the grit-size sequences, e.g., 15 μm, 9 μm, 6 μm, 3 μm and 1 μm, respectively, in descending order.

2. 3   Hardness testing

Microhardness tester equipped with Vickers indenter was used for making indentations. This hardness tester (UHL VMHT, Wateruhl Technische Mikroscopie GmbH & Co. KG, Germany) was used to measure microhardness within the load range of 0.49 N to 19.6 N using indenter speed of 50 μm/s with dwell time of 15 s. Further, macrohardness tester (Leco LV 700AT, USA) was used to generate indentations within the load range of 29.41 N to 294.19 N with indentation speed of 50 μm/s and dwell time of 15 s. At each load, 10 best indentations were chosen to generate average hardness data. Linear regression method was applied for best fitting of the hardness data in load–hardness plot. The transition load was determined from their intersecting point. Diagonal lengths of the indentations were measured using inbuilt optical microscope system of hardness tester. Great care was taken during diagonal length measurement in order to minimize the error to a least possible value. The following equation was used to calculate Vickers hardness according to ASTM C 1327:

$$HV = 1.8544(P/d^2)$$

where HV is Vickers hardness (kg/mm²); P is the indentation load (kgf); d is the average length of the two diagonals of the indentation (mm). Indentation-induced crack propagation behavior and indentation imprints were examined using image analysis software (ANALYSIS 5, Olympus, USA).

3   Results

3. 1   Load vs. hardness, indentation size

Figures 3 and 4 show the variation of hardness as a function of indentation load for β-SiAlON and α-SiC, respectively. For β-SiAlON, hardness reduced by 15.47% within the load range of 0.49 N to 29.41 N; α-SiC exhibits reduction in hardness by 31.57% within the identical load range. Hardness data are best fitted by linear regression method, which provides transition load 6.60 N with corresponding hardness of 1511.89 kg/mm² for β-SiAlON; α-SiC exhibits transition load 5.67 N with corresponding hardness of 2499.09 kg/mm². SiC exhibits brittleness value (B) of 1110.66 μm⁻¹, much higher than that of β-SiAlON (64.44 μm⁻¹). Brittleness parameter was determined using the formula

$$B = H_c E / K_{IC}^2$$

where E is the elastic modulus (GPa); H_c is the critical hardness (GPa); K_{IC} is the fracture toughness (MPa·m$^{1/2}$); B is the ratio of work of deformation to the fracture energy (μm^{-1}) [10]. β-SiAlON exhibits wider load range (i.e., 0.49–6.60 N) than that of α-SiC before transition load appears.

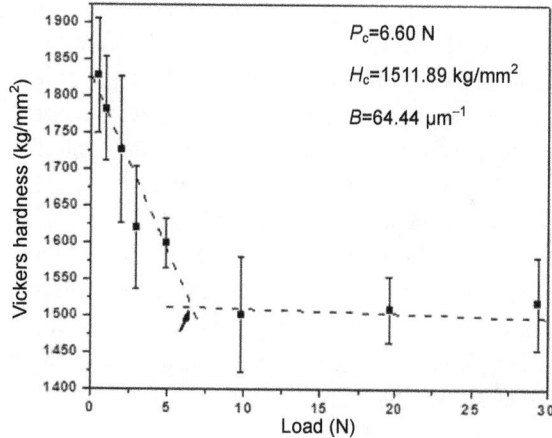

Fig. 3   Load vs. hardness for β-SiAlON.

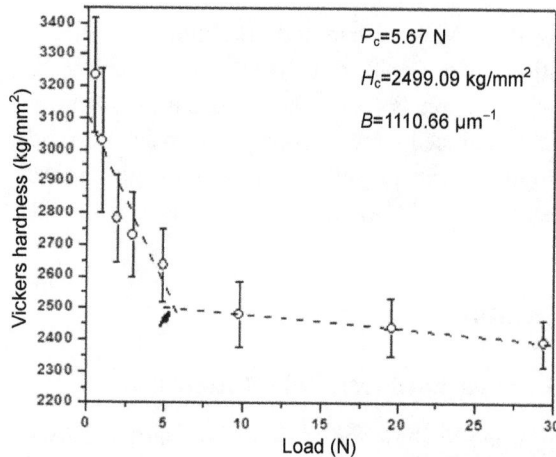

Fig. 4   Load vs. hardness for α-SiC.

Figure 5 shows the variation of indentation size as a function of indentation load for β-SiAlON and α-SiC. Within the said load range, the indentation size increases by 740% and 835% for β-SiAlON and α-SiC, respectively. At each load, the indentation size of β-SiAlON is relatively higher than that of α-SiC as shown in the figure.

3. 2   Indentation-induced cracking

β-SiAlON starts showing crack formation after 4.90 N load, whereas α-SiC exhibits crack initiation at 0.49 N

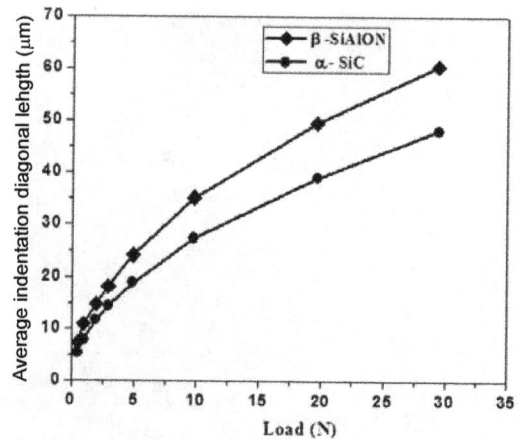

Fig. 5   Load vs. diagonal length of indentations for β-SiAlON and α-SiC.

load. Cracks are mainly radial type in nature, and no crushed zone is observed even at 294.19 N load for β-SiAlON, as shown in Figs. 6(a)–6(l). Both main and secondary radial types of cracks are found for α-SiC as shown in Figs. 6(m)–6(x). The details of indentation-induced cracks, such as average number of cracks per indentation, minimum and maximum crack sizes at each load, and types of indentation induced cracks, are provided in Table 2. It is observed from Table 2 that the average number of indentation induced cracks per indentation, and crack size are higher for α-SiC than those of β-SiAlON. As the indentation load increases, the crack length also increases significantly.

4   Discussions

In the current study, both materials are demonstrating ISE, i.e., hardness decreases with the increase of load. Among several existing models, elastic–plastic deformation (EPD) is chosen to analyse the ISE phenomena of α-SiC and β-SiAlON, because EPD is a proven technique to explain the ISE of hard and strong materials [22,23]. According to this model, the important origin of ISE in hard and strong materials is that the deformation under the indenter occurs in discrete bands rather than being continuous. Upon removal of load, the recovery of elastic increment of deformation, which precedes each new band of plastic deformation, results in the indentations appearing smaller than expected, particularly at small indentation sizes [22,23]. Thus, the hardness calculated from these residual indentation imprints increases with decreasing contact size, as well as with decreasing load. In the extreme case before the first plastic band is nucleated,

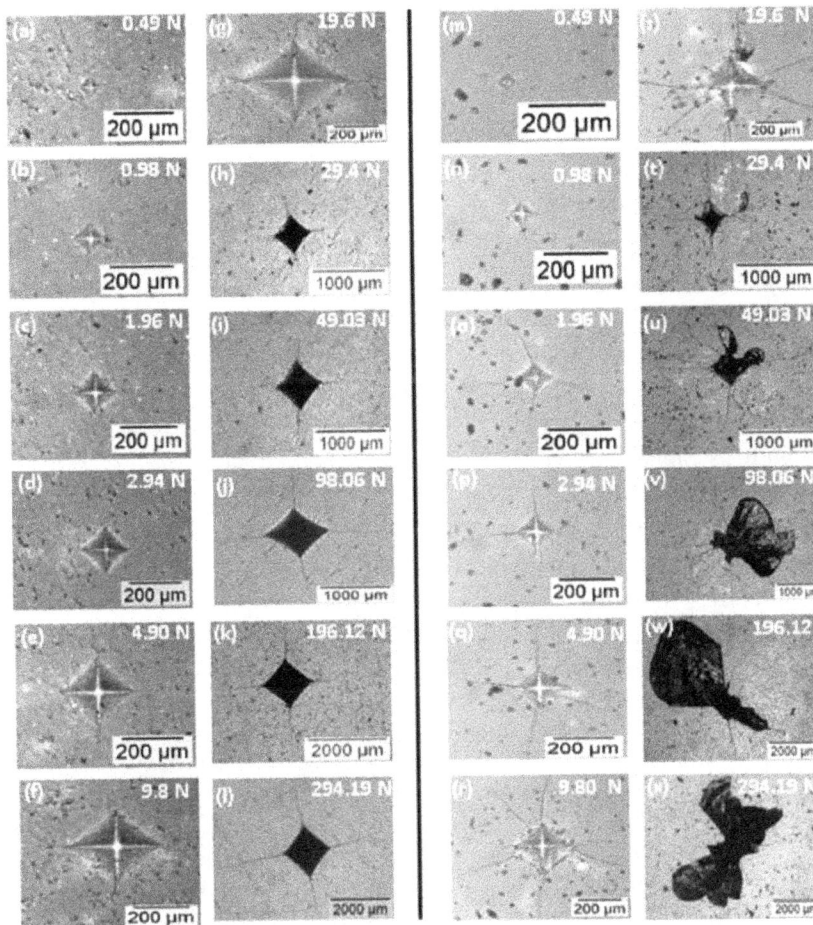

Fig. 6   (a)–(l) Indentation imprints of β-SiAlON at different loads; (m)–(x) indentation imprints of α-SiC at different loads.

Table 2   Number of indention-induced cracks, sizes and types

Sl. No.	Load (N)	Average number of cracks per indentation		Minimum crack size (μm)		Maximum crack size (μm)		Type of crack	
		β-SiAlON	α-SiC	β-SiAlON	α-SiC	β-SiAlON	α-SiC	β-SiAlON	α-SiC
1	0.49	X1	2.00	X1	11.94	X1	29.06	X1	Secondary radial
2	0.98	X1	4.00	X1	16.89	X1	55.15	X1	Main radial and secondary radial
3	1.96	X1	4.75	X1	39.00	X1	122.02	X1	Main radial and secondary radial
4	2.94	X1	6.66	X1	36.38	X1	150.06	X1	Main radial with branching and secondary radial
5	4.90	2.00	7.00	4.50	45.95	35.17	218.43	Main radial	Main radial and secondary radial
6	9.80	2.66	9.00	10.13	40.29	65.87	361.88	Main radial	Main radial with branching, secondary radial and crushed zone
7	19.61	4.00	10.00	57.43	244.93	162.16	481.28	Main radial	Crack branching associated with main and secondry radial types; crushed zone is observed
8	29.41	4.00	6.8+partial crushed zone	92.90	271.75	269.37	538.75	Main radial	Crack branching associated with main and secondry radial types; crushed zone is observed
9	49.03	4.00	X2	231.18	X2	423.64	X2	Main radial	Main and secondary radial, associated with severe crushed zone
10	98.06	4.00	X2	334.92	X2	794.33	X2	Main radial	Cracks associated with severe crushed zone
11	196.12	4.25	X2	481.71	X2	1234.41	X2	Main radial and secondary radial	Cracks associated with severe crushed zone
12	294.19	5.00	X2	1018.74	X2	2327.28	X2	Main radial and secondary radial	Severe crushed zone associated with cracks

X1: not exist; X2: crack lengths were not possible to measure because of severe crushed zone.

surface flexure will remain elastic, and fully recoverable on unloading, leaving no residual impression with which to calculate a conventional hardness value [22,23]. At higher load (generally beyond transition), indentations are associated with both elastic deformation and cracking. Fracture and cracking dominate as load increases. Thus, the recovery effect of elastic deformation becomes negligible, which eventually makes P/d^2 ratios closer at each load. Thus, load–hardness plateau becomes asymptotic particularly at higher load range. The ISE effect in α-SiC is more pronounced than that for β-SiAlON. This behavior of α-SiC and β-SiAlON is further supported by their brittleness parameter, which is much higher for α-SiC (1110.66 μm^{-1}) compared to that of β-SiAlON (64.44 μm^{-1}). The higher brittleness index of α-SiC compared to β-SiAlON, is attributed to its higher hardness at transition load and much lower fracture toughness. The higher percentages of hardness reduction can be interpreted by severe fracture, cracking of SiC combined with its higher percentage indentation size increment.

Due to less hardness and less brittleness of β-SiAlON, it deforms more than α-SiC under the application of load. Therefore, the measured indentation sizes of β-SiAlON remain higher than that of α-SiC at each load within the experimental load range. The difference between indentation sizes of β-SiAlON and α-SiC, gradually increases with load increment, as deformation is more pronounced than fracture for β-SiAlON, whereas fracture dominates over deformation for α-SiC. Since deformation leads higher indentation size than fracture, β-SiAlON exhibits relatively higher indentation size than α-SiC. Because of the similar reason, recovery effect of the incremental part of elastic deformation may persist even at higher load for β-SiAlON, but this effect is expected to be trivial for α-SiC at higher load range as is evident from severe fracture associated with its indentations. Thus, α-SiC demonstrates higher percentages of indentation size increment compared to that of β-SiAlON.

From the indentation imprints (Fig. 6), it is evident that fracture effect at each indentation load is dominant in α-SiC with comparison to β-SiAlON. Because of higher fracture toughness and less hardness, β-SiAlON withstands higher indentation load without damage around indentations due to more deformation effect; SiC is very prone to crack initiation even at low load (0.49 N) due to low fracture toughness and higher brittleness, and exhibits fragmented and irregular-shaped indentations at higher loads. The significant difference in the brittleness parameters of α-SiC and β-SiAlON agrees with the observed damage such as severe cracking and fragmented zone around the indentations of α-SiC. Further, the propensity of crack formation at lower load and development of frag-mented indentation zones for α-SiC, are related to its

extremely high brittleness; this is also the cause for the formation of different types of cracks namely main and secondary radial types in association with crack branch-ing. On the other hand, due to less brittleness of β-SiAlON, mainly radial cracks appear upon application of indentation load. A higher number of cracks is associated with each indentation for α-SiC than that of β-SiAlON, which is also attributed to higher hardness, higher brittleness and less fracture toughness of α-SiC. Due to high brittleness and transgranular fracture behavior (Fig. 7), crack propagation is much more rapid for α-SiC with comparison to β-SiAlON, which exhibits wavy fracture surface (Fig. 8). This type of fracture surface, which is observed in β-SiAlON, is the indication of relatively more deformation, which prevents rapid crack propagation and reduce indentation-induced crack length. This is the cause for larger-size indentation-induced cracks for α-SiC with comparison to β-SiAlON.

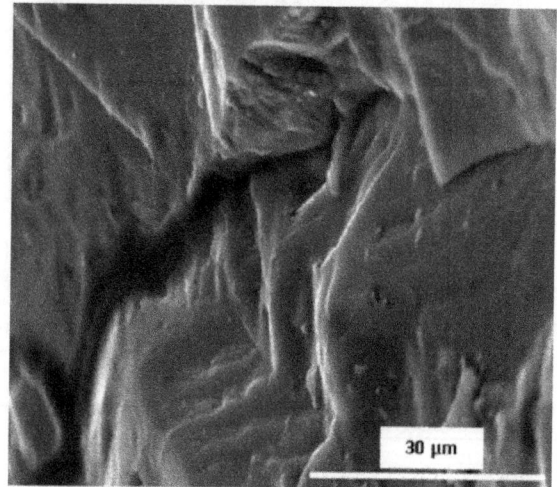

Fig. 7　Fractograph of α-SiC.

Fig. 8　Fractograph of β-SiAlON.

5 Conclusions

Based on the experimental results, it can be summarized that:

(i) ISE of both structural ceramics is explained by EPD model. Here, ISE is more pronounced for α-SiC than β-SiAlON, as dropping of hardness as a function of load is higher for α-SiC (31.57%) compared to β-SiAlON (15.47%). This effect is due to severe fracture and cracking around indentations combined with higher percentage indentation size increment for α-SiC.

(ii) The combined effect of higher hardness, lower fracture toughness and higher brittleness index is responsible for low transition load, severe cracking and fragmented zones around indentations for α-SiC, whereas effects of lower hardness, higher fracture toughness, and less brittleness make less cracking and crushed free indentations for β-SiAlON.

(iii) Main and secondary radial crack branchings are observed for α-SiC, whereas β-SiAlON exhibits mainly radial cracks. Crack branching is not evident for β-SiAlON. This effect is the cause of extreme brittleness of α-SiC with comparison to β-SiAlON.

(iv) Crack density around indentations is higher for α-SiC than that of β-SiAlON, which is also due to higher brittleness of α-SiC and its less fracture toughness.

(v) Crack sizes at each load are larger for α-SiC than that of β-SiAlON. This is due to the rapid crack propagation in α-SiC because of its transgranular fracture behavior. On the other hand, wavy fracture surface of β-SiAlON is the indication of relatively more deformation, which prevents rapid crack propagation and reduces indentation-induced crack lengths.

Acknowledgements

Authors wish to acknowledge Dr. G. Sundararajan, Director-ARCI, and Dr. Shrikant V. Joshi, Associate Director-ARCI, for their constant inspiration towards the completion of this work.

References

[1] McColm IJ. *Ceramic Hardness*. New York: Plenum Press, 1990.

[2] Rice RW, Wu CC, Boichelt F. Hardness-grain-size relations in ceramics. *J Am Ceram Soc* 1994, **77**: 2539–2553.

[3] Wilantewicz T, Cannon WR, Quinn G. The indentation size effect (ISE) for Knoop hardness in five ceramic materials. In *Advances in Ceramic Armor II: A Collection of Papers Presented at the 30th International Conference on Advanced Ceramics and Composites*. Franks LP, Wereszczak A, Lara-Curzio E, Eds. Hoboken: John Wiley & Sons, 2006: 237–250.

[4] Gong JH, Li Y. An energy-balance analysis for the size effect in low-load hardness testing. *J Mater Sci* 2000, **35**: 209–213.

[5] Dusza J, Steen M. Microhardness load size effect in individual grains of a gas pressure sintered silicon nitride. *J Am Ceram Soc* 1998, **81**: 3022–3024.

[6] Quinn GD, Green P, Xu K. Cracking and the indentation size effect for Knoop hardness of glasses. *J Am Ceram Soc* 2003, **86**: 441–448.

[7] Peng ZJ, Gong JH, Miao HZ. On the description of indentation size effect in hardness testing for ceramics: Analysis of the nanoindentation data. *J Eur Ceram Soc* 2004, **24**: 2193–2201.

[8] Sangwal K. Review: Indentation size effect, indentation cracks and microhardness measurement of brittle crystalline solids—Some basic concepts and trends. *Crys Res Technol* 2009, **44**: 1019–1037.

[9] Li H, Bradt RC. The indentation load/size effect and the measurement of the hardness of vitreous silica. *J Non-Cryst Solids* 1992, **146**: 197–212.

[10] Quinn JB, Quinn GD. Indentation brittleness of ceramics: A fresh approach. *J Mater Sci* 1997, **32**: 4331–4346.

[11] Patel PJ, Swab JJ, Staley M, *et al.* Indentation size effect (ISE) of transparent AlON and MgAl$_2$O$_4$. Available at http://www.arl.army.mil/arlreports/2006/ARL-TR-3852.pdf.

[12] Huang ZH, Jia DC, Zhou Y, *et al.* Effect of a new additive on mechanical properties of hot-pressed silicon carbide ceramics. *Mater Res Bull* 2002, **37**: 933–940.

[13] Zhang XF, Yang Q, De Jonghe LC. Microstructure development in hot-pressed silicon carbide: Effects of aluminum, boron, and carbon additives. *Acta Mater* 2003, **51**: 3849–3860.

[14] Karandikar PG, Evans G, Wong S, *et al*. A review of ceramics for armor applications. In *Advances in Ceramic Armor IV: Ceramic Engineering and Science Proceedings*. Franks LP, Obji T, Wereszczak A, Eds. Hoboken: John Wiley & Sons, 2008: 163–175.

[15] Suyama S, Kameda T, Itoh Y. Development of high-strength reaction-sintered silicon carbide. *Diam Relat Mater* 2003, **12**: 1201–1204.

[16] Fernández JM, Muñoz A, de Arellano López AR, *et al*. Microstructure–mechanical properties correlation in siliconized silicon carbide ceramics. *Acta Mater* 2003, **51**: 3259–3275.

[17] Ghosh G, Vaynman S, Fine ME, *et al*. Microstructure and ambient properties of a SiAlON composite prepared by hot pressing and reactive sintering of β-Si_3N_4 coated with Al_2O_3. *J Mater Res* 1999, **14**: 881–890.

[18] da Silva CRM, de Melo FCL, de Macedo Silva OM. Mechanical properties of SiAlON. *Mat Sci Eng A* 1996, **209**: 175–179.

[19] Abo-Naf SM, Dulias U, Schneider J, *et al*. Mechanical and tribological properties of Nd- and Yb-SiAlON composites sintered by hot isostatic pressing. *J Mater Process Tech* 2007, **183**: 264–272.

[20] Hou X-M, Chou K-C, Li F-S. Some new perspectives on oxidation kinetics of SiAlON materials. *J Eur Ceram Soc* 2008, **28**: 1243–1249.

[21] Lin MT, Shi JL, Jiang DY, *et al*. High temperature creep of a hot-pressed β-SiAlON. *Mat Sci Eng A* 2001, **300**: 61–67.

[22] Bull SJ, Page TF, Yoffe EH. An explanation of the indentation size effect in ceramics. *Phil Mag Lett* 1989, **59**: 281–288.

[23] Mukhopadhyay NK, Paufler P. Micro- and nanoindentation techniques for mechanical characterisation of materials. *Int Mater Rev* 2006, **51**: 209–245.

17

Effect of the addition of TEOS on the SiC fibers fabricated by electrospinning

Jiayan LI, Panpan CAO, Yi TAN[*], Lei ZHANG

School of Materials Science and Engineering, Dalian University of Technology, Dalian 116024, China

Abstract: Polycarbosilane (PCS) has been widely used to fabricate silicon carbide (SiC) fibers via pyrolysis. In this paper, for improving the morphology of SiC fibers, tetraethyl orthosilicate (TEOS, $m = 1$ g, 3 g and 5 g, respectively) was added into the PCS precursor solution (containing 1.5 g PCS). The continuous fibers have been prepared by electrospinning, and then the SiC fibers were synthesized by calcination at 1300 ℃, 1400 ℃ and 1600 ℃ for 4 h respectively with a heating rate of 10 ℃/min in flowing nitrogen (N_2). The morphologies of the fibers were investigated by the scanning electron microscope (SEM) and it could be seen that the crystallinity of the SiC fibers was lower, the length of the SiC fibers was increased, and the diameter was uniform with the increase of the addition amount of TEOS.

Keywords: silicon carbide (SiC); tetraethyl orthosilicate (TEOS); electrospinning; fibers

1 Introduction

In the past decade, one-dimensional (1D) nanosized semiconducting materials have attracted attention with their fascinating optical, electronic and chemical properties. The size and morphology can affect their applications as catalysts, solar cells, light-emitting diodes and biological labeling [1–3]. As a semiconducting material, silicon carbide (SiC) possesses excellent physical and electronic properties such as high mechanical strength, high thermal stability and high thermal conductivity [4]. It also has wide applications including field emission displays, nano-sensors and other nanoscale devices [5,6].

Recently, much more effort has been devoted to the synthesis process of SiC fibers with nanostructures, including carbothermal reduction of silica [7–9], chemical vapor deposition [6,10–12], carbon nanotube-templated growth [13], etc. It has been demonstrated that polycarbosilane (PCS) is used to synthesize SiC fibers by melt-spinning, curing and pyrolysis. In this paper, SiC fibers could be synthesized by a high-temperature carbothermal reduction progress using electrospun PVP (polyvinyl pyrrolidone)/PCS composite fibers as precursor. In addition, tetraethyl orthosilicate (TEOS) was added to improve the morphology of SiC fibers. Compared with the reported synthetic methods, the technique used in this work possesses the virtues including simplicity, low cost and absence of template and catalyst.

2 Experiment

As a starting material, PCS polymer was used to fabricate SiC fibers due to its rich content of silicon.

* Corresponding author.
E-mail: tanyi@dlut.edu.cn

For achieving PVP/PCS/TEOS composite, 1.5 g PCS and TEOS (1 g, 3 g and 5 g, respectively, Tianjin Kermel Chemical Reagent Ltd.) were dissolved into 11 ml solvent including 8 ml tetrahydrofuran (THF, Tianjin Kermel Chemical Reagent Ltd.) and 3 ml absolute alcohol ($\geqslant$ 99.7%, Shenyang Xinxing Chemical Reagent Ltd.). Then 0.8 g PVP (K90, Tianjin Bodi Chemical Co. Ltd.) was added in the solution to increase viscidity. Subsequently, the mixture viscous solution was magnetically stirred for 1 h at room temperature until the floccules were completely dissolved. The solution was transferred into a plastic syringe with conductive stainless steel needle and ejected with an appropriate voltage of 30 kV. The dense net of fibers was formed on the collector which was coated with an aluminium foil. The distance between the needle and the collector was about 16 cm. Whereafter, the fiber mat was immediately dried in air oven at 80 ℃ for 0.5 h, and then was heated at 190 ℃ for 6 h. The Si–H bonds in PCS were oxidized and the Si–O–Si crosslinked structure was formed. The melting of PCS was prevented and the morphology of precursor fibers was remained during the process of high-temperature pyrolysis [14,15]. Subsequently, the PCS precursor fibers were calcined respectively at 1300 ℃, 1400 ℃ and 1600 ℃ for 4 h with a heating rate of 10 ℃/min in flowing nitrogen (N_2) followed by calcination at 600 ℃ for 4 h in air to remove the remaining carbon.

X-ray diffraction (XRD) patterns were collected on a Shimadzu/XRD-6000 diffractometer with Cu Kα radiation. The morphologies of SiC fibers were characterized by JSM-5600LV scanning electron microscope (SEM, JEOL).

3 Results and discussion

The XRD patterns of the SiC fibers after calcination at 1600 ℃ for 4 h which are derived from different precursor fibers are shown in Fig. 1. The main peaks in the patterns match well with the cubic β-SiC phase (JCPDS 74-2307), and their peak positions at $2\theta = 36°$, 60° and 72° are respectively attributed to the (111), (220) and (311) planes of β-SiC. The SiC crystal is successfully synthesized at this temperature. The diffraction peaks of products become weak and broad which indicates that the crystallinity is lower and the grain size in SiC fibers is decreased with the increase

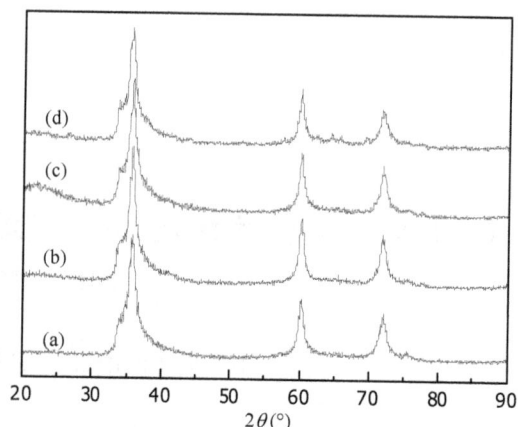

Fig. 1 XRD patterns of SiC fibers after calcination at 1600 ℃ for 4 h derived from precursor fibers with different TEOS contents: (a) 0 g, (b) 1 g, (c) 3 g, (d) 5 g.

of TEOS content. There are no other obvious diffraction peaks, which imply that the main product is cubic SiC phase and no impurity content is obtained in the fibers.

The SEM images of the PCS/TEOS precursor fibers are shown in Fig. 2. It could be observed that the surface of the precursor fibers is very smooth, which is attributed to the fine particles and the amorphous nature of the fibers. Moreover, with the increase of addition amount of TEOS, the length of the fibers is longer and the diameter is more uniform, which can also be seen in Fig. 3.

The SEM images of the SiC fibers calcined at 1300 ℃ and the corresponding columnar section of the diameter uniformity with different amounts of TEOS

(a) 0 g   (b) 1 g

(c) 3 g   (d) 5 g

Fig. 2 SEM images of the PCS precursor fibers with different TEOS contents.

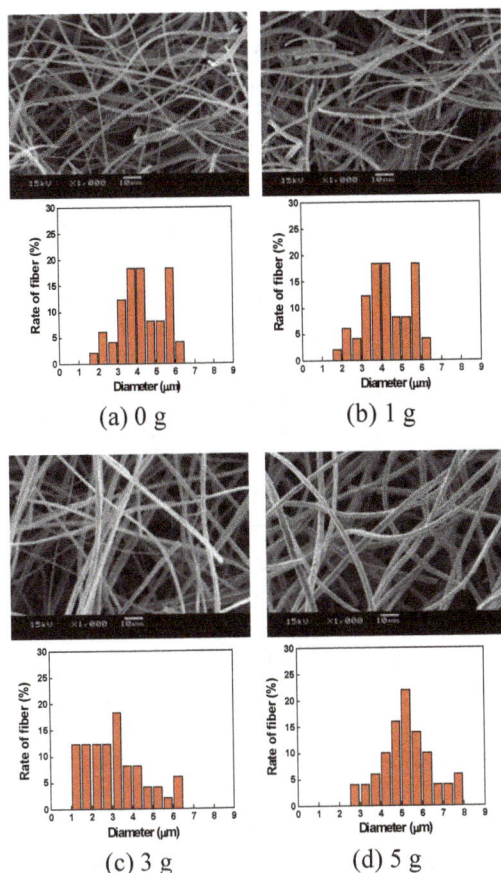

(a) 0 g   (b) 1 g

(c) 3 g   (d) 5 g

Fig. 3 SEM images of the SiC fibers calcined at 1300 ℃ and the corresponding columnar section of the diameter uniformity with different amounts of TEOS.

are shown in Fig. 3. We observe that the fibers with diameters in the range of 1–9 μm are brittle and easy to fracture when there is no or 1 g TEOS addition. However, the fibers become longer and their diameters concentrate in the range of 4–6 μm when the addition amounts of TEOS reach 3 g and 5 g.

The SEM images of SiC fibers derived from different precursor fibers are shown in Fig. 4. After calcinations at 1600 ℃ for 4 h, the SiC grains as shown on the fiber surface are averagely smaller than 500 nm, and the grains are closely connected to each other with clear grain boundaries. With the addition of TEOS, the surface of SiC fibers becomes smoother and the SiC grain sizes in the fibers are decreased, which also have been confirmed by the XRD patterns.

The SEM images of the PCS/TEOS precursor fibers and SiC fibers calcined at different temperatures with 5 g TEOS addition are shown in Fig. 5. The average diameter of the fibers is about 3 μm, and the length could reach several millimeters. The surface of SiC

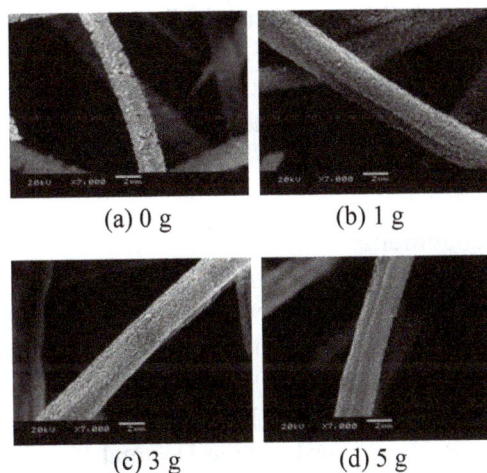

(a) 0 g   (b) 1 g

(c) 3 g   (d) 5 g

Fig. 4 SEM images of SiC fibers calcined at 1600 ℃ derived from different precursor fibers with different TEOS contents.

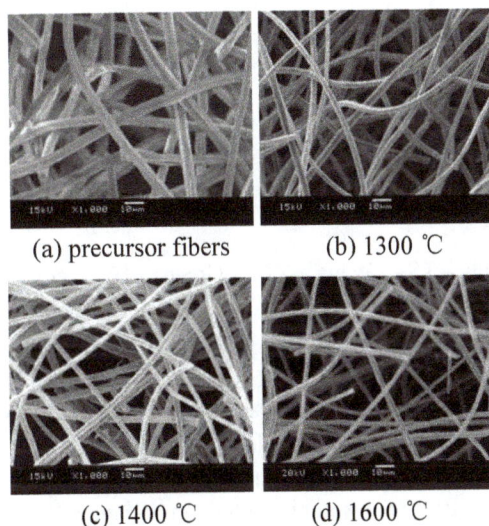

(a) precursor fibers   (b) 1300 ℃

(c) 1400 ℃   (d) 1600 ℃

Fig. 5 SEM images of the precursor fibers and SiC fibers calcined at different temperatures.

fibers after calcined for 4 h becomes rough due to the removal of organic components and the crystallization of the SiC phase, while the surface of the precursor fibers (Fig. 5(a)) is very smooth. The fibers show a slightly decrease in diameter due to the burning-off of PVP, whereas the continuous microstructure of the fibers is still maintained.

3 Conclusions

In summary, direct electrospinning was used in this work to prepare the PCS/TEOS precursor fibers, and the SiC fibers were obtained successfully after carbonization. The addition of TEOS improved the morphology of SiC fibers effectively. With the increase of addition amount of TEOS, the crystallinity was lower and the grain size in SiC fibers was decreased. The length of the fibers was longer and the diameter was more uniform.

Acknowledgements

This work was financially supported by the Fundamental Research Funds for the Central Universities (No. DUT12LAB03), the National Natural Science Foundation of China (No. 51104028), and the Specialized Research Fund for the Doctoral Program of Higher Education (20110041120031).

References

[1] Ahmadi TS, Wang ZL, Green TC, *et al*. Shape-controlled synthesis of colloidal platinum nanoparticles. *Science* 1996, **272**: 1924–1925.

[2] Huynh WU, Peng X, Alivisatos AP. CdSe nanocrystal rods/poly (3-hexylthiophene) composite photovoltaic devices. *Adv Mater* 1999, **11**: 923–927.

[3] Chan WCW, Nie S. Quantum dot bioconjugates for ultrasensitive nonisotopic detection. *Science* 1998, **281**: 2016–2018.

[4] Wang ZL, Dai ZR, Gao RP, *et al*. Side-by-side silicon carbide–silica biaxial nanowires: Synthesis, structure, and mechanical properties. *Appl Phys Lett* 2000, **77**: 3349.

[5] Zhang Y, Nishitani-Gamo M, Xiao C, *et al*. A novel synthesis method for aligned carbon nanotubes in organic liquids. *Jpn J Appl Phys* 2002, **41**: L408–L411.

[6] Choi H-J, Seong H-K, Lee J-C, *et al*. Growth and modulation of silicon carbide nanowires. *J Cryst Growth* 2004, **269**: 472–478.

[7] Bootsma GA, Knippenberg WF, Verspui G. Growth of SiC whiskers in the system SiO_2–C–H_2 nucleated by iron. *J Cryst Growth* 1971, **11**: 297–309.

[8] Wang L, Wada H, Allard LF. Synthesis and characterization of SiC whiskers. *J Mater Res* 1992, **7**: 148–163.

[9] Meng GW, Zhang LD, Mo CM, *et al*. Preparation of β-SiC nanorods with and without amorphous SiO_2 wrapping layers. *J Mater Res* 1998, **13**: 2533–2538.

[10] Zhou XT, Lai HL, Peng HY, *et al*. Thin β-SiC nanorods and their field emission properties. *Chem Phys Lett* 2000, **318**: 58–62.

[11] Leu I-C, Lu Y-M, Hon M-H. Substrate effect on the preparation of silicon carbide whiskers by chemical vapor deposition. *J Cryst Growth* 1996, **167**: 607–611.

[12] Motojima S, Hasegawa M, Hattori T. Chemical vapour growth of β-SiC whiskers from a gas mixture of Si_2Cl_6–CH_4–H_2–Ar. *J Cryst Growth* 1988, **87**: 311–317.

[13] Tang CC, Fan SS, Dang HY, *et al*. Growth of SiC nanorods prepared by carbon nanotubes-confined reaction. *J Cryst Growth* 2000, **210**: 595–599.

[14] Cheng XZ, Xiao JY, Xie ZF, *et al*. Research on curing of polycarbosilane fiber and preparation of SiC fiber. *J Mater Eng* 2004, **1**: 29–37 (in Chinese).

[15] Zheng CM, Li XD, Wang H, *et al*. Thermal stability and curing kinetics of polycarbosilane fibers. *Trans Nonferrous Met Soc China* 2006, **16**: 44–48.

18

Processing and characterization of polymer precursor derived silicon oxycarbide ceramic foams and compacts

Srinivasan NEDUNCHEZHIAN, Ravindran SUJITH, Ravi KUMAR[*]

Materials Processing Section, Department of Metallurgical and Materials Engineering, Indian Institute of Technology Madras, Chennai 600036, Tamil Nadu, India

Abstract: This work focused on the fabrication of silicon oxycarbide ceramic (SiOC) foams as well as dense compacts using poly(hydridomethylsiloxane) (PHMS) as a polymer precursor. The room-temperature cross-linking of PHMS was achieved by the addition of 1,4-diazabicyclo [2.2.2] octane (DABCO) with the release of hydrogen gas as a by-product. This resulted in self-blowing of the polymer precursor at room temperature and thereby offered the possibility of producing SiOC foams without the need of any external blowing agents. We also reported the fabrication of crack-free silicon oxycarbide compacts by cold compaction and pyrolysis route using polyvinyl alcohol (PVA) as a processing additive. Cylindrical-shaped pellets were pyrolysed at 1300 ℃ in argon atmosphere with a ceramic yield of approximately 85%. Increased resistance to phase separation and crystallization up to 1400 ℃ was attributed to the presence of large volume fraction of free carbon in the material which was confirmed by Raman spectroscopy.

Keywords: silicon oxycarbide; foams; 1,4-diazabicyclo [2.2.2] octane (DABCO); self-blown polymer; poly(hydridomethylsiloxane) (PHMS)

1 Introduction

Silicon oxycarbide ceramic (SiOC) processed through polymer precursors is one of the widely studied systems due to its attractive thermo-mechanical properties [1–6]. This is attributed to the replacement of a part of its divalent oxygen atoms by the tetravalent carbon atoms, resulting in a stronger 3-dimensional network in contrast to silica, thereby enhancing the potential of the material [7]. Polysiloxanes are widely-used polymer precursors for the production of SiOC because of the ease of availability, low cost and flexibility to handle in atmospheric ambience. Wide varieties of polysiloxanes are available in the market including poly(hydroxymethylsiloxane), poly (methylsilsesquioxane), poly(methylphenylsiloxane) and poly(hydridomethylsiloxane) (PHMS) to produce SiOC upon pyrolysis. However, the final chemical composition, ceramic yield, carbon content and thermal stability depend on the side-chain constituent (hydrogen, methyl or phenyl groups) of the polymer precursor [8]. The current study focuses on the processing of SiOC from PHMS since it contains functional hydrogen group in its side-chain which is

---
* Corresponding author.
E-mail: nvrk@iitm.ac.in

beneficial for cross-linking.

SiOC derived from PHMS requires necessarily a cross-linking step since linear polymers undergo complete thermal decomposition during pyrolysis treatment. Earlier investigation focused primarily on cross-linking of PHMS through the hydrosilylation reaction by mixing with vinyl containing precursor catalyzed by platinum-based complexes resulting in higher ceramic yield upon pyrolysis [9–12]. However, cross-linking through hydrosilylation reaction requires additional vinyl group containing precursor and platinum-based catalysts which are expensive. Here, we report rather a simple method of cross-linking PHMS at room temperature by the addition of inexpensive 1,4-diazabicyclo [2.2.2] octane (DABCO) catalyst and the ceramic yield obtained was comparable to the one achieved through cross-linking by hydrosilylation reaction. However, generation of gaseous by-product during cross-linking resulted in self-blowing, thereby producing ceramic foams. This reaction was explored to produce ceramic foams.

For the production of bulk compacts, cold compaction technique was adopted by applying a uniaxial load on cross-linked precursor powder. Fabrication of bulk compacts through polymer derived ceramic route is always a challenge because of its associated porosity and volumetric shrinkage during polymer to ceramic conversion [13]. This necessitates the use of active and passive fillers to counteract very high volumertric shrinkage during pyrolysis [14]. Here, in this paper we report an economical way of processing filler-free bulk SiOC through cold compaction using polyvinyl alcohol (PVA) as a processing additive.

2 Experiment

Commercially available PHMS $((CH_3)_3$–$Si(O)$–$[CH_3(H)Si$–$O]_n$–$Si(CH_3)_3)$ (Sigma Aldrich, India) was used as the starting polymeric precursor. The room-temperature cross-linking of precursor was achieved by the addition of 5 wt% DABCO (Sigma Aldrich, India). The solution was thoroughly mixed in a teflon beaker which was subsequently kept overnight for cross-linking. Teflon beaker was used in all the experiments to enable easy removal of the cross-linked polymer. The amount of DABCO catalyst added had a profound influence on the gelation time of the precursor (liquid to solid). The addition of 2 wt%

DABCO catalyst took almost 24 h for complete cross-linking. In order to reduce the gelation time, ~5 wt% DABCO catalyst was used for cross-linking PHMS. Subsequently, foamed polymer was subjected to pyrolysis treatment in an alumina tubular furnace by heating to 1000 °C in a flowing argon gas atmosphere (purity 99.9%, flow rate 5 L/min) at a heating rate of 5 °C/min for 2 h and the furnace cooled to produce SiOC foams.

For producing dense SiOC, cross-linked polymer was ground to fine powder using agate pestle and mortar. To improve the green strength of the pellets, 2 drops of 4% PVA solution was added to 0.75 g of polymer precursor powder prior to cold compaction. Cold uniaxial pressing was carried out in a compaction unit (Insmart XRF 40, India, capacity 40 tonnes) using zinc stearate as a lubricant by applying a compaction pressure of 440 MPa for 1 min in a cylindrical die made of cold rolled steel. It was observed that compaction carried out at a pressure greater than 440 MPa resulted in cracking of the samples during pyrolysis. Hence, all the green compacts were produced limiting the pressure to 440 MPa. Cylindrical-shaped green pellets with diameter of 12 mm and thickness of 7 mm were produced after compaction. Subsequently, cold compacted samples were pyrolyzed at 1300 °C for 2 h in argon to produce dense SiOC compacts. The elemental composition of as-thermolysed SiOC was assumed to be identical to the composition reported by Blum et al. [15].

Thermogravimetric (TG) analysis (NETZSCH STA 409 PC/PG, Germany) was performed by heating 5 mg of samples taken in an alumina crucible from room temperature to 1000 °C at a heating rate of 5 °C/min in a flowing argon atmosphere (flow rate 50 ml/min). Fourier-transformed infrared (FT-IR) spectrometer (Perkin Elmer Spectrum One, USA) was used to record transmittance spectra of samples in the range of 4000–400 cm^{-1} with a resolution of 4 cm^{-1}. X-ray diffraction (XRD) analysis was carried out on samples pyrolyzed in argon at temperatures ranging from 1000 °C to 1400 °C using D8 Discover, Bruker AXS X-ray diffractometer (USA) with Cu Kα radiation. The voltage and current settings used were 35 kV and 25 mA, respectively. All pyrolyzed ceramic samples were scanned from 5° to 90° (2θ) at a scan speed of 1 s/step. The foam morphology and presence of porosity were investigated using FEI scanning electron microscopy (SEM, Quanta 200, USA). The Raman spectra of pyrolyzed ceramic were obtained with the help of confocal Raman spectrometer (WITEC alpha 300, Germany) using Nd:YAG laser (wavelength 532 nm). The density of bulk ceramic samples produced by cold compaction and pyrolysis route were measured using both Archimedes' principle and geometrical method. The density of foam samples were calculated by measuring the weight and the volume.

3   Results and discussion

3.1   Characterization of polymer precursor

The FT-IR spectra of the as-received polymer and cross-linked polymer are exemplified in Fig. 1. The broad intense peak centred at 1100 cm^{-1} is attributed to Si–O–Si asymmetric vibration which is typical of any siloxane. The peaks observed at 1400 cm^{-1} and 1200 cm^{-1} in the as-received PHMS are attributed to Si–CH$_3$ asymmetric and symmetric bending vibrations, respectively. Moreover, the C–H stretching vibration is observed at 2966 cm^{-1} and 2901 cm^{-1}. The sharp intense peaks at 2166 cm^{-1} and 755 cm^{-1} are assigned to Si–H stretching and Si–C stretching, respectively. However, in contrast to the FT-IR spectra of the as-received PHMS, sharp decreases in the intensity of the peaks at 2166 cm^{-1} and 2966 cm^{-1} are observed in

the cross-linked PHMS attributing to the dehydrocoupling reaction. Accordingly, the most probable cross-linking mechanism is explained in Scheme 1, which is in accordance with the earlier reports [16–18]. This reaction occurs at room temperature catalyzed by the addition of DABCO catalyst. Nevertheless, it is noteworthy that even after cross-linking, Si–H peak is observed in the FT-IR spectrum of the cross-linked PHMS indicating that hydrogen is not completely consumed during cross-linking. Since the initial polymer contains reactive group (hydrogen) in its side-chain, it helps in cross-linking at room temperature by the addition of DABCO catalyst. Similar types of dehydrocoupling reactions were also reported in polysilazanes with reactive groups such as Si–H and N–H by the addition of appropriate catalyst [19]. And, also yield upon thermolysis is significantly improved because of higher dimensional cross-linked polymer molecular structure [20]. The cross-linking yield is approximately 99% and the 1% mass loss observed during cross-linking could be due to the loss of lower molecular weight groups in the polymer which is hydrogen in this case (Scheme 1).

Scheme 1   Cross-linking mechansim deduced from FT-IR spectra is:

$$Si–H + H–Si \rightarrow Si–Si + H_2$$
$$C–H + H–Si \rightarrow Si–Si + H_2$$

TG analysis of the cross-linked polymer in argon atmosphere shows that organic to inorganic transition starts at around 100 ℃ with loss of residual hydrogen and ends at ~900 ℃, resulting in SiOC with ~85% yield (Fig. 2(b)). This matches with the experimentally measured yield by taking the ratio of the weight of the

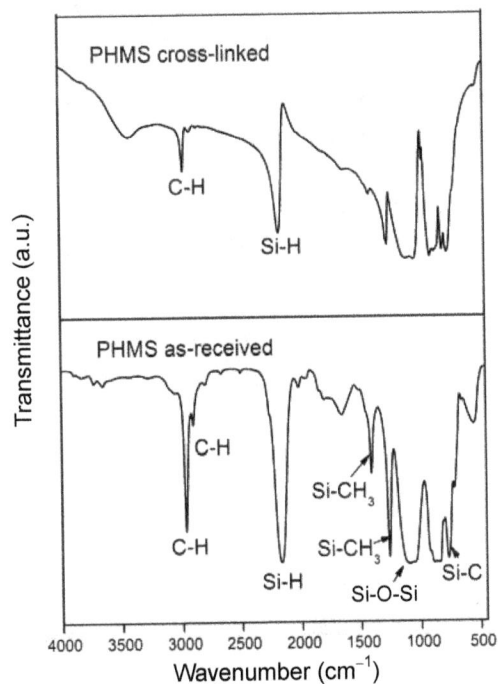

Fig. 1   FT-IR spectra of the as-received polymer and cross-linked polymer.

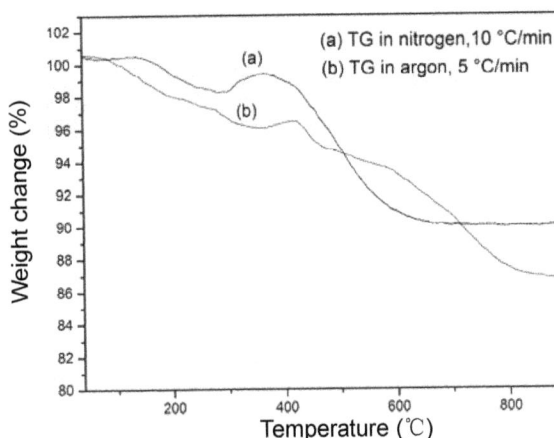

Fig. 2   TG analysis of cross-linked polymer in (a) nitrogen and (b) argon atmosphere.

ceramic obtained to the weight of the polymer taken. A small weight gain in the temperature range of around 390–410 ℃ is also observed. TG analysis in nitrogen atmosphere (Fig. 2(a)) at a heating rate of 10 ℃/min also shows a similar weight gain of ~1% but at a different temperature range of 320–400 ℃ which could be due to the difference in heating rate and change in atmosphere. Moreover, from the TG data it could be inferred that the ceramic yield is higher in nitrogen atmosphere (~88%) in comparison to argon atmosphere, implying the significance of processing atmosphere.

3. 2 XRD and spectroscopic characterization of SiOC

The XRD of the SiOC samples pyrolysed between 1000 ℃ and 1400 ℃ are shown in Fig. 3. Complete absence of phase separation and crystallization up to 1300 ℃ and only a broad amorphous hump at $2\theta = 21°$ corresponding to the most intense peak of silicon dioxide (cristobalite) are observed. However, samples exposed to 1400 ℃ for longer duration (10 h) result in the phase separation of silicon dioxide and β-silicon carbide from the amorphous SiOC network (Fig. 3).

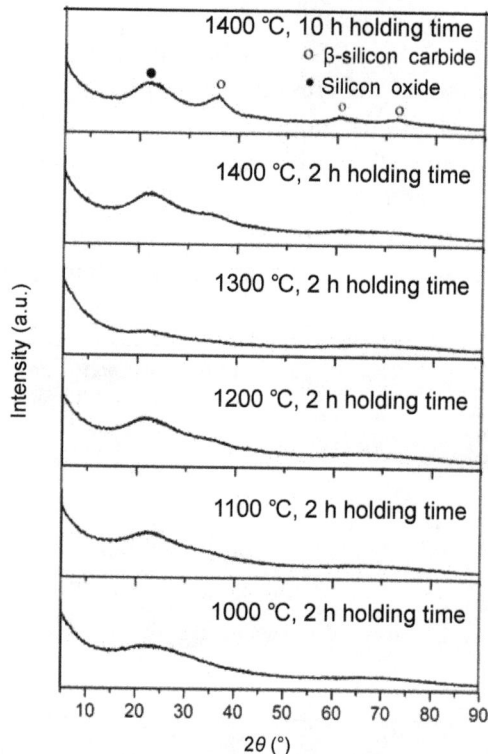

Fig. 3   X-ray   diffractograms   of   the as-thermolysed and heat-treated SiOC samples.

Silicon carbide peaks ($2\theta = 35.5°$, 61° and 72°) start to appear for the samples pyrolysed at 1400 ℃ for 10 h. However, the peaks are broad indicating the nanocrystalline nature of SiC. The crystallite size of β-SiC measured by Scherrer method is ~2.5 nm inferring the insignificant crystallite coarsening. This could be in accodance with the method suggested by Walter et al. [21], in which the presence of free carbon was reported to restrict the mobility of SiC in the SiOC network. However, the presence of graphite-like carbon/free carbon is not revealed in the X-ray diffractograms. Nevertheless, Raman spectrum of SiOC sample heat treated at 1400 ℃ shows characteristic peaks (D and G bands) of free carbon phase (Fig. 4). D band observed at 1344 cm^{-1} corresponds to disorderness in the sp^2 carbon (breaking of translational symmetry) and G band observed at 1605 cm^{-1} corresponds to tangential vibration mode of graphite [22]. In addition, the limited crystallite growth of β-SiC could also be attributed to the high viscosity of SiOC ($\sim 10^{14}$–10^{16} Pa·s) as proposed by Schiavon et al. [23] and Rouxel et al. [24].

3. 3 Ceramic foams

The hydrogen evolution during cross-linking results in self-blowing of the polymer as explained schematically in Fig. 5. Bubbles of hydrogen nucleate at the bottom of the container and move upwards resulting in the formation of localized colonies followed by the growth of existing bubbles at the expense of newly formed bubbles. The continous bubble formation followed by stabilization results in blowing of the entire polymer. Elongated pores oriented along the foaming direction are observed in a foamed precursor. The foam structure is   retained   during   the   pyrolysis   treatment.   The

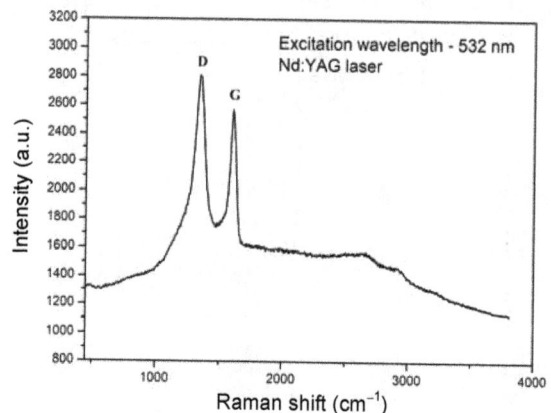

Fig. 4   Raman spectrum of the sample heat treated at 1400 ℃ for 10 h.

self-blowing of poly(phenylmethylsilsesquioxane) at 300 ℃ due to release of water and ethanol was reported in the literatures by Gambaryan-Roisman et al. [25] and Zeschky et al. [26–28]. This method of producing SiOC foams is promising, since the removal of sacrificial phase during pyrolysis is not required. However, it is often difficult to control the density of ceramic foams processed through this method. The photographs and SEM images of pyrolyzed ceramic foams is shown in Figs. 6(a) and 6(b) indicating that the processed foams are relatively free of cracks.

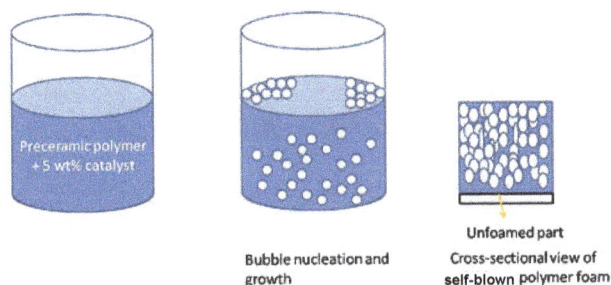

Fig. 5 Schematic chart of self-blowing phenomenon in PHMS.

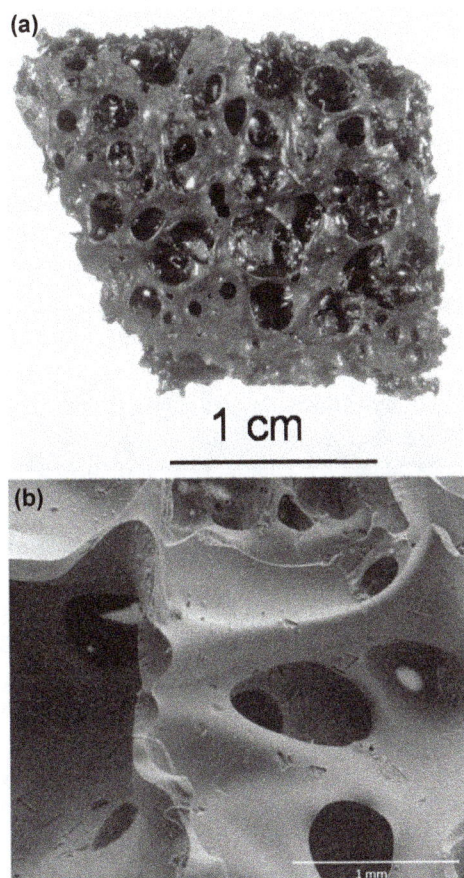

Fig. 6 (a) Photograph of SiOC foams; (b) SEM showing interconnected porosity in SiOC foams.

3. 4 Bulk compacts

The photographs of the cold compacted pellet and the SiOC compact after pyrolysis are shown in Fig. 7. An optimized compaction pressure of 440 MPa was used for the processing of bulk compacts, since residual porosity in the green stage for gaseous by-products to escape during pyrolysis is always desired. It is observed that the addition of PVA reduces the free flowability of powder thereby retaining the shape after pressing. However, in contrast to warm pressing where welding and plastic deformation occur at the pressing temperature, only particle rearrangement occurs in cold pressing and PVA plays a vital role in producing bulk compacts. This is supported by the SEM images of cross-linked polymer powder particles and pyrolyzed SiOC as shown in Figs. 8(a) and 8(b), respectively. Individual particles could be seen even after the pyrolysis with residual irregular shaped pores (Fig. 8(b)). The density of pyrolyzed samples is 1.4–1.5 g/cm^3 and geometric density is close to 1.5–1.6 g/cm^3. Compared to the density values reported in the existing literatures for SiOC [29–31], the observed density is lower for the cold pressed and pyrolysed SiOC. This could be attributed to the presence of residual porosity due to the escapement of volatile matter during the thermolysis. However, the linear shrinkages in axial and radial directions are ~26% and ~27% respectively and the volumetric shrinkage is ~67%. Harshe et al. [7] and Weisbarth and Jansen [32] reported axial shrinkage of ~16% and ~19% for bulk SiOC and $SiBN_3C$, respectively. Henceforth, the shrinkage occuring in these ceramics could be considered as a useful indication of densification. Despite the fact that very high linear shrinkage is observed, crack-free SiOC is obtained after pyrolysis and is mainly attributed to isotropic shrinkage and high ceramic yield of polymer precursor.

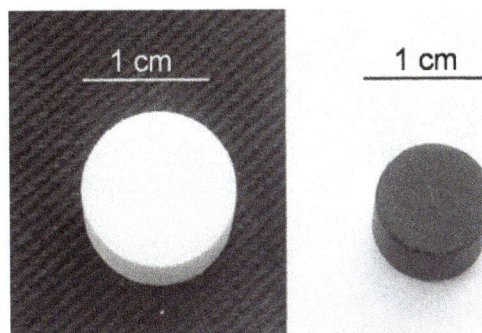

Fig. 7 Photographs of cold compacted SiOC samples before and after pyrolysis.

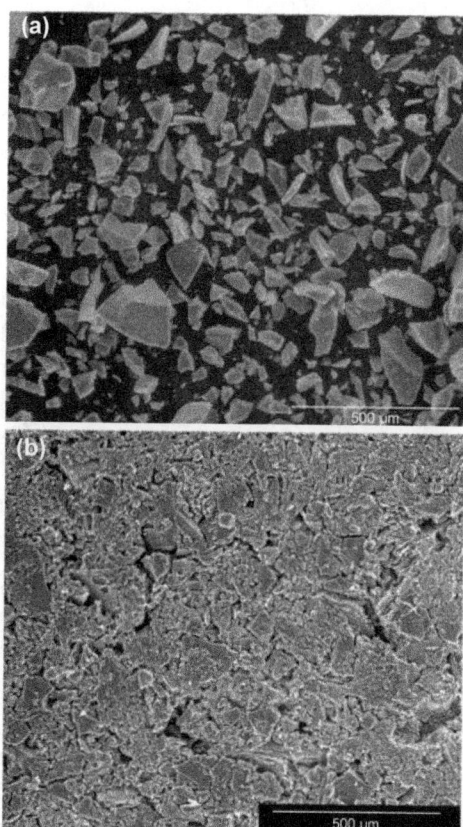

Fig. 8 SEM images of cross-linked polymer powder and pyrolyzed ceramic.

4 Conclusions

Room-temperature cross-linking of PHMS was achieved by the addition of DABCO catalyst with a ceramic yield of approximately 85%. SiOC was thermally stable (absence of phase separation) up to 1300 ℃ in argon atmosphere. Crystalline peaks corresponding to β-SiC were observed only after prolonged exposure at 1400 ℃. The presence of free carbon was confirmed by Raman spectroscopy. SEM micrographs revealed the presence of porosity in optimized SiOC compacts and interconnected pores in self-blown ceramic foams. Filler-free monolithic SiOC compacts using PVA as a binder were found to be promising and economical. SiOC foams through self-blowing mechanism were found to be attractive because of the ease in processability in contrast to other processing routes.

Acknowledgements

This work was supported financially by Indian Space Research Organisation and their support is gratefully acknowledged.

References

[1] Zhang H, Pantano CG. Synthesis and characterization of silicon oxycarbide glasses. *J Am Ceram Soc* 1990, **73**: 958–963.

[2] Renlund GM, Prochazka S, Doremus RH. Silicon oxycarbide glasses: Part II. Structure and properties. *J Mater Res* 1991, **6**: 2723–2734.

[3] Bois L, Maquet J, Babonneau F, *et al*. Structural characterization of sol–gel derived oxycarbide glasses. 1. Study of the pyrolysis process. *Chem Mater* 1994, **6**: 796–802.

[4] Corriu RJP, Leclercq D, Mutin PH, *et al*. ^{29}Si nuclear magnetic resonance study of the structure of silicon oxycarbide glasses derived from organosilicon precursors. *J Mater Sci* 1995, **30**: 2313–2318.

[5] Mutin PH. Control of the composition and structure of silicon oxycarbide and oxynitride glasses derived from polysiloxane precursors. *J Sol–Gel Sci Technol* 1999, **14**: 27–38.

[6] Kleebe HJ, Turquat C, Sorarù GD. Phase separation in an SiCO glass studied by transmission electron microscopy and electron energy-loss spectroscopy. *J Am Ceram Soc* 2001, **84**: 1073–1080.

[7] Harshe R, Balan C, Riedel R. Amorphous Si(Al)OC ceramic from polysiloxanes: Bulk ceramic processing, crystallization and applications. *J Eur Ceram Soc* 2004, **24**: 3471–3482.

[8] Colombo P, Mera G, Riedel R, *et al*. Polymer-derived ceramics: 40 years of research and innovation in advanced ceramics. *J Am Ceram Soc* 2010, **93**: 1805–1837.

[9] Radovanovic E, Gozzi MF, Gonçalves MC, *et al*. Silicon oxycarbide glasses from silicone networks. *J Non-Cryst Solids* 1999, **248**: 37–48.

[10] Modena S, Sorarù GD, Blum Y, *et al*. Passive oxidation of an effluent system: The case of polymer-derived SiCO. *J Am Ceram Soc* 2005, **88**: 339–345.

[11] Liu X, Li YL, Hou F. Fabrication of SiOC ceramic microparts and patterned structures from polysiloxanes via liquid cast and pyrolysis. *J Am*

Ceram Soc 2009, **92**: 49–53.

[12] Su D, Li YL, An HJ, *et al.* Pyrolytic transformation of liquid precursors to shaped bulk ceramics. *J Eur Ceram Soc* 2010, **30**: 1503–1511.

[13] Jiang T, Hill A, Fei W, *et al.* Making bulk ceramics from polymeric precursors. *J Am Ceram Soc* 2010, **93**: 3017–3019.

[14] Greil P. Active-filler-controlled pyrolysis of preceramic polymers. *J Am Ceram Soc* 1995, **78**: 835–848.

[15] Blum YD, MacQueen DB, Kleebe H-J. Synthesis and characterization of carbon-enriched silicon oxycarbides. *J Eur Ceram Soc* 2005, **25**: 143–149.

[16] Lavedrine A, Bahloul D, Goursat P, *et al.* Pyrolysis of polyvinylsilazane precursors to silicon carbonitride. *J Eur Ceram Soc* 1991, **8**: 221–227.

[17] Yvie NSCK, Corriu RJP, Leclerq D, *et al.* Silicon carbonitride from polymeric precursors: Thermal cross-linking and pyrolysis of oligosilazane model compounds. *Chem Mater* 1992, **4**: 141–146.

[18] Wen G, Bai H, Huang X, *et al.* Lotus-type porous SiOCN ceramic fabricated by undirectional solidification and pyrolysis. *J Am Ceram Soc* 2011, **94**: 1309–1313.

[19] Kroke E, Li Y-L, Konetschny C, *et al.* Silazane derived ceramics and related materials. *Mat Sci Eng R* 2000, **26**: 97–199.

[20] Blum YD, Schwartz KB, Laine RM. Preceramic polymer pyrolysis. *J Mater Sci* 1989, **24**: 1707–1718.

[21] Walter S, Soraru GD, Bréquel H, *et al.* Microstructural and mechanical characterization of sol–gel-derived Si–O–C glasses. *J Eur Ceram Soc* 2002, **22**: 2389–2400.

[22] Wang Y, Alsmeyer DC, McCreery RL. Raman spectroscopy of carbon materials: Structural basis of observed spectra. *Chem Mater* 1990, **2**: 557–563.

[23] Schiavon MA, Gervais C, Babonneau F, *et al.* Crystallization behavior of novel silicon boron oxycarbide glasses. *J Am Ceram Soc* 2004, **87**: 203–208.

[24] Rouxel T, Soraru G-D, Vicens J. Creep viscosity and stress relaxation of gel-derived silicon oxycarbide glasses. *J Am Ceram Soc* 2001, **84**: 1052–1058.

[25] Gambaryan-Roisman T, Scheffler M, Buhler P, *et al.* Processing of ceramic foam by pyrolysis of filler containing phenylmethyl polysiloxane precursor. *Ceram Trans* 2000, **108**: 121–130.

[26] Zeschky J, Goetz-Neunhoeffer F, Neubauer J, *et al.* Preceramic polymer derived cellular ceramics. *Compos Sci Technol* 2003, **63**: 2361–2370.

[27] Zeschky J, Höfner T, Arnold C, *et al.* Polysilsesquioxane derived ceramic foams with gradient porosity. *Acta Mater* 2005, **53**: 927–937.

[28] Zeschky J, Lo J, Höfner T, *et al.* Mg alloy infiltrated Si–O–C ceramic foams. *Mat Sci Eng A* 2005, **403**: 215–221.

[29] Mazo MA, Palencia C, Nistal A, *et al.* Dense bulk silicon oxycarbide glasses obtained by spark plasma sintering. *J Eur Ceram Soc* 2012, **32**: 3369–3378.

[30] Sujith R, Srinivasan N, Kumar R. Small-scale deformation of pulsed electric current sintered silicon oxycarbide polymer derived ceramics. *Adv Eng Mater* 2013, DOI: 10.1002/adem.201300146.

[31] Konetschny C, Galusek D, Reschke S, *et al.* Dense silicon carbonitride ceramics by pyrolysis of cross-linked and warm pressed polysilazane powders. *J Eur Ceram Soc* 1999, **19**: 2789–2796.

[32] Weisbarth R, Jansen M. SiBN$_3$C ceramic workpieces by pressureless pyrolysis without sintering aids: Preparation, characterization and electrical properties. *J Mater Chem* 2003, **13**: 2975–2978.

19

Mechanical and dielectric behaviors of perovskite (Ba,Sr)TiO$_3$ borosilicate glass ceramics

Avadhesh Kumar YADAV[a,*], C. R. GAUTAM[a], Abhinay MISHRA[b]

[a]*Department of Physics University of Lucknow, Lucknow-226007, India*
[b]*School of Materials Science and Technology, Indian Institute of Technology, Banaras Hindu University, Varanasi-221005, India*

Abstract: Perovskite (Ba,Sr)TiO$_3$ glass ceramics were crystallized in the presence of La$_2$O$_3$ for glass ceramic system [(Ba$_{1-x}$Sr$_x$)TiO$_3$]–[2SiO$_2$–B$_2$O$_3$]–[K$_2$O]–[La$_2$O$_3$] ($x = 0.0$ and 0.4). The formation of major crystalline phase of BaTiO$_3$ along with secondary phase of Ba$_2$TiSi$_2$O$_8$ was confirmed by X-ray diffraction (XRD) studies. Major crystalline phase was clearly seen in the micrographs of (Ba,Sr)TiO$_3$ borosilicate glass ceramic samples. The prepared glass ceramic samples showed very high values of toughness and elastic modulus. Barium strontium titanate (BST) glass ceramics are used in barrier layer capacitors for storage of high energy due to their very high dielectric constant and low dielectric loss.

Keywords: (Ba,Sr)TiO$_3$; X-ray diffraction (XRD); scanning electron microscopy (SEM); elastic modulus; dielectric constant

1   Introduction

Glass ceramics are polycrystalline materials formed by controlled crystallization of glasses, and they possess very high mechanical strength [1]. The mechanical strength of glass ceramics depends on structural composition and processing variables, and it has complex inter-relationship with microstructure [2,3]. The size, shape, natural wetting and presence or absence of internal cracks are also affecting the microhardness of glass ceramics [4]. Gomma *et al.* [5] reported the electrical and mechanical properties of alkali BaTiO$_3$ alumino–borosilicate glass ceramics containing Sr or Mg. In this glass ceramic system, it was seen that microhardness depends on SrO/BaO or MgO/Al$_2$O$_3$. The measurement of mechanical properties such as hardness and elastic modulus of sintered BaTiO$_3$ ceramic shows the grain size dependence of densification, and it was found that the microhardness increases with decrease in grain size [6]. The carbon nanotube in the glass matrix enhances the toughness of composite materials.

The variation in electric behavior is due to a sequence of phase transitions and the space charge polarization between glass matrix, crystalline phases and inter grain boundary. At high temperature, BaTiO$_3$ is paraelectric with cubic structure. On cooling, this material undergoes successive structural phase transitions from cubic, tetragonal, orthorhombic and rhombohedral [7]. Such glass ceramics are mainly used in high energy storage device such as barrier layer capacitors. When some of Ba^{2+} ions are replaced by Sr^{2+} ions, electrical properties are extensively changed due to space charge polarization. BST (barium strontium titanate) glass ceramics have perovskite

* Corresponding author.
E-mail: yadav.av11@gmail.com

structure and show ferroelectric behavior. Divya *et al.* [8,9] have investigated the crystallization and dielectric properties of BST borosilicate glass ceramics. This group has also reported dielectric behavior of BST with addition of bariumaluminosilicate. They found that the dielectric constant is 10 000 in bariumaluminosilicate system, but the replacement of aluminosilicate with borosilicate enhances the dielectric constant up to 12 000. BST glass ceramics in Al_2O_3 addition make the dielectric constant of 1000 and breakdown strength of 800 kV/cm [10,11]. The replacement of AlF_3 by Ga_2O_3 in glass ceramic system $60(Ba_{0.7}Sr_{0.3})TiO_3–25SiO_2–15AlF_3$ have enhanced the crystallization, while Bi_2O_3 retards the crystallization. $Ba_{0.4}Sr_{0.6}TiO_3$ in the presence of 5 vol% $BaO–SiO_2–B_2O_3$ glass has the highest energy density of 0.89 J/cm^3, which is 2.4 times higher than that of pure $Ba_{0.4}Sr_{0.6}TiO_3$ (0.37 J/cm^3). The improvement of energy density could be mainly due to the increase of the breakdown strength and the decrease of the remnant polarization. Glass ceramics barium/lead based sodium niobates and barium titanate based silicates are investigated, whose dielectric constant are ranging from 20 to 700 [12–14]. The breakdown strength of 1400 kV/cm and the energy storage density of 4.0 J/cm^3 show strontium barium niobate based glass ceramics as strong candidate for high energy density storage capacitors for portable or pulsed power applications, and the breakdown strength increases with crystallization temperature [15]. The increase of Sr/Ba ratio supports the formation of uniform dense and fine-grained structure in glass ceramics, suppresses the crystallization of strontium barium niobate based borate glass, and leads to raised breakdown strength but decreased dielectric constant [16]. Dielectric properties of BST glass ceramics depend upon crystallization, composition of glass ceramic system and doping elements. Recently, our group reported the crystallization and dielectric behavior of BST borosilicate glass ceramics with addition of La_2O_3. La_2O_3 in BST glass ceramics enhances the crystallization and it acts as nucleating agent for crystallization [17,18]. In our previous reported works, the dielectric constant was found to be about 350. Mechanical behavior of BST borosilicate glass ceramics was not reported so far in literature as per best of our knowledge. So, we are trying for enhancing the dielectric constant and mechanical strength of glass ceramics in system $[(Ba_{1−x}Sr_x)TiO_3–$

$[2SiO_2–B_2O_3]–[K_2O]–[La_2O_3]$ ($x = 0.0$ and 0.4).

2 Experimental

2.1 Glass and glass ceramic sample preparation

High purity inorganic ingradients $BaCO_3$ (99%), $SrCO_3$ (99%), TiO_2 (99%), SiO_2 (99.5%), H_3BO_3 (99.8%), K_2CO_3 (99.9%) and La_2O_3 (99.9%) of HiMedia Laboratories Pvt. Limited, India were used for the preparation of glass samples in composition system $[(Ba_{1−x}Sr_x)TiO_3]–[2SiO_2–B_2O_3]–[K_2O]–[La_2O_3]$ ($x = 0.0$ and 0.4). The appropriate amounts of these ingredients were mixed in a mortar with pestle in acetone medium. Well mixed and dried powder of raw materials was melted in platinum crucible with high temperature SiC programmable electric furnace at 1200 ℃. Finally, well melted ingredients were poured into an aluminum mould and pressed by a thick aluminum plate. Then, the quenched glass samples were immediately transferred into a preheated programmable muffle furnace for annealing at 450 ℃ for 3 h. Now, the prepared glass samples were grinded in mortar with pestle to form the fine powder, and this glass powder was calcined at 600 ℃ for 3 h with a heating rate of 10 ℃/min. The calcined powder was again grinded using 2.5% of polyvinyl alcohol (PVA) as binder, and pellets (1 cm in diameter) of this powder were formed at load of 10 ton by BEST hydraulic press machine (New Delhi). These pellets were sintered with the heating rate of 5 ℃/min at 850 ℃ for 3 h and 6 h, respectively. Nomenclature of these glass samples was described as: first three alphabets BST denote barium strontium titanate, 5K denotes 5% of K_2O, 1L means 1 mol% of La_2O_3, numeric digit is the fractional content of Sr, T and S denote the sintering time of 3 h and 6 h respectively, and 850 means the sintering temperature (e.g., BST5K1L0.4T850).

2.2 X-ray diffraction

X-ray diffraction (XRD) data of powdered glass ceramic samples was recorded by a Rigaku Miniflex-II X-ray Diffractometer using Cu Kα radiation with step size of 0.02°. XRD patterns of these glass ceramic samples were compared with standard JCPDS files for different constituting phases. The particle size was calculated by Debye–Scherer (DS) method and Williamson and Hall (WH) plots.

2.3 Scanning electron microscopy

Scanning electron microscopy (SEM) is one of the most versatile instruments available for the examination and analysis of the microstructural morphology of solid objects. The glass ceramic samples were polished by using SiC powder of 100 mesh, 600 mesh and 800 mesh to attain the smooth surfaces. These smooth glass ceramic samples were further polished by emery paper of different grades of 1/0, 2/0, 3/0 and 4/0, and the final polishing was done on a blazer cloth using diamond paste (1 μm and 6 μm) with Hifin fluid "OS". Glass ceramic samples were polished and etched using a solution of 30%HNO$_3$ + 20%HF. Au–Pd coating was done by sputtering method (Poraton Sc-7640 Sputter) on the etched surface of various glass ceramic samples for SEM. SEM images were recorded for surface morphological studies of crystalline phases (voltage of 15 kV and magnification of 5000× by using Model LEO 430 Cambridge Instruments Ltd., UK).

2.4 Density measurements

The geometrical density of green pellets and sintered glass ceramic pellets was measured by using formula of mass unit volume. The geometrical density of these samples has been listed in Table 1.

2.5 Mechanical measurements

Mechanical characterizations were carried out in a universal testing machine (Instron 3639) in compression mode of cylindrical glass ceramic sample having dimensions of 1 cm in diameter and 2.5 cm in length. The toughness of BST borosilicate glass ceramic samples was measured as area under stress–strain curve within limit of initial point to elastic limit point.

2.6 Dielectric measurements

For dielectric measurements of glass ceramic samples, the both surfaces of these samples were made smooth and electroded by applying the silver paint (Code No. 1337-A, Elteck Corporation, India). The electroded samples were cured at 600 ℃ for 5 min. The capacitance C and dissipation factor tanδ were measured at frequencies of 20 Hz, 100 Hz, 1 kHz, 10 kHz and 100 kHz in equal intervals of temperature during heating in the range of room temperature to 150 ℃ by Wayne Kerr 6500 P (high frequency LCR meter, frequency: 20 Hz–5 MHz). The samples were heated in the heating chamber to the required temperature at a rate of 2 ℃/min.

Dielectric constant ε_r was calculated from the measured capacitance C using the following equation:

$$\varepsilon_r = \frac{Cd}{\varepsilon_0 A} \tag{1}$$

where C is the capacitance in farad; ε_0 is the permittivity of free space (8.854×10^{-12} F/m); d is the thickness (m); and A is the area of the samples (m^2). The dielectric constant and dissipation factor were plotted as a function of temperature at a few selected frequencies.

3 Results and discussion

3.1 XRD analysis

XRD patterns of BST borosilicate glass ceramic samples BT5K1L0.0T850, BT5K1L0.0S850, BST5K1L0.4T850 and BST5K1L0.4S850 are shown in Figs. 1–4. XRD pattern of sintered glass ceramic sample BT5K1L0.0T850 shows major crystalline phase of BaTiO$_3$ (BT, JCPDS Card No. 34-0129) and secondary phase of Ba$_2$TiSi$_2$O$_8$ (BTS, JCPDS Card No. 22-0513) (Fig. 1) [19–22]. Figure 2 depicts the XRD pattern of glass ceramic sample BT5K1L0.0S850. This pattern is found to be similar as BT5K1L0.0T850, only has a little difference in low intensity peaks of secondary phase. The comparative studies of these XRD patterns for 3 h and 6 h show that 6 h sintering time gives better crystallization. In glass ceramic

Table 1 Sample code, density, crystallite size, breaking strength, elastic modulus, strain and toughness of glass ceramic samples

Glass ceramic sample code	Density (g/cm^3)		Crystallite size (nm)		Breaking strength (MPa)	Elastic modulus (GPa)	Strain	Toughness (MJ/m^3)
	Green pellets	Sintered pellets	L_{DS}	L_{WH}				
BT5K1L0.0T850	3.07	3.36	46.34	54.03	757.85	3.041±0.001	2.39×10^{-3}	131
BT5K1L0.0S850	3.07	3.38	44.43	50.56	2204.15	5.664±0.012	2.64×10^{-3}	614
BST5K1L0.4T850	2.95	3.13	35.62	42.89	4849.44	5.587±0.001	2.00×10^{-3}	3142
BST5K1L0.4S850	2.95	3.22	27.99	40.95	11294.49	7.070±0.001	2.80×10^{-3}	11824

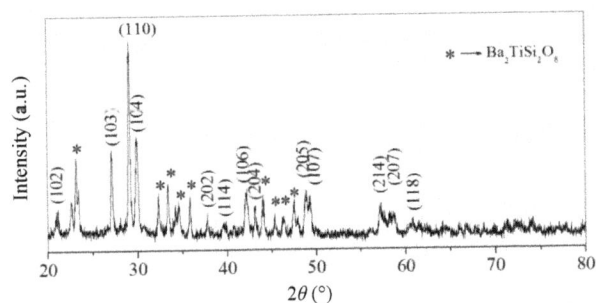

Fig. 1   XRD   pattern   of   glass   ceramic   sample BT5K1L0.0T850.

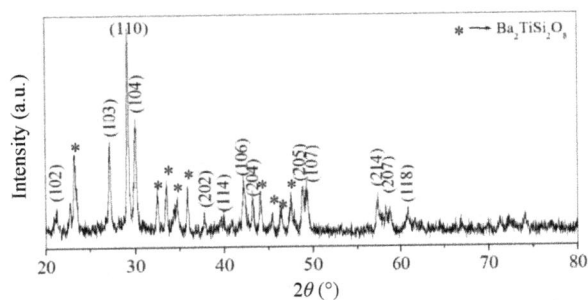

Fig. 2   XRD   pattern   of   glass   ceramic   sample BT5K1L0.0S850.

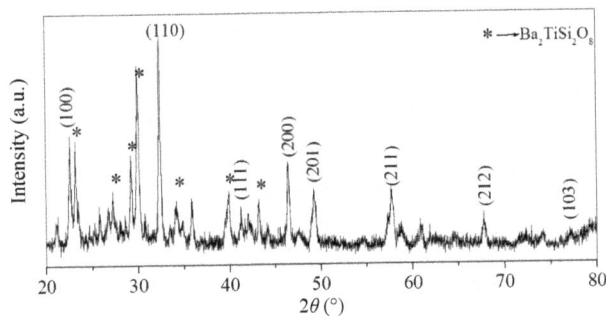

Fig. 3   XRD   pattern   of   glass   ceramic   sample BST5K1L0.4T850.

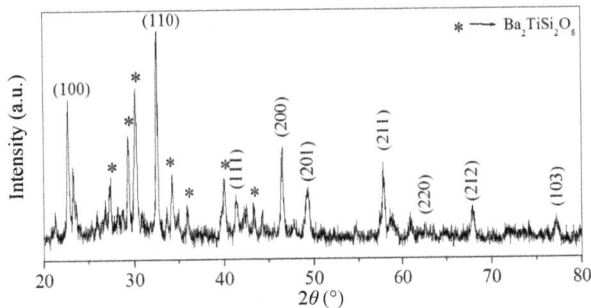

Fig. 4   XRD   pattern   of   glass   ceramic   sample BST5K1L0.4S850.

samples BT5K1L0.0T850 and BT5K1L0.0S850, there is slight shifting of high intensity peak near 29.5° to a smaller angle due to the tetragonal distortion or crystal clamping. The crystal structure of BT is hexagonal. The lattice parameters for glass ceramic sample BT5K1L0.0T850 are $a = 5.75$ Å, $b = 13.87$ Å and $c = 2.38$ Å. XRD pattern of glass ceramic sample BST5K1L0.4T850 ($x = 0.4$) is shown in Fig. 3, and it confirms the formation of major crystalline phase of BST along with secondary phase of BTS which is similar to XRD pattern of glass ceramic sample BST5K1L0.4S850 sintered for 6 h (Fig. 4) [23]. The crystallite size decreases with increasing the sintering time for the crystallization of the calcined glass samples.

WH plots for all glass ceramic samples are shown in Figs. 5–8. The particle size and strain are obtained by comparing the trend line equation from WH plots and listed in Table 1. The particle sizes obtained from both methods (Debye–Scherer formula and WH plots) show quite different values for corresponding samples. This difference in particle size obtained from two methods occurs due to strain broadening in the samples. The strain for glass ceramic sample BT5K1L0.0T850 is observed to be 2.39×10^{-3}, while it is 2.64×10^{-3} for glass ceramic sample BT5K1L0.0S850. Thus, the strain broadening between the particles increases with

sintering time which is clear from Table 1. The positive value of strain between the particles shows the tensile strain in BST glass ceramics.

3.2   Microstructural analysis

SEM micrographs of BST borosilicate glass ceramic samples BT5K1L0.0T850, BT5K1L0.0S850, BST5K1L0.4T850 and BST5K1L0.4S850 are shown in Figs. 9–12. Figure 9 shows the SEM image of glass ceramic sample BT5K1L0.0T850. The interconnected grains with non uniform grain distribution of the major phase of BST inside the glassy matrix are observed. Figure 10 depicts the micrograph of glass ceramic sample BT5K1L0.0S850, and it shows the dense microstructure of grains of smaller size and this confirms nucleation of crystallites in glassy network. Well developed and interconnected grain crystallite phase in the glass matrix of BST is found in SEM image of glass ceramic sample BST5K1L0.4T850 as shown in Fig. 11. The better nucleation is seen in the glass ceramic sample due to heat treatment of 6 h instead of 3 h (Fig. 12). Finally, these micrographs show almost the same grain geometry, only with the difference observed in the densification of the phase formation of crystalline phase. The grain size and porosity in glassy matrix depend upon the sintering as well as glass composition.   The grain size is decreased

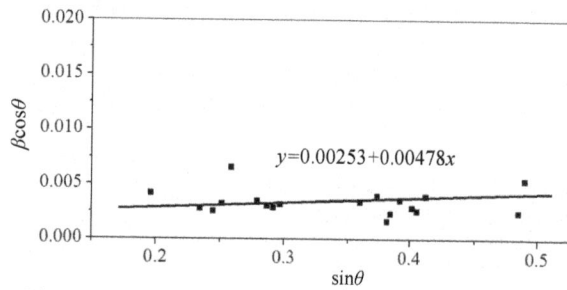

Fig. 5   WH plot of glass ceramic sample BT5K1L0.0T850.

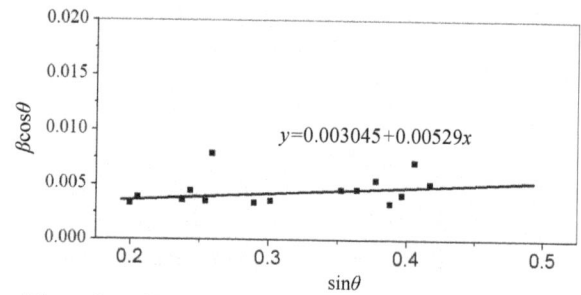

Fig. 6   WH plot of glass ceramic sample BT5K1L0.0S850.

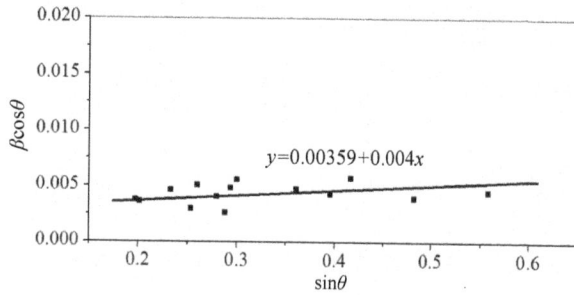

Fig. 7   WH plot of glass ceramic sample BST5K1L0.4T850.

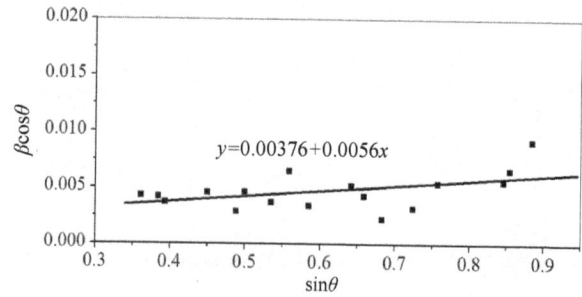

Fig. 8   WH plot of glass ceramic sample BST5K1L0.4S850.

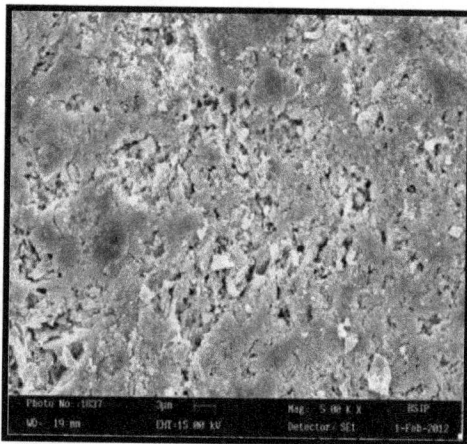

Fig. 9   SEM micrograph of glass ceramic sample BT5K1L0.0T850.

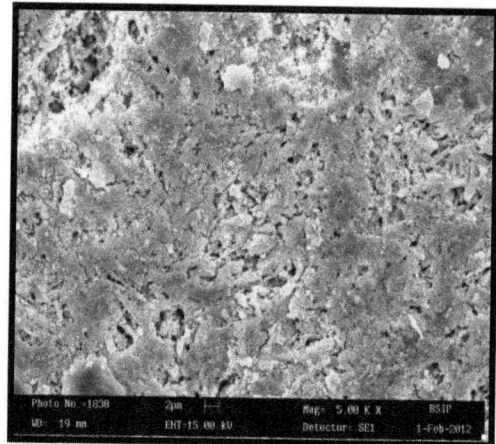

Fig. 10   SEM micrograph of glass ceramic sample BT5K1L0.0S850.

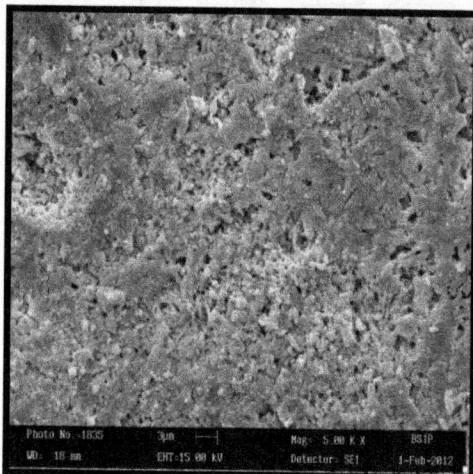

Fig. 11   SEM micrograph of glass ceramic sample BST5K1L0.4T850.

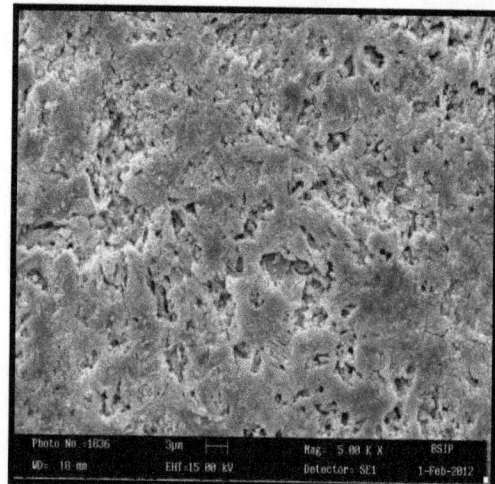

Fig. 12   SEM micrograph of glass ceramic sample BST5K1L0.4S850.

with increase of sintering time and increase of the content of SrO in the glass ceramic samples [5].

3.3 Density measurements

The geometrical densities of green and sintered pellets of samples BT5K1L0.0T850, BT5K1L0.0S850, BST5K1L0.4T850 and BST5K1L0.4S850 have been listed in Table 1. The density of green pellets of samples BT5K1L0.0 is found to be 3.07 g/cm^3, and that of sintered pellets of sample BT5K1L0.0T850 is 3.36 g/cm^3 while for BT5K1L0.0S850 is 3.38 g/cm^3. The density of sintered glass ceramic samples is found to be higher than the reported value of density of glass ceramic samples formed by controlled crystallization of glass for the same system [24]. This shows that densification of glass ceramic samples on sintering increases with increasing the sintering time. This is also confirmed by XRD and SEM analyses. Similarly, the densities of glass ceramic samples BST5K1L0.4T850 and BST5K1L0.4S850 also increase with sintering time. Not only the sintering effect on density but the compositional effects are observed in these glass ceramic samples. With addition of Sr content, the density of glass ceramic samples decreases due to less density of Sr (2.6 g/cm^3) than Ba (3.5 g/cm^3).

3.4 Mechanical behavior

Mechanical properties depend upon crystallize size and porosity, and all such parameters depend on soaking time of sintering as well as compositions of the glass ceramics. Stress vs. strain measurements were carried out in compressive mode of uniform cylindrical glass ceramic samples BT5K1L0.0T850, BT5K1L0.0S850, BST5K1L0.4T850 and BST5K1L0.4S850 as shown in Figs. 13 and 14. The compressive study of glass ceramic sample BT5K1L0.0T850 reveals that Hooke's law of elasticity follows up to load of 757.85 MPa (Fig. 13(a)). Figure 13(b) shows the variation of strain with compressive stress and it varies directly up to load 2204.15 MPa. The glass ceramic sample BT5K1L0.0S850 withstands more loads as compared to glass ceramic sample BT5K1L0.0T850. Mechanical strength of glass ceramic samples increases with decreasing grain size or crystallite size, as clear from Table 1 [5]. Thus, compressive mechanical strength increases with increasing the sintering time for crystallization. This high elastic modulus may be attributed to van der Waals–London attractive

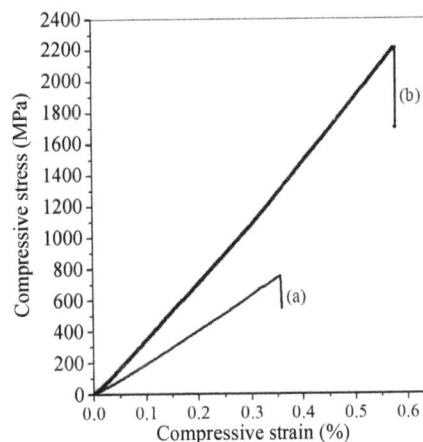

Fig. 13 Variations of compressive strain vs. compressive stress: (a) BT5K1L0.0T850 and (b) BT5K1L0.0S850.

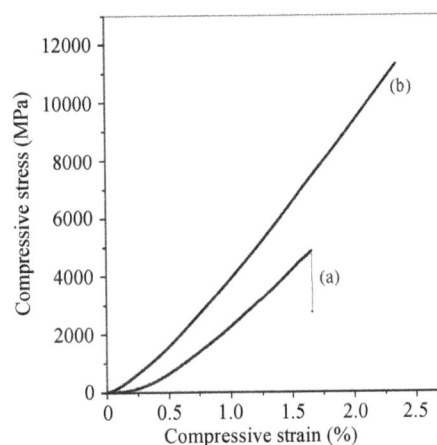

Fig. 14 Variations of compressive strain vs. compressive stress: (a) BST5K1L0.4T850 and (b) BST5K1L0.4S850.

interactions and it varies inversely with size of particles. Elastic modulus is determined by slope of stress–strain curve at any point in linear portion of the curve. The value of elastic modulus is found to be 5.664±0.012 GPa for BT5K1L0.0S850, while it is 3.041±0.001 GPa for BT5K1L0.0T850. Elastic property of glass ceramic sample BST5K1L0.4T850 describes that the Hooke's law does not be obeyed initially, and it shows the ductile nature which is due to 40% addition of Sr content in glass ceramic sample BST5K1L0.4T850. Compressive stress follows linearly with further increase of load up to yield point 4849.44 MPa (Fig. 14(a)). Thus, addition of Sr content increases the mechanical strength of glass ceramic sample. Figure 14(b) shows that ductileness of glass ceramic sample BST5K1L0.4S850 decreases with increasing sintering time for crystallization. The compressive strength is found to be 11 294.49 MPa. This high value of strength is due to addition of Sr content as well as soaking time for sintering. The glass ceramic sample BST5K1L0.4S850 is found more elastic than others and its elastic modulus is 7.070±0.001 GPa. This value of elastic modulus is much higher than those of other reported glass ceramics [25–27]. The compressive strength, elastic modulus and toughness of these glass ceramic samples have been listed in Table 1.

The toughness of glass ceramic sample BT5K1L0.0T850 is found to be 131 MJ/m^3, whereas its value for glass ceramic sample BT5K1L0.0S850 is 614 MJ/m^3. Thus, with increasing the soaking time of sintering, the ability to resist the crack in glass ceramic sample increases, and it confirms the sintering time dependence on densification. Dependence of densification is also observed from SEM micrographs and XRD analysis. When 40 mol% of Sr content is added in the glass ceramic system of BST borosilicate, then toughness is increased up to value of 3142 MJ/m^3 for glass ceramic sample BST5K1L0.4T850 and 11 824 MJ/m^3 for sample BST5K1L0.4S850. Such type of glass ceramics can be used for high load bearing applications. These glass ceramics are having high resistance to corrosion and also withstand high temperature.

3.5　Dielectric behavior

The dielectric behavior of glass ceramic samples depends on various factors such as the nature and amount of crystallinity of crystalline phases, crystallite size, grain size and grain boundary morphology, pyrochlore phases, crystal clamping and the connectivity of the high permittivity perovskite crystals in the low permittivity glassy matrix. The nature of crystalline phases and microstructure of glass ceramics can be controlled by heat treatment conditions during the crystallization of glass to glass ceramics. The solid solution of barium titanate and strontium titanate ceramic to form $(Ba_{1-x}Sr_x)TiO_3$ have significant importance for technological point of view, as such glass ceramics have very high dielectric constant. The ferroelectric phase transition in these glass ceramics has been controlled by composition as well as heat treatment schedule.

Figures 15 and 16 depict the variations of dielectric constant and dielectric loss with temperature for glass ceramic samples BT5K1L0.0T850 and

BST5K1L0.4T850 at frequencies 20 Hz, 100 Hz, 1 kHz, 10 kHz and 100 kHz. The dielectric behavior of glass ceramic sample BT5K1L0.0T850, sintered for 3 h at 850 ℃, is shown in Fig. 15. The double phase transition is observed at temperatures of 60 ℃ and 125 ℃, and it is similar to previously reported results on (BaSr)TiO$_3$ by Zhang et al. [28]. The double anomaly occurs due to successive transitions from orthorhombic to tetragonal and tetragonal to cubic crystal structures. The ferroelectric features are confirmed by symmetry breaking in dielectric constant. Here, the second phase transition is also attributed to the transition of ferroelectric phase to paraelectric at Curie temperature of 125 ℃. The highest value of dielectric constant is found to be 31 497 at frequency of 20 Hz. The dielectric loss at low frequencies increases gradually with increasing the temperature, and its value on room temperature at 20 Hz is found to be 0.01. The peak position variation with temperature is non-monotonic. This may be due to non uniform distribution of secondary phase formation in glassy matrix or defects produced by crystal clamping. Figure 16 shows the dielectric behavior of glass ceramic sample BST5K1L0.4T850 which is derived by 60/40 (in mol%) of Ba/Sr ratio for the same glass system, and it is sintered for 3 h. The dielectric constant for this glass ceramic sample at 20 Hz is observed to be 32 349. This value of dielectric constant is greater than the value of dielectric constant at the same frequency for Sr free glass ceramic sample BT5K1L0.0T850. This increase in the value of dielectric constant is attributed to the higher conductivity difference among the crystalline phase, pyrochlore phase and glass interface. The dielectric loss increases gradually with increasing the temperatures for frequencies of 20 Hz, 10 kHz and 100 kHz, and its value for 20 Hz at room temperature is 0.01.

The variations of dielectric constant and dielectric loss with temperature at above mentioned frequencies for glass ceramic samples BT5K1L0.0S850 and BST5K1L0.4S850 are shown in Figs. 17 and 18, which are derived by sintering of 6 h soaking time. The sintering time for the crystallization affects the value of dielectric constant but it does not change the dielectric behavior of the glass ceramic samples. Curie temperature of 6 h sintered glass ceramics is decreased, and this may lead to reduction of residual glass and pyrochlore phases in the glassy matrix. Figure 17 depicts the dielectric behavior of glass ceramic sample

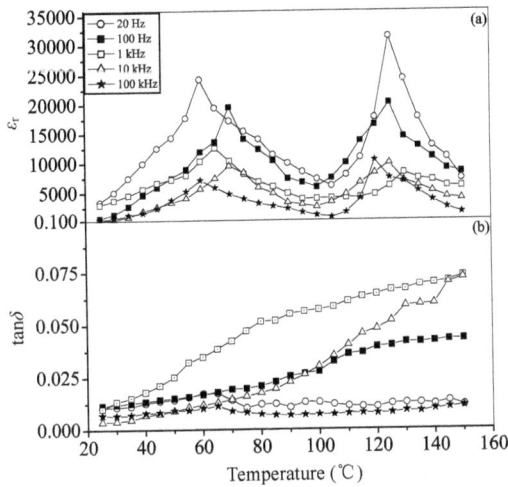

Fig. 15 Variations of (a) dielectric constant ε_r and (b) dissipation factor tanδ with temperature at different frequencies for glass ceramic sample BT5K1L0.0T850.

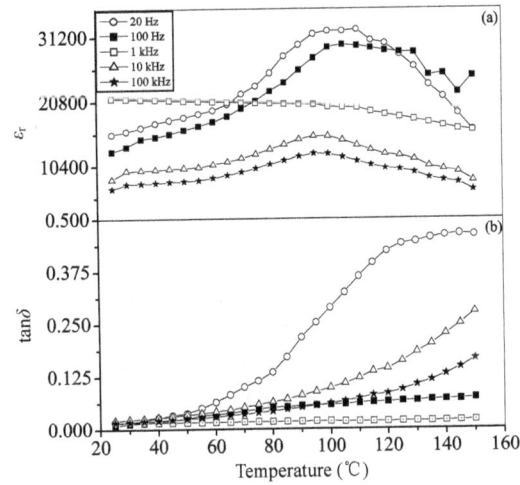

Fig. 16 Variations of (a) dielectric constant ε_r and (b) dissipation factor tanδ with temperature at different frequencies for glass ceramic sample BST5K1L0.4T850.

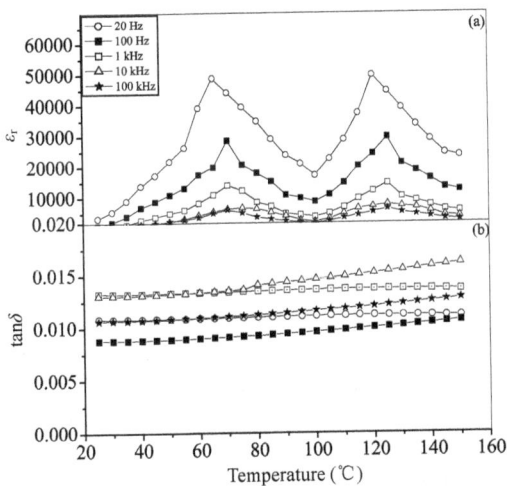

Fig. 17 Variations of (a) dielectric constant ε_r and (b) dissipation factor tanδ with temperature at different frequencies for glass ceramic sample BT5K1L0.0S850.

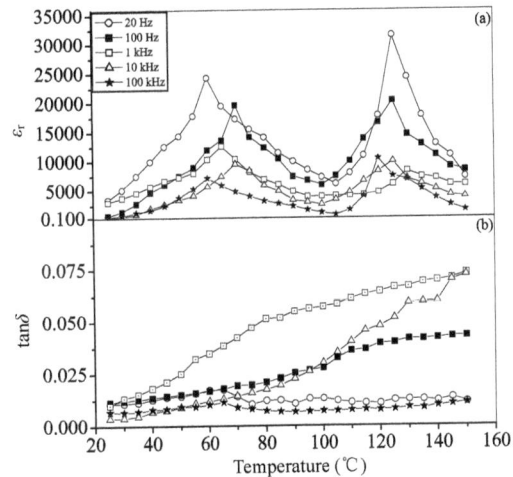

Fig. 18 Variations of (a) dielectric constant ε_r and (b) dissipation factor tanδ with temperature at different frequencies for glass ceramic sample BST5K1L0.4S850.

BT5K1L0.0S850, and this pattern shows that the dielectric constant is increased up to 49 612 from 31 497 with increase in sintering time. This high value of dielectric constant is attributed to interfacial or space charge polarization which is arising from the conductivity difference among the various phases such as major BT/BST phase, secondary phase and glass matrix interface [29]. In 6 h sintered glass ceramic samples, the amount of secondary phase is less as compared to that of 3 h, and hence, there is large conductivity difference between semiconducting grains and insulating glassy matrix. Curie temperature for

glass ceramic sample BT5K1L0.0S850 is found to be 120 ℃. The dielectric loss for this glass ceramic sample is almost independent of temperature which is good indication for high energy storage device applications such as capacitors and barrier layer capacitors. The value of dielectric loss for 20 Hz frequency at room temperature is 0.01. The variation of dielectric constant and dielectric loss with temperature at frequencies 20 Hz, 100 Hz, 1 kHz, 10 kHz and 100 kHz of glass ceramic sample BST5K1L0.4S850 is shown in Fig. 18. The peak value of dielectric constant is observed at 20 Hz and it is found to be 39 532. The dielectric loss for glass ceramic samples is not much influenced by the temperature variations, and its value at room temperature corresponding to 20 Hz is found to be 0.01. The values of dielectric constant and dielectric loss at various frequencies have been listed in Table 2. These glass ceramic samples show the ferroelectric behavior due to addition of La_2O_3. The solid solution crystallites of BT/BST become semiconducting due to diffusion of La ions in the solution. Strontium titanate leads to the formation of electronic defects/vacancies of titanium cation with addition of La_2O_3 and depends on the doping level and processing condition as per the following equations, where all symbols are in Kröger–Vink notations.

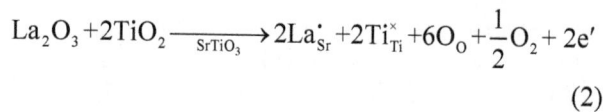

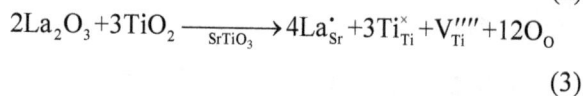

$$La_2O_3 + 2TiO_2 \xrightarrow{SrTiO_3} 2La_{Sr}^{\cdot} + 2Ti_{Ti}^{x} + 6O_O + \frac{1}{2}O_2 + 2e' \tag{2}$$

$$2La_2O_3 + 3TiO_2 \xrightarrow{SrTiO_3} 4La_{Sr}^{\cdot} + 3Ti_{Ti}^{x} + V_{Ti}'''' + 12O_O \tag{3}$$

The doping of La_2O_3 in small concentration in perovskite ceramics induces n-type semiconductivity by electronic compensation which is given by Eq. (4), but higher concentration of La^{3+} creates titanium vacancies given by Eq. (5).

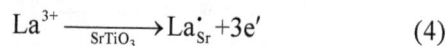

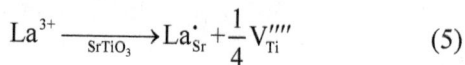

$$La^{3+} \xrightarrow{SrTiO_3} La_{Sr}^{\cdot} + 3e' \tag{4}$$

$$La^{3+} \xrightarrow{SrTiO_3} La_{Sr}^{\cdot} + \frac{1}{4}V_{Ti}'''' \tag{5}$$

4 Conclusions

Bulk transparent BST borosilicate glasses were prepared successfully by melt quench method, and their glass ceramics were formed by solid state sintering route. XRD patterns of these glass ceramic samples confirm the formation of the major perovskite phase of $BaTiO_3/(Ba,Sr)TiO_3$ along with trace amount of secondary phase of $Ba_2TiSi_2O_8$. The crystallite size decreases with increasing the sintering time or

increasing the content of SrO. The mechanical strength of glass ceramic samples increases with decreasing the crystallite size. The value of dielectric constant is found to maximum (49 612) for glass ceramic sample BT5K1L0.0S850. The dielectric constant is increased with increasing the soaking time for sintering of glass ceramics. These glass ceramic samples show the ferroelectric behavior in barium rich glass ceramic samples.

Acknowledgements

The authors gratefully acknowledge the University Grant Commission (UGC), New Delhi, India for financial support under major research project F.No.37-439/2009 (SR).

References

[1] Beall GH, Pinckney LR. Nanophase glass-ceramics. *J Am Ceram Soc* 1999, **82**: 5–16.

[2] Thakur OP, Kumar D, Parkash O, *et al.* Electrical characterization of strontium titanate borosilicate glass ceramics system with bismuth oxide addition using impedance spectroscopy. *Mater Chem Phys* 2003, **78**: 751–759.

[3] Henry J, Hill RG. The influence of lithia content on the properties of fluorphlogopite glass-ceramics. II. Microstructure hardness and machinability. *J Non-Cryst Solids* 2003, **319**: 13–30.

[4] Schneider SJ. *Engineered Materials Handbook Volume 4: Ceramics and Glasses.* Materials Park, Ohio: ASM International, 1991.

[5] Gomaa MM, Abo-Mosallam HA, Darwish H. Electrical and mechanical properties of alkali barium

Table 2 **Dielectric characteristics of glass ceramic samples BT5K1L0.0T850, BT5K1L0.0S850, BST5K1L0.4T850 and BST5K1L0.4S850**

Glass ceramic sample code	Curie temperature T_C (°C)	Room temperature dielectric parameters at 20 Hz		Value of maximum dielectric constant ε_r				
		ε_r	$\tan\delta$	20 Hz	100 Hz	1 kHz	10 kHz	100 kHz
BT5K1L0.0T850	125	3448	0.01	31497	20712	9897	8366	10420
BT5K1L0.0S850	120	3370	0.01	49612	29956	14707	8142	6267
BST5K1L0.4T850	110	15695	0.01	32349	30032	21441	15195	12276
BST5K1L0.4S850	100	20758	0.01	39532	32310	22177	18365	23447

titanium alumino borosilicate glass-ceramics containing strontium or magnesium. *J Mater Sci: Mater El* 2009, **20**: 507–516.

[6] Haung Q, Gao L, Sun J. Effect of adding carbon nanotubes on microstructure, phase transformation, and mechanical property of BaTiO$_3$ ceramics. *J Am Ceram Soc* 2005, **88**: 3515–3318.

[7] Jana F, Shirane J. *Ferroelecrtrics Crystals*. Oxford: Peragamon Press, 1962.

[8] Divya PV, Kumar V. Crystallization studies and properties of (Ba$_{1-x}$Sr$_x$)TiO$_3$ in borosilicate glass. *J Am Ceram Soc* 2007, **90**: 472–476.

[9] Divya PV, Vignesh G, Kumar V. Crystallization studies and dielectric properties of (Ba$_{0.7}$Sr$_{0.3}$)TiO$_3$ in bariumaluminosilicate glass. *J Phys D: Appl Phys* 2007, **40**: 7804–7810.

[10] Gorzkowski EP, Pan M-J, Bender B. Glass-ceramics of barium strontium titanate for high energy density capacitors. *J Electroceram* 2007, **18**: 269–276.

[11] Gorzkowski EP, Pan M-J, Bender BA, *et al.* Effect of additives on the crystallization kinetics of barium strontium titanate glass–ceramics. *J Am Ceram Soc* 2008, **91**: 1065–1069.

[12] Zhang Q, Wang L, Luo J, *et al.* Improved energy storage density in barium strontium titanate by addition of BaO–SiO$_2$–B$_2$O$_3$ glass. *J Am Ceram Soc* 2009, **92**: 1871–1873.

[13] Isaka N, Ohkawa K, Kiyono H, *et al.* Effects of glass components on crystallization and dielectric properties of BST glass–ceramics. *J Mater Sci: Mater El* 2008, **19**: 1233–1239.

[14] Rangarajan B, Jones B, Shrout T, *et al.* Barium/lead-rich high permittivity glass–ceramics for capacitor applications. *J Am Ceram Soc* 2007, **90**: 784–788.

[15] Chen G, Zhang W, Liu X, *et al.* Preparation and properties of strontium barium niobate based glass-ceramics for energy storage capacitors. *J Electroceram* 2011, **27**: 78–82.

[16] Song J, Chen G, Yuan C, *et al.* Effect of the Sr/Ba ratio on the microstructures and dielectric properties of SrO–BaO–Nb$_2$O$_5$–B$_2$O$_3$ glass–ceramics. *Mater Lett* 2014, **117**: 7–9.

[17] Yadav AK, Gautam C, Singh P. Crystallization kinematics and dielectric behavior of (Ba,Sr)TiO$_3$ borosilicate glass ceramics. *New Journal of Glass and Ceramics* 2012, **2**: 126–131.

[18] Yadav AK, Gautam CR. A review on crystallisation behaviour of perovskite glass ceramics. *Adv Appl Ceram* 2014, **113**: 193–207.

[19] Gautam CR, Yadav AK, Singh P. Synthesis, crystallisation and microstructural study of perovskite (Ba,Sr)TiO$_3$ borosilicate glass ceramic doped with La$_2$O$_3$. *Mater Res Innov* 2013, **17**: 148–153.

[20] Arend H, Kihlborg L. Phase composition of reduced and reoxidized barium titanate. *J Am Ceram Soc* 1969, **52**: 63–65.

[21] Alfors JT, Stinson MC, Matthew RA, *et al.* Seven new barium minerals from eastern Fresno County, California. *Am Mineral* 1965, **50**: 314–340.

[22] Yadav AK, Gautam CR, Gautam A, *et al.* Structural and crystallization behavior of (Ba,Sr)TiO$_3$ borosilicate glasses. *Phase Transitions* 2013, **86**: 1000–1016.

[23] Joseph J, Vimala TM, Raja J, *et al.* Structural investigations on the (Ba,Sr)(Zr,Ti)O$_3$ system. *J Phys D: Appl Phys* 1999, **32**: 1049–1054.

[24] Gautam C, Yadav AK, Mishra VK, *et al.* Synthesis, IR and Raman spectroscopic studies of (Ba,Sr)TiO$_3$ borosilicate glasses with addition of La$_2$O$_3$. *Open Journal of Inorganic Non-metallic Material* 2012, **2**: 47–54.

[25] Dupen B. Measuring Young's modulus with metal flatstock. *The Technology Interface Journal* 2007, available at http://technologyinterface.nmsu.edu/Fall07/10_Dupen/index.pdf.

[26] Ryu S-S, Kim H-T, Kim H-J, *et al.* Characterization of mechanical properties of BaTiO$_3$ ceramic with different types of sintering aid by nanoindentation. *J Ceram Soc Jpn* 2009, **117**: 811–814.

[27] Yadav AK. Synthesis, microstructure and dielectric properties of barium strontium titanate borosilicate glass ceramics. PhD Thesis. Lucknow, India: University of Lucknow, 2013: 294–312.

[28] Zhang Y, Ma T, Wang X, *et al.* Two dielectric relaxation mechanisms observed in lanthanum doped barium strontium titanate glass ceramics. *J Appl Phys* 2011, **109**: 084115.

[29] Thakur OP, Kumar D, Parkash O, *et al.* Crystallization, microstructure development and dielectric behaviour of glass ceramics in the system [SrO·TiO$_2$]–[2SiO$_2$·B$_2$O$_3$]–La$_2$O$_3$. *J Mater Sci* 2002, **37**: 2597–2606.

20

Frequency and temperature dependent electrical characteristics of CaTiO$_3$ nano-ceramic prepared by high-energy ball milling

Subhanarayan SAHOO[a,*], Umasankar DASH[b], S. K. S. PARASHAR[c], S. M. ALI[a]

[a]School of Electrical Engineering, KIIT University, Bhubaneswar 751024, India
[b]Center for Nanotechnology, KIIT University, Bhubaneswar 751024, India
[c]School of Applied Sciences, KIIT University, Bhubaneswar 751024, India

Abstract: Nanocrystalline calcium titanate (CT) ceramic has been synthesized by a combination of solid-state reaction and high-energy ball milling. This nano-ceramic is characterized by X-ray diffraction (XRD), dielectric study and impedance spectroscopy. The XRD pattern shows single phase ceramic of orthorhombic symmetry. The frequency-dependent dielectric study shows that the dielectric constant is maximized at low frequencies and decreases with an increase in frequency. Impedance spectroscopy analyses reveal a non-Debye type relaxation phenomenon. A significant shift in impedance loss peaks toward the higher-frequency side indicates conduction in the material favoring the long-range motion of mobile charge carriers. The grain conduction effect is observed from the complex impedance spectrum by the appearance of one semicircular arc in Nyquist plot. It is also observed that the resistance decreases with an increase in temperature showing a negative temperature coefficient of resistance (NTCR). Various thermistor parameters have been calculated by fitting with Steinhart–Hart equation. The modulus plots represent the presence of temperature-dependent electrical relaxation phenomenon with the material. The frequency-dependent AC conductivity at different temperatures indicates that the conduction process is thermally activated. The activation energy has been calculated from an Arrhenius plot of DC conductivity and relaxation frequency.

Keywords: high-energy ball milling; dielectric study; impedance spectroscopy; thermistor; conductivity

1 Introduction

CaTiO$_3$ (CT) belongs to the important group of compounds with a perovskite-type structure which has been widely used in electronic devices, and it is a key component of synroc (a type of synthetic rock used to store nuclear waste) [1]. It has high dielectric constant, low dielectric loss and large temperature coefficient of resonant frequency, making it a promising component in the production of communication equipment operating at microwave frequencies (ultra high frequency (UHF) and super high frequency (SHF)), which in turn are used in microwave dielectric applications (as resonators and filters) [2–4]. Also, it is a material that can be employed as a thermally sensitive resistor element due to its negative temperature coefficient, and for the immobilization of highly radioactive wastes. Such unique properties give this material much attention, and many investigations regarding its many uses have been carried out in recent

---

* Corresponding author.
E-mail: subhanarayannanotech@gmail.com

years [5]. Recently, visible photoluminescence properties at room temperature in disordered structurally perovskite titanates (CaTiO$_3$) and highly emissive red-emitting phosphors have been reported in the literature [6–11].

Different methods have been reported in the literature for the synthesis of CaTiO$_3$ powders. This perovskite was initially prepared by conventional solid-state reaction between TiO$_2$ and CaCO$_3$ or CaO at temperature of approximately 1623 K [12]. However, CaTiO$_3$ powders obtained by this method present several problems, such as high-processing temperature, inhomogeneity and contamination by impurities with a non-uniform particle size distribution [5]. To minimize these problems, wet chemical methods have been employed to synthesize CaTiO$_3$ powders with desired stoichiometry, such as sol–gel [2], co-precipitation [13], combustion method [14], organic–inorganic solution technique [15] and hydrothermal process [16].

High-energy ball milling has been used for many years in producing ultra fine powders of nano and submicron sizes. The severe and intense mechanical action on the solid surfaces leads to physical and chemical changes in the near surface region where the solids come into contact under mechanical forces. These mechanically initiated chemical and physicochemical effects in solids are generally termed as the mechanochemical effect. This route is currently being used to synthesize inorganic materials as it exhibits some advantages, such as the reduction in sintering temperature [17,18]. The study of the mechanochemical effect on fine particles has created much interest among researchers because of its several advantages to downstream processes like reducing annealing and sintering temperature, reducing phase transformation temperature, enhancing leaching process, decreasing thermal decomposition temperature, and increasing particle reactivity [19,20]. The mechanochemical synthesis process is carried out in high-intensity grinding mills such as vibro mills, planetary mills and oscillating mills. It has been noticed that the size reduction process and the microstructural evolution of CaTiO$_3$ during milling process are mainly influenced by the type of impulsive stress applied by the grinding media, which can either be an impact or shear type. Moreover, other parameters such as atmosphere composition and presence of different liquid media inside the grinding mill affect the mechanochemical process. In fact, when the mechanochemical synthesis of CaO and TiO$_2$ is carried out in planetary mills at higher rotational speeds to produce CaTiO$_3$, the impact stress is dominant, and not much attention is given on the mechanochemical mechanism itself.

The aim of this work, therefore, is to give an additional contribution in understanding the mechanochemical synthesis of CaTiO$_3$ nanoparticles without the deleterious phase and to study their electrical behavior via impedance and dielectric investigation.

2 Experiment

CaTiO$_3$ ceramic powders were prepared by high-energy ball milling technique. In this technique, CaO (99.9%) and TiO$_2$ (99.95%) were used as starting materials. Stoichiometric amounts of these materials were weighed. The high-energy ball milling of these weighed powders was done in Retsch High Energy ball mill. Initially, the weighed powders were mixed in tungsten carbide vial with tungsten carbide balls (10 mm in diameter) as milling media for 30 min at 100 rpm. The ball-to-powder ratio (BPR) was maintained at 20:1 by weight. Then it was increased to 300 rpm and high-energy milling was started. It was observed that the dry milling was not suitable for these materials because of sticking of powders to the walls of vials. To avoid this problem, toluene was added as a medium which also acted as a coolant and facilitated proper mixing of the powders during milling. Sampling was done every 5 h of milling. It was observed that CT phase was not formed even after 15 h of milling, the milled powders required further heat treatment to realize CT phase formation. To avoid the contamination, milling time was restricted to 15 h. After proper mixing, mixed powders were calcined at 600 ℃ for 2 h by an indigenous programmable furnace with intermediate grinding to avoid agglomeration of the particles. The calcined powders were used for the study of their phase formation as well as their reaction mechanism. After studying the phase formation, the powders were again ground and mixed with poly vinyl alcohol (PVA) (which acted as a binder) to reduce the brittleness, and have better compactness amongst the granules of the material. The green pellets with dimensions of 10 mm in diameter and 1 mm in thickness were made using a uniaxial press with the help of tungsten carbide dye.

The synthesized powders were structurally characterized by X-ray diffraction (XRD) using a Philips diffractometer model PW-1830 with Cu Kα radiation ($\lambda = 1.5418$ Å) in a wide range of 2θ ($20° < 2\theta < 70°$) at a scanning rate of 2 (°)/min. In order to measure the electrical properties of CT ceramic, a disc was pressed uniaxially at 200 MPa with 2 wt% PVA added as a binder. Afterwards, this disc was sintered at 1200 ℃ for 2 h. Silver contacts were deposited on the opposite disc faces and heated at 700 ℃ for 15 min. The frequency (100 Hz–1 MHz) and temperature (300–500 ℃) dependent dielectric measurements were carried out using an LCR meter (HIOKI 3532 LCR HiTester) connected to a computer.

3 Results and discussion

3.1 XRD analysis

Figure 1 shows the XRD plots of $CaTiO_3$ powders ball-milled for different time durations at 300 rpm. It is observed that the powders ball-milled for 15 h show almost a single phase compound with a very small amount of impurity. The 15-h ball-milled powders were calcinied at 600 ℃ for 2 h. The XRD of the clacinied powders is shown in Fig. 2. The XRD figure shows a single phase material with orthorhombic structure in accordance with the ICDD No. 89-56. The average crystallite size is estimated by Scherrer's equation using the full width at half maximum (FWHM) of the most intense peak and found to be 90 nm. The lattice parameters are calculated to be $a = 5.3914(0)$ Å, $b = 5.4425(4)$ Å, and $c = 7.6526(4)$ Å.

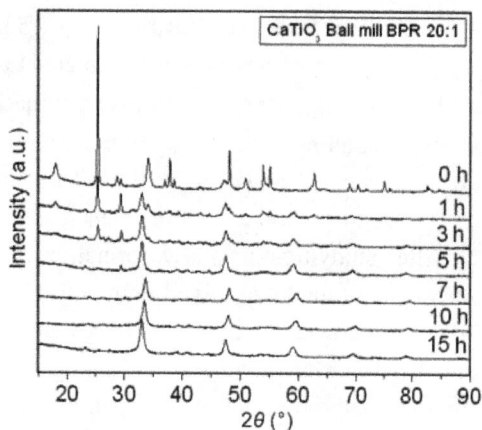

Fig. 1 XRD plots of $CaTiO_3$ powders ball-milled for different time durations.

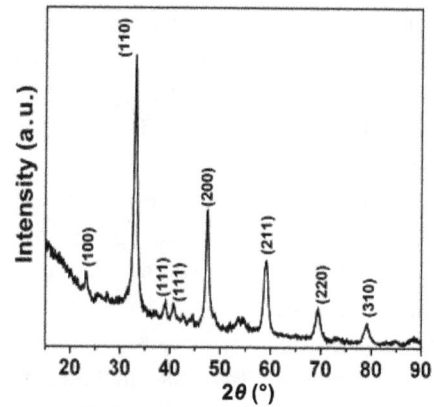

Fig. 2 XRD plot of $CaTiO_3$ powders ball-milled for 15 h and calcinied at 600 ℃ for 2 h.

3.2 Dielectric analysis

Figure 3(a) shows the variation of dielectric constant with respect to frequency at various temperatures. The nature of the dielectric permittivity for free dipoles oscillating in an alternating field may be described in the following way. At very low frequencies ($\omega \ll 1/\tau$, τ is the relaxation time), dipoles follow the field and we have $\varepsilon' = \varepsilon_s$ (value of the dielectric constant at quasistatic fields). As the frequency increases (with $\omega < 1/\tau$), dipoles begin to lag behind the field and ε' slightly decreases. When frequency reaches the characteristic frequency ($\omega = 1/\tau$), the dielectric constant drops (relaxation process). At very high frequencies ($\omega \gg 1/\tau$), dipoles can no longer follow the field and $\varepsilon' \approx \varepsilon_\infty$ (high-frequency value of ε'). Qualitatively this behavior has been observed in Fig. 3(a). The dielectric constant at low frequency is rather high, and is found to decrease with frequency at first and then becomes more or less stabilized. The high value of ε' at frequencies lower than 1 kHz, which increases with decreasing frequency and increasing temperature, corresponds to bulk effect of the system.

Figure 3(b) plots the angular frequency dependence of dielectric loss of $CaTiO_3$ at various temperatures. The dielectric loss is rather high at low frequency but falls quickly with rising frequency. Similar to the dependence of dielectric constant in temperature, the dielectric loss increases with increasing temperature. This indicates the thermally activated nature of the dielectric relaxation of the system. The fast rising trend of $\tan\delta$ at low frequencies is a representative of the presence of DC conductivity in CT ceramic. Higher values of dielectric constant interestingly observed only at very high temperature and very low frequencies may be attributed to free charge buildup at the interfaccs within the bulk of the sample (interfacial Maxwell–Wagner (MW) polarization [21]) and at the interface between the sample and the electrodes (space–charge polarization) [22].

3.3 Impedance spectroscopy study

Figure 4(a) shows the variation of the real part of impedance (Z') with frequency at various temperatures. It is observed that the magnitude of Z' decreases with increase in both frequency and temperature, indicating an increase in AC conductivity with rise in temperature and frequency. The Z' values for all temperatures merging at high frequency may be due to the release of space charges, as a result of reduction in barrier properties of the material with rise in temperature. Further, at low frequency Z' values decreases with increase in temperature show

negative temperature coefficient of resistance (NTCR)-type behavior similar to that of semiconductors.

Figure 4(b) shows the variation of the imaginary part of impedance (Z'') with frequency at different temperatures. The curves show that Z'' values reach a maxima peak (Z''_{max}) and the value of Z''_{max} shifts to higher frequencies with increasing temperature. A typical peak broadening which is slightly asymmetrical in nature can be observed with the rise in temperature. The broadening of peaks in frequency explicit plots of Z'' suggests that there is a spread of relaxation time, i.e., the existence of a temperature-dependent electrical relaxation phenomenon in the material. The merger of Z'' values in the high frequency region may possibly be an indication of the accumulation of space charges in the material. The relaxation process may be due to the presence of immobile species at low temperature and defects at higher temperature.

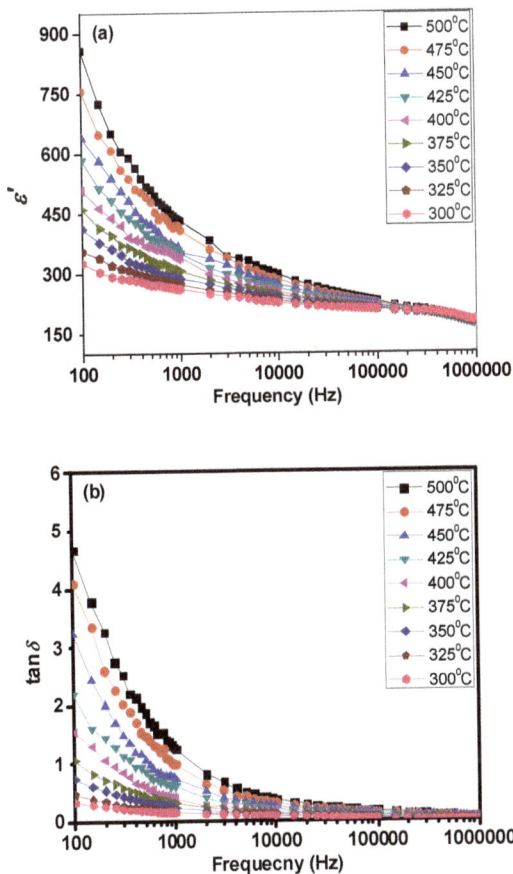

Fig. 3 Frequency-dependent (a) dielectric constant and (b) dielectric loss of CaTiO₃ at different temperatures.

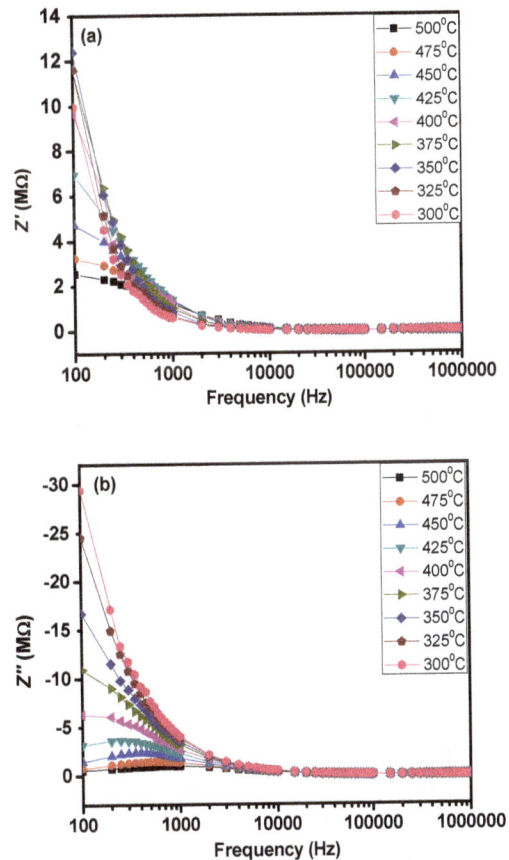

Fig. 4 Frequency-dependent variations of (a) real part Z' and (b) imaginary part Z'' for CT ceramic at different temperatures.

Figure 5 shows the complex impedance plots (Nyquist plots) of CT ceramic at different temperatures between 300 ℃ and 500 ℃. Single depressed semicircle is observed within the studied temperature range, representing the grain effect in the material. It is also observed that with the increase in temperature the radius of the semicircles decreases representing decrement in the resistivity. All the semicircles exhibit some depression degree instead of a semicircle centered at the real axis Z' due to a distribution of relaxation time. The bulk effect of the material is obtained by fitting the experimental response to that of an equivalent circuit, which is usually considered to comprise of one parallel resistor–CPE (constant phase element) elements. The circuit fitting parameter was done by ZView software with the modeled circuits. The experimental value of bulk resistance (R_b) at different temperatures has been obtained from the intercept of the semicircular arc on the real axis (Z'). The depression of the semicircle is considered further evidence of polarization phenomena with a distribution of relaxation time. This can be referred as the non-Debye type relaxation in which there is a distribution of relaxation time. This non-ideal behavior can be correlated to several factors, such as grain orientation, grain boundary, stress–strain phenomena and atomic defect distribution. This modification leads to Cole–Cole empirical behavior described by the following equation:

$$Z^* = R / (1 + (j\omega / \omega_0)^{1-n}) \qquad (1)$$

where n represents the magnitude of the departure of the electrical response from an ideal condition and can be determined from the location of the center of the Cole–Cole circles. Least-squares fitting to the complex

impedance data gives the value of $n > 0$ at all the temperatures, suggesting the dielectric relaxation to be of poly-dispersive non-Debye type.

After performing the impedance spectroscopy data simulation, the calculated values of the equivalent electric circuit parameters were then related to the characteristic parameters. Results of the data simulation in a form of estimated parameters of the relaxation processes are given in Table 1.

Figure 6 shows both the experimental and simulated resistance values within the temperature range of 300–500 ℃. The simulated data has been obtained by performing Steinhart–Hart equation within the temperature range of 300–500 ℃.

From the graph we can conclude the experimental data are nearly equal to the simulated data. The experimental and simulated data both show NTCR property within the material, which signifies that nanocrystalline CaTiO$_3$ can be used as a promising environment-friendly material for high-temperature thermistor application. The Steinhart–Hart equation with the squared term eliminated is the most common

Table 1 Parameters obtained from temperature dependent impedance spectroscopy data for CT ceramic

Temperature (℃)	$R\,(\Omega)$	$C\,(\text{F})$
300	9.6525E7	8.2146E-12
325	7.2310E7	1.0928E-11
350	3.6844E7	2.1410E-11
375	2.5996E7	3.0043E-11
400	1.4624E7	5.3835E-11
425	9.0277E6	8.6176E-11
450	5.8032E6	6.7179E-11
475	3.7285E6	6.7649E-11
500	7.2358E6	7.2358E-11

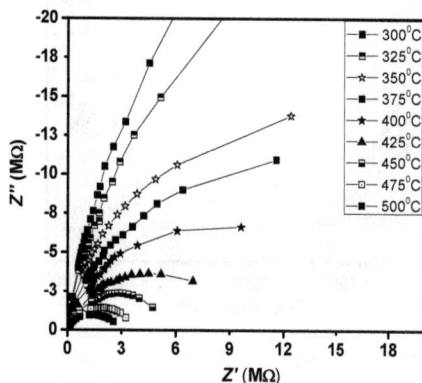

Fig. 5 Cole–Cole plots between Z' and Z'' for CT ceramic measured at various temperatures.

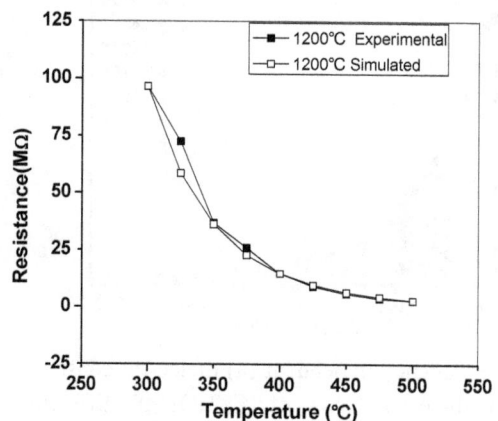

Fig. 6 Variations of both the experimental and simulated resistances.

form used and is usually found explicit in temperature T:

$$1/T = A + B\ln R + C(\ln R)^3 \qquad (2)$$

where T is in kelvin unit; A, B and C are the curve fitting coefficients; and $\ln R$ is the natural logarithm of a resistance in ohm. Solving this equation set we obtain following values for fitting coefficients of Steinhart–Hart equation: $A = 0.000416221$, $B = 0.000034033$ and $C = 0.000000113$. Putting these results back into the above equation, we obtain the simulated value of resistances at corresponding temperatures. The A, B and C values obtained from the fitting are well within standard acceptable value of thermistor.

The β (known as sensitivity index of thermistor materials) value is a very important parameter in the description and specification of thermistor materials and thermistor components. The thermistor characteristic parameter can be expressed as

$$\beta = \left(\frac{TT_N}{T_N - T}\right)\ln\frac{R_T}{R_N} \qquad (3)$$

where R_T is the resistance at temperature T; R_N is the resistance at temperature T_N known. Figure 7 shows β value in the temperature range of 300–500 ℃.

α is a material characteristic which shows the resistance change percentage per degree centigrade. The temperature coefficient is a basic concept in thermistor calculation.

$$\alpha = (1/R)[\mathrm{d}(R)/\mathrm{d}T] = -\beta/T^2 \qquad (4)$$

Because the resistance of NTCR thermistor is a function of temperature, the α value of a particular thermistor material is also nonlinear across the relevant temperature range. The variation of α parameter with temperature is shown in Fig. 8. For the usual thermistor materials β constant is closely related to

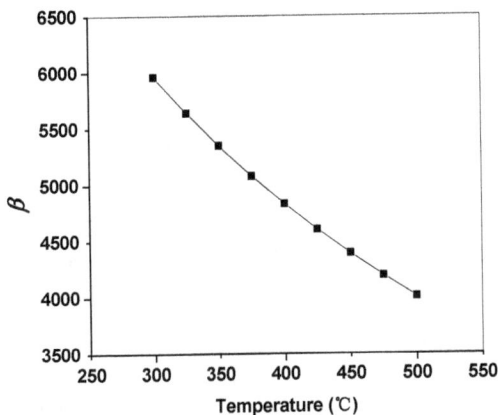

Fig. 7   Variation of β value with temperature.

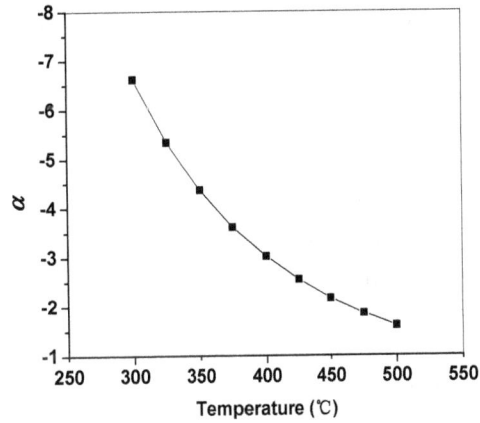

Fig. 8   Variation of α value with temperature.

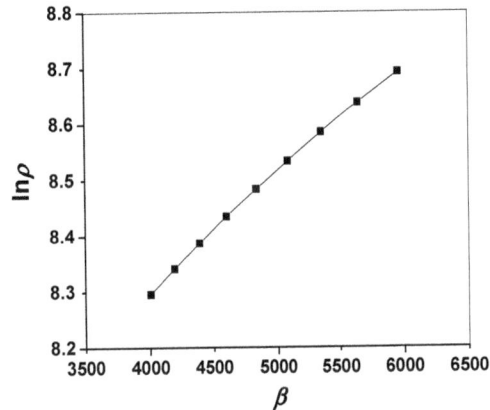

Fig. 9   Variation of $\ln\rho$ with β.

resistivity as shown in Fig. 9. The general information on sensitivity of material resistivity to temperature can be interpreted from the β value indicated in Fig. 9. It is well known that material having high β value has high temperature stability. Here we got β value in the range of 4000–6000 K, which means CT ceramic is capable to withstand temperature surges and very suitable in high temperature electronic applications.

3.4   Electrical modulus

In order to clarify whether the multiple relaxations and/or conduction processes are involved in the complex dielectric response of $CaTiO_3$, ceramics are given with the dielectric modulus formalism in addition to the complex permittivity analysis. The complex dielectric modulus corresponds to the relaxation of the electric field in the material when the electrical induction D is maintained as constant and defined as

$$M^*(f) = \frac{1}{\varepsilon^*} = M'(f) + iM''(f) \qquad (5)$$

The frequency dependence of the imaginary part of the

dielectric modulus $M''(f)$ at various temperatures generally provides information concerning the charge transport mechanism such as electrical transport and conductivity relaxation, and has been successfully used to distinguish localized dielectric relaxation processes from long-range conductivity and short/long-range polaron hoping in ceramics. A conductivity relaxation is indicated by the presence of a peak in the imaginary part of the dielectric modulus spectrum not accompanied by a peak in the imaginary part of the permittivity, while a dielectric relaxation phenomenon gives maxima both in the imaginary parts of permittivity and of dielectric modulus spectra.

Figures 10(a) and 10(b) display the (angular) frequency dependence of $M'(\omega)$ and $M''(\omega)$ for CT ceramic at different temperatures. $M'(\omega)$ shows a dispersion tendency toward M_∞ (the asymptotic value of $M'(\omega)$ at higher frequencies. In the low-temperature region, the value of $M'(\omega)$ increases with the increase in frequency and decreases with rise in temperature with slow rate, while in the high-temperature region, the value of $M'(\omega)$ increases rapidly with the increase in both temperature and frequency. It may be contributing to the conduction phenomena due to the short range mobility of charge carriers. This implies the lack of a restoring force for flow of charges under the influence of a steady electric field.

Figure 10(b) shows the frequency dependence of the imaginary part $M''(\omega)$ of the electric modulus at different temperatures. The plots show an asymmetric behavior with respect to peak maxima whose positions are frequency and temperature dependent. These spectra also reflect the motions of the ions in the material by exhibiting two apparent relaxation regions. The left region of the peak indicates the conduction process, while region on the right of the peak is associated to the relaxation process where the ions can make localized motion within the well. The asymmetric modulus peak shifts towards higher-frequency side exhibiting the correlation between the motions of mobile charge carriers. The asymmetry in peak broadening shows the spread of relaxation time with different time constant, and hence the relaxations of non-Debye type. The existence of low-frequency peaks suggests that the ions can move over long distances, whereas high-frequency peaks suggest about

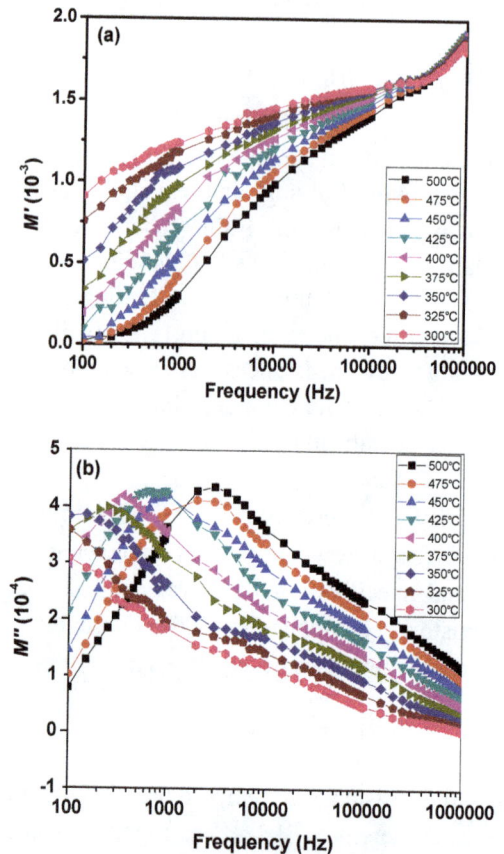

Fig. 10   Frequency dependent variations of (a) real part $M'(\omega)$ and (b) imaginary part $M''(\omega)$ for the CT ceramic with at different temperatures.

the confinement of ions in their potential well. The nature of modulus spectrums confirms the existence of hopping mechanism in the electrical conduction of the materials.

The frequency ω_m (corresponding to Z''_{max} and M''_{max}) gives the most probable relaxation time τ_m from the condition $\omega_m \tau_m = 1$. The most probable relaxation time follows the Arrhenius law, given by

$$\omega_m = \omega_0 \exp\left(\frac{-E_a}{k_B T}\right) \quad (6)$$

where ω_0 is the pre-exponential factor; k_B is Boltzman constant; and E_a is the activation energy. Figure 11 shows a plot of $\log \omega_m$ versus $1/T$, where the dots are the experimental data and the solid lines are the least-squares straight-line fit from the relaxation of both Z'' and M''. The activation energy E_a were calculated from the least-squares fit to the points. From Fig. 11 it can be seen that the activation energy calculated from Arrhenius relation

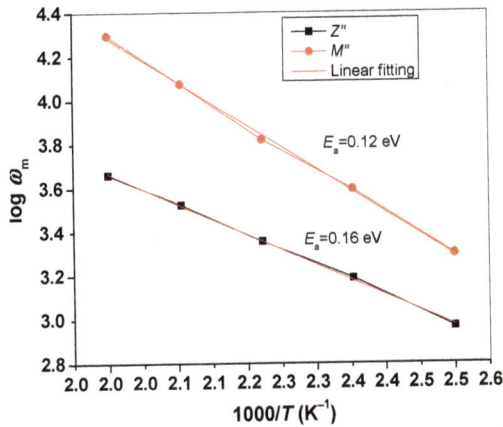

Fig. 11   Arrhenius plots of $\log \omega_\mathrm{m}$ from imaginary part of (a) impedance and (b) modulus.

are $E_\mathrm{a} = 0.16$ eV for Z'' and $E_\mathrm{a} = 0.12$ eV for M''.

3.5   Conductivity

The AC electrical conductivity was obtained in accordance with the relation $\sigma_{AC} = d/(A \cdot Z')$, where d is the thickness and A is the surface area of the specimen. Figure 12 shows the variation of AC electrical conductivity σ_{AC} of CT ceramic as a function of frequency at different temperatures. The conductivity plot possesses the following characteristics: (i) dispersion at lower and merging at higher frequencies of conductivity spectra with the increase in temperature. The plot shows conductivity increases with increase of temperature. Frequency independent behavior of the conductivity in the low-frequency region is observed but that becomes sensitive at high-frequency region, which is generally known as hopping frequency, shifted towards higher-frequency side with increase of temperature. In the higher-frequency region, the conductivity increases due to the hopping of charge carriers in finite clusters. Frequency independent AC conductivity observed in the high temperature indicates the long-range movement of mobile charge carriers. The high frequency variation of σ_{AC} is found to obey universal Jonscher's power law behavior, $\sigma_{AC} = K\omega^s$, with $0 \leqslant s \leqslant 1$, where ω is the angular frequency of AC field, in the frequency sensitive region.

Figure 13 shows the variation of σ_{DC} against the $10^3/T$. The value of bulk conductivity of the material is evaluated from the AC conductivity plot of the sample at different temperatures by theoretical fitting using Joncher's power law. At higher temperature, the conductivity versus temperature response is more or less a straight line and can be explained by a thermally activated transport of Arrhenius type:

$$\sigma_{DC} = \sigma_0 \exp\left(\frac{-E_\mathrm{a}}{k_\mathrm{B}T}\right) \qquad (7)$$

where σ_0, E_a and k_B represent the pre-exponential term, the activation energy of the mobile charge carriers and Boltzmann's constant, respectively. At lower temperature, a small deviation from the linear behavior of conductivity has been noticed and can be attributed to Mott's hopping type phenomena. The DC activation energy of the material has been estimated to 0.12 eV in the temperature. The activation energy values for the electric modulus (0.11 eV) and for DC conductivity (0.11 eV) are almost identical suggesting a hopping mechanism for CT ceramic.

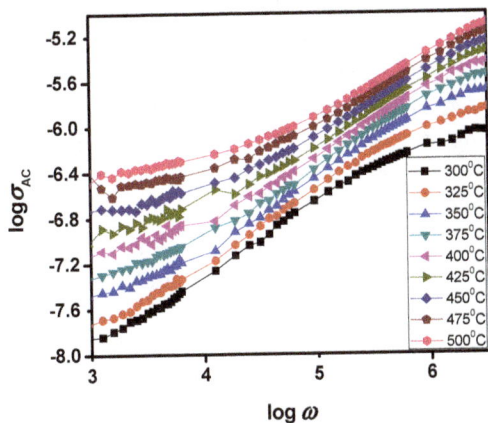

Fig. 12   Frequency dependence of AC conductivity of CT ceramic at different temperatures.

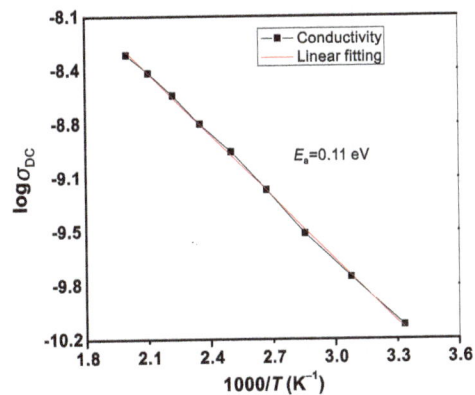

Fig. 13   Temperature dependence of DC conductivity for CT ceramic. The dots are the experimental points and the solid lines are the least-squares straight line fit.

4 Conclusions

Nanocrystalline calcium titanate ceramic has been prepared by a high-energy ball milling technique. X-ray analysis confirms the orthorhombic structure. The temperature-dependent dielectric study reveals a normal ferroelectric behavior in the material. Electrical parameters such as the real part of impedance (Z'), the imaginary part of impedance (Z'') and AC/DC conductivity as functions of both frequency and temperature have been studied through impedance spectroscopy. Nyquists plots show bulk effect and the resistance decreases with rise in temperature, which indicates the NTCR behavior of the sample. The thermistor parameter has been obtained by the standard equations. It is found that the values of the parameters obtained from the experimental results are in accordance with the other thermistor used in industries. The electrical relaxation process occurring in the material has been found to be temperature dependent. Modulus analysis has established the possibility of hopping mechanism for electrical transport processes in the system. The AC conductivity spectrum is found to obey Jonscher's universal power law. The frequency-dependent AC conductivity at different temperatures indicates that the conduction process is thermally activated. In particular, a high-performance NTCR thermistor is important because the thermistor should detect temperature over a wide range and under very severe condition. The thermistor obeying Steinhart–Hart equation means all the prepared sample useful for high temperature applications with rapid thermal response.

References

[1] Lutze W, Ewing RC, Eds. *Radioactive Waste Forms for the Future*. Amsterdam: Elsevier, 1988: 233–334.

[2] Pfaff G. Synthesis of calcium titanate powders by the sol–gel process. *Chem Mater* 1994, **6**: 58–62.

[3] Cavalcante LS, Marques VS, Sczancoski JC, *et al*. Synthesis, structural refinement and optical behavior of $CaTiO_3$ powders: A comparative study of processing in different furnaces. *Chem Eng J* 2008, **143**: 299–307.

[4] Cavalcante LS, Simões AZ, Santos LPS, *et al*. Dielectric properties of $Ca(Zr_{0.05}Ti_{0.95})O_3$ thin films prepared by chemical solution deposition. *J Solid State Chem* 2006, **179**: 3739–3743.

[5] Evans IR, Howard JAK, Sreckovic T, *et al*. Variable temperature in situ X-ray diffraction study of mechanically activated synthesis of calcium titanate, $CaTiO_3$. *Mater Res Bull* 2003, **38**: 1203–1213.

[6] Marques VS, Cavalcante LS, Sczancoski JC, *et al*. Synthesis of $(Ca,Nd)TiO_3$ powders by complex polymerization, Rietveld refinement and optical properties. *Spectrochim Acta A* 2009, **74**: 1050–1059.

[7] de Lazaro S, Milanez J, de Figueiredo AT, *et al*. Relation between photoluminescence emission and local order–disorder in the $CaTiO_3$ lattice modifier. *Appl Phys Lett* 2007, **90**: 111904.

[8] Pan Y, Su Q, Xu H, *et al*. Synthesis and red luminescence of Pr^{3+}-doped $CaTiO_3$ nanophosphor from polymer precursor. *J Solid State Chem* 2003, **174**: 69–73.

[9] Haranath D, Khan AF, Chander H. Bright red luminescence and energy transfer of Pr^{3+}-doped $(Ca,Zn)TiO_3$ phosphor for long decay applications. *J Phys D: Appl Phys* 2006, **39**: 4956.

[10] Zhang X, Zhang J, Nie Z, *et al*. Enhanced red phosphorescence in nanosized $CaTiO_3:Pr^{3+}$ phosphors. *Appl Phys Lett* 2007, **90**: 151911

[11] Okamoto S, Kobayashi H, Yamamoto H. Enhancement of characteristic red emission from $SrTiO_3:Pr^{3+}$ by Al addition. *J Appl Phys* 1999, **86**: 5594.

[12] Kay HF, Bailey PC. Structure and properties of $CaTiO_3$. *Acta Cryst* 1957, **10**: 219–226.

[13] Zhang X, Zhang J, Ren X, *et al*. The dependence of persistent phosphorescence on annealing temperatures in $CaTiO_3:Pr^{3+}$ nanoparticles prepared by a coprecipitation technique. *J Solid State Chem* 2008, **181**: 393–398.

[14] Muthuraman M, Patil KC, Senbagaraman S, *et al*. Sintering, microstructural and dilatometric studies of combustion synthesized synroc phases. *Mater Res Bull* 1996, **31**: 1375–1381.

[15] Lee SJ, Kim YC, Hwang JH. An organic–inorganic solution technique for fabrication of nano-sized $CaTiO_3$ powder. *J Ceram Process Res* 2004, **5**: 223–226.

[16] Kutty TRN, Vivekanandan R, Murugaraj P. Precipitation of rutile and anatase (TiO_2) fine powders and their conversion to $MTiO_3$ (M = Ba, Sr, Ca) by the hydrothermal method. *Mater Chem Phys*

1988, **19**: 533–546.

[17] Kong LB, Zhu W, Tan OK. Preparation and characterization of Pb(Zr$_{0.52}$Ti$_{0.48}$)O$_3$ ceramics from high-energy ball milling powders. *Mater Lett* 2000, **42**: 232–236.

[18] German RM. *Sintering Theory and Practice*. New York: Wiley, 1996.

[19] Wang J, Xue JM, Wan DM, *et al.* Mechanically activating nucleation and growth of complex perovskites. *J Solid State Chem* 2000, **154**: 321–328.

[20] Siqueira JRR, Simões AZ, Stojanovic BD, *et al.* Influence of milling time on mechanically assisted synthesis of Pb$_{0.91}$Ca$_{0.1}$TiO$_3$ powders. *Ceram Int* 2007, **33**: 937–941.

[21] Bidault O, Goux P, Kchikech M, *et al.* Space charge relaxation in perovskites. *Phys Rev B* 1994, **49**: 7868–7873.

[22] Kyritsis A, Pissis P, Grammatikakis J. Dielectric relaxation spectroscopy in poly(hydroxyethy acrylate)/water hydrogels. *J Polym Sci Pol Phys* 1995, **33**: 1737–1750.

21

Synthesis and optical characterization of porous ZnO

K. SOWRI BABU[*], A. RAMACHANDRA REDDY, Ch. SUJATHA,
K. VENUGOPAL REDDY, A. N. MALLIKA

Department of Physics, National Institute of Technology Warangal, Warangal-506 004, Andhra Pradesh, India

Abstract: In this paper, a simple and cheap method to prepare porous ZnO by using zinc nitrate, ethanol and triethanolamine (TEA) is reported. The as-prepared sample consisted of nano and micro pores. The sample was calcined at 300 ℃, 400 ℃ and 500 ℃ with different heating rates. At 500 ℃, the nano pores disappeared but the sample maintained its micro porosity. Field emission scanning electron microscopy (FE-SEM) pictures confirmed that the size and growth of ZnO nanoparticles depended on the heating conditions. The infrared (IR) absorption peak of Zn–O stretching vibration positioned at 457 cm^{-1} was split into two peaks centered at 518 cm^{-1} and 682 cm^{-1} with the change of morphology. These results confirmed that Fourier transform infrared (FT-IR) spectrum was sensitive to variations in particle size, shape and morphology. The photoluminescence (PL) spectrum of porous ZnO contained five emission peaks at 397 nm, 437 nm, 466 nm, 492 nm and 527 nm. Emission intensity enhanced monotonously with increase of temperature and the change was rapid between temperatures of 300 ℃ and 500 ℃. This was due to the elimination of organic species and improvement in the crystallanity of the sample at 500 ℃.

Keywords: semiconductors; porous ZnO; optical properties

1 Introduction

Materials exhibit novel and fascinating properties with reduction of size from bulk to nanometer scale. As far as the semiconductor nanoparticles are concerned, the increase in band gap has been observed when the particle size is comparable to or smaller than the excitonic Bohr radius. Consequently, the emission spectrum shows blue shift (shift towards lower wavelength) and it is called quantum confinement effect. Among the various known semiconductors, ZnO has several advantages compared to other wide-band-gap semiconductors. It has a direct band gap of 3.37 eV and large exciton binding energy of 60 meV at room temperature [1]. Due to these unique properties, it finds applications in antireflection coatings, transparent electrodes in solar cells, ultraviolet (UV) light emitters, diode lasers, varistors, piezoelectric devices, spin-electronics, surface acoustic wave propagation, and also in sensing of gas [2]. It is well known that nanostructures have very good sensitivity with respect to size, shape, type of material and preparation techniques. In the case of ZnO, it is able to satisfy above properties because it can be easily synthesized in various shapes like particles, rods, needles, shells, tubes, wires, nails, belts, combs, propellers, springs, tetrapods, etc. [3–8].

Apart from the above variety of structures, porous

* Corresponding author.
E-mail: sowribabuk@gmail.com

ZnO can also be prepared. Porous ZnO has some specific advantages such as high surface area, chemical and photochemical stability, uniformity in pore size, shape selectivity, and rich surface chemistry [9]. The high surface area of porous ZnO makes its surface more active. The highly active surface would increase the probability of interaction of gases with the semiconductor, which in turn increases the sensitivity of the material [10]. So, the material has found a variety of promising applications such as catalysts, nano-sieve filters, dye sensitized solar cells, bio- and electrochemical sensors, bone-replacement materials and also in gas sensors [9,11,12]. For example, for dye sensitized solar cells, ZnO thin films should be porous and have high specific surface area for obtaining high conversion efficiency of light into current [13]. Recently, some attention has been paid to the synthesis of porous ZnO by using various methods such as thermal decomposition method, template assisted methods, hydrothermal method, solid-vapor process, radio frequency sputtering, electrochemical deposition, etc. [13–15]. To the best of our knowledge, an easy and economical method to prepare porous ZnO is not addressed so far. So, because of the importance of porous ZnO in various applications mentioned above, the following work is taken up.

In this work, a simple and cheap method to prepare nanoporous ZnO using zinc nitrate, ethanol and triethanolamine (TEA) is reported. The as-prepared sample was calcined at 300 ℃, 400 ℃ and 500 ℃ and studied with their structural and optical properties. The sample was characterized with X-ray diffraction (XRD), field emission scanning electron microscopy (FE-SEM), ultraviolet–visible (UV–Vis) spectrophotometer, and Fourier transform infrared (FT-IR) and photoluminescence (PL) spectrometers to study their structural and optical properties.

2 Experimental procedures

The preparation of porous ZnO is as follows. 0.53 g of $Zn(NO_3)_2 \cdot 6H_2O$ was dissolved in ethanol and stirred on magnetic hot plate until it became clear and stable. To this solution, a few drops of TEA were added and stirring was continued for one more hour at room temperature. In the next step, the temperature of the hot plate was increased to 80 ℃ to evaporate the solvent. Then, the solution was converted into a gel. The gel was subsequently swelling into foam and

underwent a self-propagating combustion reaction on the hot plate itself to give fine powder. The combustion took place within few seconds and the resulted powder was of dark brown color. Afterwards, the sample was calcined at 300 ℃ and 500 ℃ with a heating rate of 2 ℃/min in a programmable furnace. As the temperature was increased to 500 ℃, the color of the sample changed from brown to white. To know the influence of heating conditions on the size of the particles and growth, the sample was also calcined at 400 ℃ with a heating rate of 1 ℃/min.

XRD measurements were taken on Inel XRG 3000 diffractometer equipped with Cobalt Kα ($\lambda = 1.78897$ Å) radiation (15 mA, 30 kV). The morphology of the sample was studied by using ZEISS FE-SEM. Absorption spectra were acquired on Analyticazena (Specord 205) UV–Vis spectrophotometer. FT-IR spectra of the sample were recorded on Bruker Optics FT-IR Spectrometer Model: Tensor 27. PL measurements were performed on Jobin Yuon spectrofluorometer with 450 W xenon arc lamp.

3 Results and discussion

3.1 XRD analysis

Figure 1 shows the XRD patterns for the as-prepared porous ZnO and porous ZnO calcined at 300 ℃ and 500 ℃. This picture depicts the characteristic peaks (100), (002), (101), (102), (110), (103), etc., which correspond to the wurtzite ZnO structure. The broadening of these peaks decreases continuously, which is an indication of the increase in particle size and improved crystallanity with a rise of temperature. The change is drastic as the temperature increases from 300 ℃ to 500 ℃. The crystallite size of the sample is calculated using Debye–Scherrer formula:

$$D = \frac{0.9\lambda}{\beta \cos\theta}$$

where D is the particle diameter; λ is the wavelength of the X-ray used; β is the FWHM (full width at half maximum); and θ is the diffraction angle. The crystallite size values for the as-prepared porous ZnO and the samples calcined at 300 ℃ and 500 ℃ are 3 nm, 6 nm and 43 nm, respectively. This drastic change in crystallite size indicates that up to 300 ℃ only the progress of crystallization from amorphous state occurs and further growth takes place above

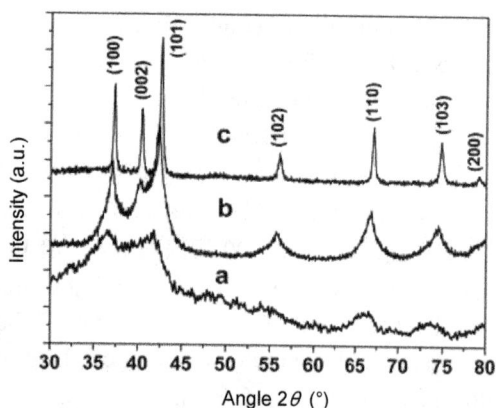

Fig. 1 Typical XRD patterns of (a) the as-prepared porous ZnO and porous ZnO annealed at (b) 300 ℃ and (c) 500 ℃.

300 ℃. Hence well crystallized ZnO is formed at 500 ℃ and the color of the sample turns out from brown to white. It can also be seen from Fig. 1 that diffraction peaks of ZnO shift towards the higher angle side compared to the as-prepared porous ZnO with a rise of temperature. This implies that ZnO nanoparticles are under compressive stress.

3.2 FE-SEM analysis

The FE-SEM micrographs of the as-prepared porous ZnO and porous ZnO annealed at 500 ℃ are presented in Fig. 2. The as-prepared sample contains nano and micro pores with diameters ranging from a few nanometers to micrometers. The porous ZnO is shown in Fig. 2(b) representing the nano pores. But, such nano pores are not detected when it is calcined at 500 ℃. From the FE-SEM micrograph in Fig. 2(d), the average particle size of the porous ZnO calcined at 500 ℃ comes out to be 47 nm. This value is in good agreement with the value obtained from the XRD data.

3.3 UV–Vis analysis

The UV–Vis absorbance spectra of the as-prepared porous ZnO and samples annealed at 300 ℃ and 500 ℃ are shown in Fig. 3. The as-prepared porous ZnO depicts an absorption at 347 nm, which is blue shifted compared to the bulk ZnO whose absorption edge is observed at 380 nm. With an increase in temperature, there is a systematic shift in the absorption edge towards higher wavelength side but the maximum shift occurrs between temperatures of 300 ℃ and 500 ℃. The absorption edge values corresponding to the samples annealed at 300 ℃ and

500 ℃ are 355 nm and 370 nm. This gradual red shift of absorption edge with increase in annealing temperature is obviously due to the increase of particle size. This indicates the strong quantum confinement effects of the porous ZnO nanoparticles. The particle size has also been calculated from the absorption edge values of all the three samples according to Ref. [2]. The effective mass model has been used for finding particle size of ZnO nanoparticles. This model utilizes the fact that the wavelength of absorption depends on particle size. The particle sizes obtained for all the three samples are 5 nm, 7 nm and 13 nm, repectively. The particle size values for the as-prepapred sample and sample calcined at 300 ℃ are in good agreement with the values measured from XRD. But in the case of

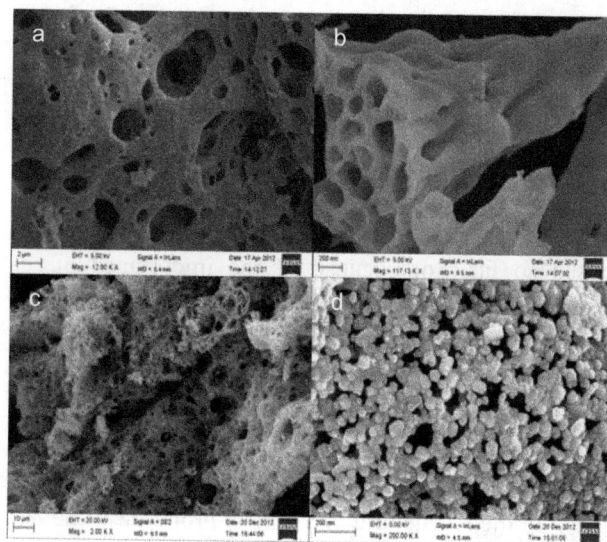

Fig. 2 FE-SEM pictures of the as-prepared porous ZnO ((a) and (b)) and porous ZnO annealed at 500 ℃ ((c) and (d)).

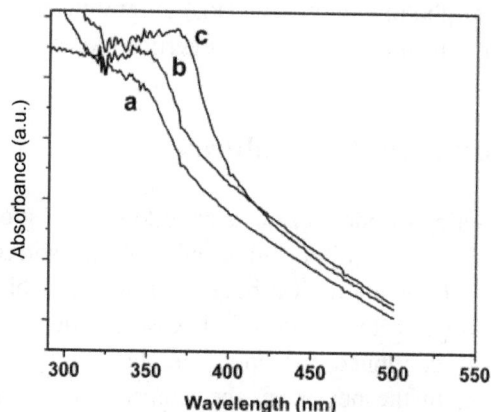

Fig. 3 UV–Vis absorption spectra of (a) the as-prepared porous ZnO and porous ZnO calcined at (b) 300 ℃ and (c) 500 ℃.

the sample calcined at 500 ℃, the particle size calculated from UV–Vis absorbance spectrum is not coinciding with the results obtained from FE-SEM and XRD.

3.4 FT-IR analysis

The FT-IR spectra of the samples measured in the range of 4000–400 cm^{-1} are presented in Fig. 4. The as-prepared ZnO shows IR peaks at 457 cm^{-1}, 1063 cm^{-1}, 1390 cm^{-1}, 1602 cm^{-1} and 3423 cm^{-1}. The sharp peak positioned at 457 cm^{-1} is attributed to the Zn–O stretching bonds. The IR bands shown in Fig. 4 are in the region of 1700–600 cm^{-1} and correspond to C=O, C–O and C–H vibrations respectively [16]. Remaining peaks are due to the O–H stretching vibrations and bending modes of the adsorbed water. With the increase of temperature, the intensity of peaks centered at 1063 cm^{-1}, 1390 cm^{-1} and 1602 cm^{-1} are deteriorated and at 500 ℃ these peaks are almost disappeared. It indicates that organic species are completely removed at 500 ℃. It is observed that the position of 457 cm^{-1} peak remains the same when annealed at 300 ℃ but shifts to 439 cm^{-1} at 500 ℃. The shift of the XRD peaks is also observed between 300 ℃ and 500 ℃, which is attributed to the compressive stress acting on ZnO nanoparticles. So, the broadening and shift of the IR peak to lower wavenumber could be due to the stress acting on ZnO nanoparticles and variation in morphology. Figure 4(b) shows the FT-IR spectrum of porous ZnO calcined at

400 ℃ at a heating rate of 1 ℃/min along with its FE-SEM micrograph. It is apparent from the micrograph that the initial particles are spherical in shape with an approximate size of 9 nm. This size seems to be critical and if once reaching that size, they tend to agglomerate to bigger particles. These nanoparticles grow to larger sizes at the expense of other small paraticles through Ostwald ripening process. Interestingly, the sharp peak at 457 cm^{-1} splits into two peaks positioned at 518 cm^{-1} and 682 cm^{-1}. Similar results are observed in the case of ZnO microcrystals and the position of the peaks is dependent on axial ratio (c/a) of the crystals [17]. The FE-SEM picture of this sample shows the change of morphology of ZnO nanoparticles from spherical to cylinder shape, so the splitting of Zn–O stretching band in FT-IR spectrum can be assigned to the change in morphology of ZnO nanoparticles. It is obvious from FT-IR spectrum of porous ZnO calcined at 400 ℃ at a heating rate of 1 ℃/min that organic species are not eliminated at this temperature.

3.5 PL analysis

Room-temperature PL spectra of the samples recorded at an excitation wavelength of 320 nm are shown in Fig. 5. All the samples exhibit near UV emission at around 397 nm and four defect related peaks at 437 nm, 466 nm, 492 nm and 527 nm. The intensity of all these emission peaks is enhanced with the rise of temperature. It can be seen from the figure that there is no significant change in the emission intensities of the

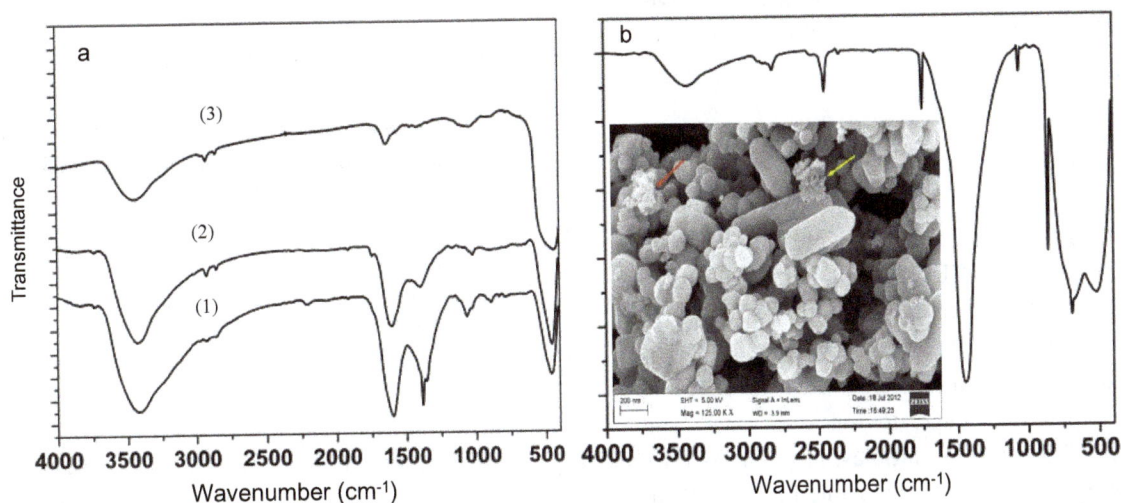

Fig. 4   (a) FT-IR spectra of (1) the as-prepared porous ZnO and porous ZnO calcined at (2) 300 ℃ and (3) 500 ℃; (b) FT-IR spectrum of the sample calcined at 400 ℃ with a heating rate of 1 ℃/min and inset shows the FE-SEM picture of the sample.

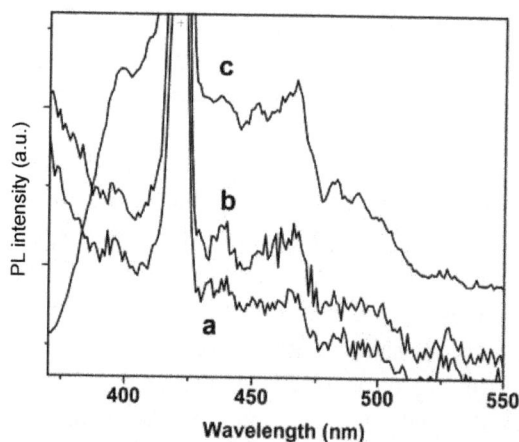

Fig. 5   PL spectra of (a) the as-prepared porous ZnO and porous ZnO calcined at (b) 300 ℃ and (c) 500 ℃.

as-prepared porous ZnO and porous ZnO calcined at 300 ℃. But when the temperature is increased to 500 ℃, the intensities of emission peaks are enhanced remarkably. Very recently, it was proved that 398 nm peak is originated from the electron transition from the localized level, slightly below conduction band to the valence band [18]. The enhanced intensity of these emission peaks is obviously due to the improved crystallanity of the ZnO with rise of temperature as confirmed from XRD, FE-SEM and UV–Vis results. The enhancement of the defect related emission intensities along with the UV emission indicates that these defects are intrinsic in nature. The emission peak at 437 nm can be attributed to the zinc vacancy [19]. The other visible emission peaks at 466 nm, 492 nm and 527 nm have been assigned to various deep level defect states originated from the zinc interstitials and/or oxygen vacancies [5,19].

4   Conclusions

Porous ZnO was synthesized by a simple method using zinc nitrate, ethanol and TEA. The as-prepared sample contained both nano and micro pores. The nano pores were disappeared but microporosity of the sample was retained even at 500 ℃. It was found that the shape of the FT-IR spectrum depends on the size and morphology of ZnO nanoparticles. The stress and change in morphology were responsible for the shift of Zn–O stretching mode to lower wavenumber side. The absorption spectra showed continuous red shift from 347 nm to 370 nm with rise of temperature. PL

emission intensity enhanced with increase of temperature. The near UV emission at 397 nm has been attributed to the electron transition from the localized level slightly below conduction band to the valence band. The visible emission at 437 nm was attributed to the zinc vacancy. The other visible emissions at 466 nm, 492 nm and 527 nm were assigned to the zinc interstitials and/or oxygen vacancies.

Acknowledgements

The authors thank the dean of School of Physics, University of Hyderabad, for providing FE-SEM facility generously, and are also grateful to Y. B. Ravi Shankar for his continuous support in XRD analysis of the samples.

References

[1]   Service RF. Will UV lasers beat the blues? *Science* 1997, **276**: 895.

[2]   Singh AK, Viswanath V, Janu VC. Synthesis, effect of capping agents, structural, optical and photoluminescence properties of ZnO nanoparticles. *J Lumin* 2009, **129**: 874–878.

[3]   Koch U, Fotik A, Weller H, *et al.* Photochemistry of semiconductor colloids. Preparation of extremely small ZnO particles, fluorescence phenomena and size quantization effects. *Chem Phys Lett* 1985, **122**: 507–510.

[4]   Djurišić AB, Leung YH, Tam KH, *et al.* Green, yellow, and orange defect emission from ZnO nanostructures: Influence of excitation wavelength. *Appl Phys Lett* 2006, **88**: 103107.

[5]   Jamali-Sheini F. Chemical solution deposition of ZnO nanostructure films: Morphology substrate angle dependency. *Ceram Int* 2012, **38**: 3649–3657.

[6]   Wei A, Sun XW, Xu CX, *et al.* Growth mechanism of tubular ZnO formed in aqueous solution. *Nanotechnology* 2006, **17**: 1740–1744.

[7]   Wu L, Wu Y, LÜ W. Preparation of ZnO nanorods and optical characterizations. *Physica E* 2005, **28**:

76–82.

[8] Roy VAL, Djurišić AB, Chan WK, *et al.* Luminescent and structural properties of ZnO nanorods prepared under different conditions. *Appl Phys Lett* 2003, **83**: 141–143.

[9] Kılıç B, Gür E, Tüzemen S. Nanoporous ZnO photoelectrode for dye-sensitized solar cell. *J Nanomater* 2012, DOI: 10.1155/2012/474656.

[10] Li B, Wang Y. Hierarchically assembled porous ZnO microstructures and applications in a gas sensor. *Superlattice Microst* 2011, **49**: 433–440.

[11] Jeon SM, Kim MS, Cho MY, *et al.* Fabrication of porous ZnO nanorods with nano-sized pores and their properties. *J Korean Phys Soc* 2010, **57**: 1477–1481.

[12] Dai Z, Liu K, Tang Y, *et al.* A novel tetragonal pyramid-shaped porous ZnO nanostructure and its application in the biosensing of horseradish peroxidase. *J Mater Chem* 2008, **18**: 1919–1926.

[13] Liu Z, Jin Z, Li W, *et al.* Preparation of porous ZnO plate crystal thin films by electrochemical deposition using PS template assistant. *Mater Lett* 2006, **60**: 810–814.

[14] Li S, Zhang X, Jiao X, *et al.* One-step large-scale synthesis of porous ZnO nanofibers and their application in dye-sensitized solar cells. *Mater Lett* 2011, **65**: 2975–2978.

[15] Ching CG, Ooi PK, Ng SS, *et al.* Fabrication of porous ZnO via electrochemical etching using 10 wt% potassium hydroxide solution. *Mat Sci Semicon Proc* 2013, **16**: 70–76.

[16] Xiong H-M, Shchukin DG, Möhwald H, *et al.* Sonochemical synthesis of highly luminescent zinc oxide nanoparticles doped with magnesium(II). *Angew Chem Int Edit* 2009, **48**: 2727–2731.

[17] Andrés Vergés M, Mifsud A, Sernad CJ. Formation of rod-like zinc oxide microcrystals in homogeneous solutions. *J Chem Soc Faraday Trans* 1990, **86**: 959–963.

[18] Sowri Babu K, Ramachandra Reddy A, Sujatha Ch, *et al.* Optimization of UV emission intensity of ZnO nanoparticles by changing the excitation wavelength. *Mater Lett* 2013, **99**: 97–100.

[19] Giri PK, Bhattacharyya S, Singh DK, *et al.* Correlation between microstructure and optical properties of ZnO nanoparticles synthesized by ball milling. *J Appl Phys* 2007, **102**: 093515.

Permissions

List of Contributors

Maryam HOSSEINI-ZORI
Department of Inorganic Pigments and Glazes, Institute for Color Science and Technology (ICST), P.O. Box 1668814811, Tehran, Iran

QU Lijie
State Key Laboratory of Advanced Welding and Joining, Harbin Institute of Technology, Harbin 150001, China
Department of Materials Science and Engineering, Jiamusi University, Jiamusi 154007, China

LI Muqin
State Key Laboratory of Advanced Welding and Joining, Harbin Institute of Technology, Harbin 150001, China
Department of Materials Science and Engineering, Jiamusi University, Jiamusi 154007, China

LIU Miao
Department of Stomatology, Jiamusi University, Jiamusi 154007, China

Erlin ZHANG
Department of Materials Science and Engineering, Jiamusi University, Jiamusi 154007, China

MA Chen
Department of Materials Science and Engineering, Jiamusi University, Jiamusi 154007, China

Yongshang TIAN
Faculty of Material Science and Chemistry, China University of Geosciences, Wuhan 430074, People's Republic of China
Engineering Research Center and Application of Nano-Geomaterials of Ministry of Education,

Yansheng GONG
Faculty of Material Science and Chemistry, China University of Geosciences, Wuhan 430074, People's Republic of China
Engineering Research Center and Application of Nano-Geomaterials of Ministry of Education,

LI Zhaoying
Faculty of Material Science and Chemistry, China University of Geosciences, Wuhan 430074, People's Republic of China

Feng JIANG
Faculty of Material Science and Chemistry, China University of Geosciences, Wuhan 430074, People's Republic of China

Hongyun JIN
Faculty of Material Science and Chemistry, China University of Geosciences, Wuhan 430074, People's Republic of China
Engineering Research Center and Application of Nano-Geomaterials of Ministry of Education,
China University of Geosciences, Wuhan 430074, People's Republic of China

Pavel CTIBOR
Institute of Plasma Physics, ASCR, v.v.i., Za Slovankou 3, Prague, Czech Republic

Vaclav STENGL
Institute of Inorganic Chemistry, ASCR, v.v.i., Husinec-Rez, Czech Republic

Zdenek PALA
Institute of Plasma Physics, ASCR, v.v.i., Za Slovankou 3, Prague, Czech Republic

Venkata Ramana MUDINEPALLI
Department of Physics, "National Taiwan Normal University", Taipei 11677, Taiwan, China

Leng FENG
Department of Chemistry, Shenzhen Graduate School, Peking University, Shenzhen 518055, China

Wen-Chin LIN
Department of Physics, "National Taiwan Normal University", Taipei 11677, Taiwan, China

B. S. MURTY
Nanotechnology Laboratory, Department of Metallurgical and Materials Engineering, Indian Institute of Technology-Madras, Chennai 600036, India

Juliana Kelmy M. F. DAGUANO
aUniversidade de São Paulo - Escola de Engenharia de Lorena, USP-EEL - Pólo Urbo-Industrial, s/n, Gleba AI-6, Lorena-SP, CEP 12600-000, Brazil

Paulo A. SUZUKI
Universidade de São Paulo - Escola de Engenharia de Lorena, USP-EEL - Pólo Urbo-Industrial, s/n, Gleba AI-6, Lorena-SP, CEP 12600-000, Brazil

Kurt STRECKER
Universidade Federal de São João del-Rei, – UFSJ-CENEN, Campus Sto Antônio - Praça Frei Orlando 170 – Centro, S. J. del-Rei-MG. CEP 36307-352, Brazil

José Martinho Marques de OLIVEIRA
Escola Superior Aveiro Norte, Edifício Rainha, 3720-232
O. Azeméis, Portugal

Maria Helena Figueira Vaz FERNANDES
Universidade de Aveiro, Campus Universitário de
Santiago 3810-193 Aveiro, Portugal

Claudinei SANTOS
Universidade de São Paulo - Escola de Engenharia de
Lorena, USP-EEL - Pólo Urbo-Industrial, s/n, Gleba AI-6,
Lorena-SP, CEP 12600-000, Brazil
Universidade do Estado do Rio de Janeiro – Faculdade
de Tecnologia de Resende – UERJ-FAT – Rod. Presidente
Dutra, km, 298, Resende-RJ, CEP 27537-000, Brazil

Shaohua QIAN
State Key Laboratory of Mechanics and Control of
Mechanical Structures, Nanjing University of Aeronautics
and Astronautics, Nanjing 210016, China

Kongjun ZHU
State Key Laboratory of Mechanics and Control of
Mechanical Structures, Nanjing University of Aeronautics
and Astronautics, Nanjing 210016, China

Xuming PANG
State Key Laboratory of Mechanics and Control of
Mechanical Structures, Nanjing University of Aeronautics
and Astronautics, Nanjing 210016, China

Jing WANG
State Key Laboratory of Mechanics and Control of
Mechanical Structures, Nanjing University of Aeronautics
and Astronautics, Nanjing 210016, China

Jinsong LIU
State Key Laboratory of Mechanics and Control of
Mechanical Structures, Nanjing University of Aeronautics
and Astronautics, Nanjing 210016, China

Jinhao QIU
State Key Laboratory of Mechanics and Control of
Mechanical Structures, Nanjing University of Aeronautics
and Astronautics, Nanjing 210016, China

Xuegang HUANG
Hypervelocity Aerodynamics Institute, China
Aerodynamics Research and Development Center,
Mianyang 621000, China

Jie HUANG
Hypervelocity Aerodynamics Institute, China
Aerodynamics Research and Development Center,
Mianyang 621000, China

Zhongmin ZHAO
Department of Vehicle and Electrical Engineering,
Mechanical Engineering College, Shijiazhuang 050003,
China

Long ZHANG
Department of Vehicle and Electrical Engineering,
Mechanical Engineering College, Shijiazhuang 050003,
China

Junyan WU
Department of Vehicle and Electrical Engineering,
Mechanical Engineering College, Shijiazhuang 050003,
China

Pil-Gyu CHOI
Department of Applied Chemistry, Faculty of Engineering,
Osaka University, 2-1 Yamadaoka, Suita, Osaka 565-0871,
Japan

Takanobu OHNO
Department of Applied Chemistry, Faculty of Engineering,
Osaka University, 2-1 Yamadaoka, Suita, Osaka 565-0871,
Japan

Nashito FUKUHARA
Department of Applied Chemistry, Faculty of Engineering,
Osaka University, 2-1 Yamadaoka, Suita, Osaka 565-0871,
Japan

Toshiyuki MASUI
Department of Applied Chemistry, Faculty of Engineering,
Osaka University, 2-1 Yamadaoka, Suita, Osaka 565-0871,
Japan

Nobuhito IMANAKA
Department of Applied Chemistry, Faculty of Engineering,
Osaka University, 2-1 Yamadaoka, Suita, Osaka 565-0871,
Japan

Manju PANDEY
Nano-Sensor Research Laboratory, F/O Engineering and
Technology, Jamia Millia Islamia (Central University),
New Delhi, India

Prabhash MISHRA
Nano-Sensor Research Laboratory, F/O Engineering and
Technology, Jamia Millia Islamia (Central University),
New Delhi, India

Debdulal SAHA
Sensors and Actuators Division, Central Glass & Ceramic
Research Institute, 196 Raja S. C. Mullick Road, Kolkata
700032, India

S. S. ISLAM
Nano-Sensor Research Laboratory, F/O Engineering and Technology, Jamia Millia Islamia (Central University), New Delhi, India

L. Jay DEINER
Department of Chemistry, New York City College of Technology, City University of New York, 300 Jay St., Brooklyn, NY 11201, USA

Michael A. ROTTMAYER
The Air Force Research Labs, Wright-Patterson Air Force Base, OH 45433, USA

Bryan C. EIGENBRODT
Department of Chemistry, Villanova University, 800 E. Lancaster Ave., Villanova, PA 19085, USA

P. N. MEDEIROS
Department of Materials Engineering, Federal University of Rio Grande do Norte, Campus Lagoa Nova, CEP 59078-900-Natal/RN, Brazil

Y. F. GOMES
Department of Materials Engineering, Federal University of Rio Grande do Norte, Campus Lagoa Nova, CEP 59078-900-Natal/RN, Brazil

M. R. D. BOMIO
Department of Materials Engineering, Federal University of Rio Grande do Norte, Campus Lagoa Nova, CEP 59078-900-Natal/RN, Brazil

I. M. G. SANTOS
Department of Chemistry, Federal University of Paraíba, Cidade Universitária, CEP 58051-900-João Pessoa/PB, Brazi

M. R. S. SILVA
Department of Chemistry, Federal University of Paraíba, Cidade Universitária, CEP 58051-900-João Pessoa/PB, Brazi

C. A. PASKOCIMAS
Department of Materials Engineering, Federal University of Rio Grande do Norte, Campus Lagoa Nova, CEP 59078-900-Natal/RN, Brazil

M. S. LI
Institute Physics of São Carlos, USP, CEP 13566-590, São Carlos, São Paulo, Brazil

F. V. MOTTA
Department of Materials Engineering, Federal University of Rio Grande do Norte, Campus Lagoa Nova, CEP 59078-900-Natal/RN, Brazil

Nobuhito IMANAKA
Department of Applied Chemistry, Faculty of Engineering, Osaka University, 2-1 Yamadaoka, Suita, Osaka 565-0871, Japan

Toshiyuki MASUI
Department of Applied Chemistry, Faculty of Engineering, Osaka University, 2-1 Yamadaoka, Suita, Osaka 565-0871, Japan

Kazuya JYOKO
Department of Applied Chemistry, Faculty of Engineering, Osaka University, 2-1 Yamadaoka, Suita, Osaka 565-0871, Japan

Pattem Hemanth KUMAR
Department of Ceramic Engineering, Indian Institute of Technology (BHU), Varanasi, India

Abhinav SRIVASTAVA
Department of Ceramic Engineering, Indian Institute of Technology (BHU), Varanasi, India

Vijay KUMAR
Department of Ceramic Engineering, Indian Institute of Technology (BHU), Varanasi, India

Nandini JAISWAL
Department of Ceramic Engineering, Indian Institute of Technology (BHU), Varanasi, India

Pradeep KUMAR
Department of Chemical Engineering, Indian Institute of Technology (BHU), Varanasi, India

Vinay Kumar SINGH
Department of Ceramic Engineering, Indian Institute of Technology (BHU), Varanasi, India

Prasenjit BARICK
Centre for Non-Oxide Ceramics, International Advanced Research Centre for Powder Metallurgy and New Materials, PO: Balapur, RCI Road, Hyderabad 500005, Andhra Pradesh, India

Dulal Chandra JANA
Centre for Non-Oxide Ceramics, International Advanced Research Centre for Powder Metallurgy and New Materials, PO: Balapur, RCI Road, Hyderabad 500005, Andhra Pradesh, India

Bhaskar Prasad SAHA
Centre for Non-Oxide Ceramics, International Advanced Research Centre for Powder Metallurgy and New Materials, PO: Balapur, RCI Road, Hyderabad 500005, Andhra Pradesh, India

Jiayan LI
School of Materials Science and Engineering, Dalian University of Technology, Dalian 116024, China

Panpan CAO
School of Materials Science and Engineering, Dalian University of Technology, Dalian 116024, China

Yi TAN
School of Materials Science and Engineering, Dalian University of Technology, Dalian 116024, China

Lei ZHANG
School of Materials Science and Engineering, Dalian University of Technology, Dalian 116024, China

Srinivasan NEDUNCHEZHIAN
Materials Processing Section, Department of Metallurgical and Materials Engineering, Indian Institute of Technology Madras, Chennai 600036, Tamil Nadu, India

Ravindran SUJITH
Materials Processing Section, Department of Metallurgical and Materials Engineering, Indian Institute of Technology Madras, Chennai 600036, Tamil Nadu, India

Ravi KUMAR
Materials Processing Section, Department of Metallurgical and Materials Engineering, Indian Institute of Technology Madras, Chennai 600036, Tamil Nadu, India

Avadhesh Kumar YADAV
Department of Physics University of Lucknow, Lucknow-226007, India

C. R. GAUTAM
Department of Physics University of Lucknow, Lucknow-226007, India

Abhinay MISHRA
School of Materials Science and Technology, Indian Institute of Technology, Banaras Hindu University, Varanasi-221005, India

Subhanarayan SAHOO
School of Electrical Engineering, KIIT University, Bhubaneswar 751024, India

Umasankar DASH
Center for Nanotechnology, KIIT University, Bhubaneswar 751024, India

S. K. S. PARASHAR
School of Applied Sciences, KIIT University, Bhubaneswar 751024, India

S. M. ALI
School of Electrical Engineering, KIIT University, Bhubaneswar 751024, India

K. SOWRI BABU
Department of Physics, National Institute of Technology Warangal, Warangal-506 004, Andhra Pradesh, India

A. RAMACHANDRA REDDY
Department of Physics, National Institute of Technology Warangal, Warangal-506 004, Andhra Pradesh, India

Ch. SUJATHA
Department of Physics, National Institute of Technology Warangal, Warangal-506 004, Andhra Pradesh, India

K. VENUGOPAL REDDY
Department of Physics, National Institute of Technology Warangal, Warangal-506 004, Andhra Pradesh, India

A. N. MALLIKA
Department of Physics, National Institute of Technology Warangal, Warangal-506 004, Andhra Pradesh, India

www.ingramcontent.com/pod-product-compliance
Lightning Source LLC
Chambersburg PA
CBHW050501200326
41458CB00014B/5256